**6**

# Maths
## Teacher's Guide

Tony Cotton

Ray Huntley

**OXFORD**
UNIVERSITY PRESS

**OXFORD**
UNIVERSITY PRESS

Great Clarendon Street, Oxford, OX2 6DP, United Kingdom

Oxford University Press is a department of the University of Oxford. It furthers the University's objective of excellence in research, scholarship, and education by publishing worldwide. Oxford is a registered trade mark of Oxford University Press in the UK and in certain other countries.

© Ray Huntley and Tony Cotton 2021

The moral rights of the author have been asserted.

First published in 2014



British Library Cataloguing in Publication Data

Data available

ISBN 9781382017312

15

Paper used in the production of this book is a natural, recyclable product made from wood grown in sustainable forests. The manufacturing process conforms to the environmental regulations of the country of origin.

Printed in Great Britain by CPI Group (UK) Ltd., Croydon CR0 4YY

**Acknowledgements**

The publisher and authors would like to thank the following for permission to use photographs and other copyright material:

**Cover:** Artwork by Peskimo. **Photos: pv(t):** foto-bee/Alamy Stock Photo; **pv(m):** Alistair McDonald/Shutterstock; **pv(b):** 2R fotografia/Shutterstock; **px:** Monkey Business Images/ Shutterstock; **pxviii:** monkeybusinessimages/iStockphoto.

Artwork by Q2A Media Services Pvt. Ltd.

Every effort has been made to contact copyright holders of material reproduced in this book. Any omissions will be rectified in subsequent printings if notice is given to the publisher.

The manufacturer's authorised representative in the EU for product safety is Oxford University Press España S.A. of El Parque Empresarial San Fernando de Henares, Avenida de Castilla, 2 – 28830 Madrid (www.oup.es/en or product.safety@oup.com). OUP España S.A. also acts as importer into Spain of products made by the manufacturer.

# Contents

# Introduction

## The joy of learning maths

We are living in an ever-changing world, where the way we work, live, learn, communicate and relate to one another is constantly shifting. In this climate, we need to instill in our learners the skills to equip them for every eventuality so they are able to overcome challenges, adapt to change and have the best chance of success. To do this, we need to evolve beyond traditional teaching approaches and foster an environment where students can start to build lifelong learning skills for success. Students need to learn how to learn, how to problem solve, be agile and work flexibly. Going hand-in-hand with this is the development of self-awareness and mindfulness through the promotion of wellbeing to ensure that students learn the socio-emotional skills to succeed.

With *Oxford International Primary Maths*, students develop lifelong learning skills as well as mathematical skills. The course promotes the development of real-world skills including financial literacy. The activities in the Student Books and Practice Books offer numerous opportunities to think creatively and develop interpersonal skills. Fundamentally, *Oxford International Primary Maths* promotes students' self-development as critical thinking and motivation are at the heart of the problem-solving approach in the course.

This series is based on the English National Curriculum Programme of Study for Primary Maths. The *Oxford International Primary Maths* books for each stage meet all the learning objectives from the curriculum. Each lesson includes the learning objectives and a summary of the key teaching points. A full mapping grid identifying the unit and lesson where each objective can be found is available online at www.oxfordowl.co.uk

## Oxford International Primary Maths: A problem-solving approach

In this second edition of *Oxford International Primary Maths*, there is a strong focus on using a problem-solving approach. While mathematical facts are important, it is unlikely that simply giving students the information they need will result in them understanding the mathematics and being able to apply their learning in new problem-solving situations. This is often described as a move from 'surface learning' to 'deep learning'.

Many people remember mathematics lessons as places where the teacher stood at the front of the class writing on the board. Students wrote down the information, maybe worked through a couple of examples with the teacher and then proceeded to complete a series of exercises to practise the skill that they had been taught. This can be described as a *didactic* approach and it relies on the idea that direct instruction is the appropriate strategy to adopt. The authors of this series would argue that *heuristic* strategies encourage students to explore the mathematics for themselves supported by the teacher. 'Heuristic' derives from the Greek word meaning to discover, and in mathematics learning, heuristic strategies are ones where students engage in exploration and discovery to solve a problem. Heuristic strategies include making a visual representation of a problem, making a calculated guess or estimate, simplifying a problem or following a known method. This results in a deeper understanding.

When faced with any problem in mathematics, there are recognised stages to go through in order to solve the problem, and these have been developed and agreed by many researchers. One version that summaries the problem-solving process comes from Georg Polya.

1. Understand the problem.

2. Devise a plan.

3. Carry out the plan.

4. Check the reasoning.

In following these stages, students use a number of skills that support problem solving, such as using trial and improvement, working systematically, pattern spotting, visualising, conjecturing and generalising.

### Embedding a mastery approach

In recent years, the term 'mastery' has been used in conjunction with mathematics learning. It has been drawn from teaching approaches in countries where mathematics performance is deemed to be very high. The essence of mastery is to produce students who have deep conceptual understanding and procedural fluency through learning in a collaborative and problem-solving context. Mastery learning incorporates use of manipulatives, exposure to different methods of solving a problem, dialogue and explanation.

### Following a Concrete Pictorial Abstract (CPA) approach

One of the more successful approaches to learning was provided by Jerome Bruner in his model of enactive, iconic and symbolic modes. This has been developed in recent years to form the CPA approach. CPA stands for concrete, pictorial and abstract, each of which aligns with Bruner's modes. The concrete phase involves students making use of physical manipulatives to help understand the learning, before moving to record the learning in pictorial form as individuals. As the learning develops, students will begin to recognise how to record their learning in a more general and abstract way. The CPA approach is not necessarily sequential, and students might move between the different modes as they work through a problem.

## *Oxford International Primary Maths* and the use of manipulatives

Throughout the series, students are encouraged to use manipulatives, or concrete objects, to model addition, subtraction, multiplication and division. These manipulatives include:

- base-ten equipment (ones-cubes, tens-rods, hundreds-flats and thousands-cubes)

- place-value counters

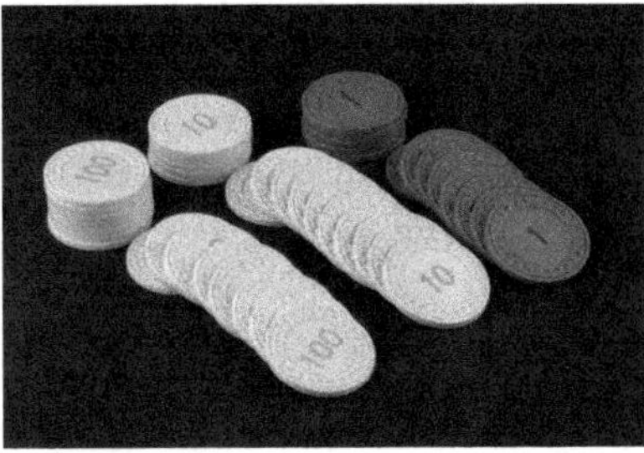

- number rods.

Such manipulatives are used to explain to students how the written methods 'work', for example by modelling exchanging 10 ones-cubes for 1 tens-rod in an addition.

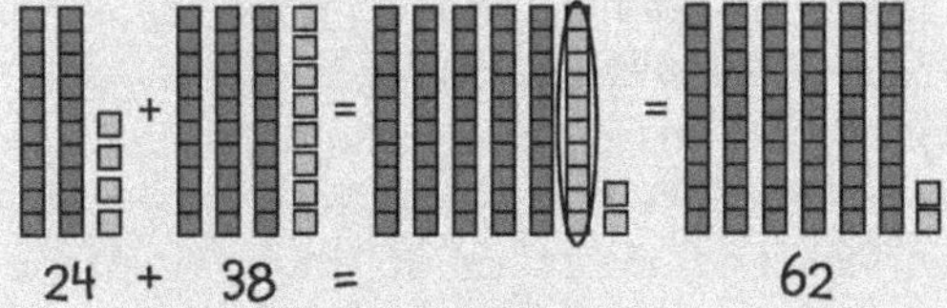

## Differentiation

There are several ways that you can differentiate learning in the classroom. These include differentiation by:

- task
- outcome
- support
- grouping.

It has been traditional in some schools to offer up to three different levels of tasks for each lesson. This is differentiation by task. It is important that all students are exploring the same area of mathematics as they can collaborate and discuss their mathematics in a way that is not possible if students are engaged on different activities. This approach has been extensively researched and published by Jo Boaler of Stanford University, California. For example, she has outlined projects that gave students in different schools either a differentiated approach in lessons, or lessons where everyone worked on the same task (Boaler, J., 2005. *The Elephant in the Classroom.* Souvenir Press). Where all abilities worked on the same task, every student made and sustained 'better than expected' progress, and performed better on statutory tests and exams. The Education Endowment Foundation teacher's toolkit suggests that collaborative learning can result in a five-month acceleration in students' learning. (See https://educationendowmentfoundation.org.uk/resources/teaching-learning-toolkit.)

The expectation in this series is that all students will be offered the same starting point. The activities are carefully designed to be accessible to all students in your class and the teacher's notes for the activity offer differentiated outcomes for students. It is also important that you offer differentiated support to different students. You will mainly do this through the sort of questioning that you engage in and support you offer. You will ask challenging questions and supporting questions to help all students access the task. For example, when engaging in a simple counting activity with some students you might model the action of counting by placing a finger on each object as you count and emphasise the last number you say to model that the last number you say gives the number of objects in the set. You might ask other students engaged in the same activity to compare two sets, or to find one more or one less than the set they are counting.

## Grouping students to promote a growth mindset

When engaging in learning mathematics, it is expected that you will use a variety of student groupings. This may be a change for some teachers who have previously grouped students by prior attainment in their classroom. Research has shown that grouping students 'by ability', which usually means grouping students using test results, can have a negative impact on their future attainment. It is more effective to use a range of ways of grouping students. You will decide on the most appropriate way of grouping depending on the activity. You are also given advice in the teacher's notes. It is important that the teacher is active in deciding which form of grouping is appropriate. It is also important that students learn how to operate in a range of different groups and with a range of different students so that they get used to working in a variety of ways and with different people.

The three main ways of grouping students are based on:

- friendship
- ability/prior experience
- mixed attainment.

**Friendship groups** are most appropriate for activities in which the students have been given some element of choice. Perhaps they are carrying out some research for a data handling project or exploring data on animals to develop their understanding of measurement. This grouping is the default if teachers do not actively group students.

**Ability groups**, or groups based on students' prior experience, may be helpful if the lesson requires a very specific prior knowledge. You can group together the students you know have this knowledge and they can then work with minimal guidance from you, which allows you to focus on groups who need additional support.

**Mixed-attainment groups** are encouraged for the majority of the activities. This form of grouping is also favoured by those following a mastery approach. Working in collaborative, all-attainment groups also supports students' wellbeing and promotes a growth mindset, as described in research by Carol Dweck. She found that students who were grouped by ability tended to stay in those groupings throughout their school life, and regard themselves as having a fixed ability that could not be changed. This has dire consequences for students in middle or lower sets. When placed in mixed-ability groups, all students can develop a growth mindset which enables them to believe they can learn and improve, whatever their starting point (Dweck, C., 2007. 'The Perils and Promise of Praise'. *Educational Leadership*. October 2007, 65(2), 34–39). A growth mindset is promoted when students do not feel that their future success is predicated on prior achievement. This kind of grouping is particularly helpful for students new to English. Mixed-attainment groups allow students who are less confident in English to hear more-confident peers using mathematical vocabulary. Research has shown that mixed-attainment groups benefit both high attainers, who become more secure in their mathematics knowledge through explaining their thinking to peers, and those less secure in their mathematical knowledge as peer teaching has been shown to be effective.

Whatever form of grouping you choose, it is helpful to assign roles to individuals in the group. Some teachers use 'role cards' to remind members of the group of the role they should play. Here are some examples of roles.

- Leader: You should make sure everyone has a chance to speak and focus the discussion around the task.
- Time keeper: You should encourage the group to stay on task. Announce when the time is half way through and when time is nearly up.
- Recorder: You should write down group members' ideas or draw a collective graphic. You will write on the board during the presentation.
- Presenter: You will present the group's findings to the whole class at the end of the session.
- Resource organiser: You will make sure that group members have all the resources they need during the task.



Assessment is the process of establishing how individual students are progressing and what they have achieved, or a means of measuring their learning. Assessment is usually carried out in two main ways – assessment of learning and assessment for learning.

Assessment of learning is sometimes called summative assessment, and takes place at the end of a lesson, a unit, a term or even a year. It measures what students know at that point as a summary of their learning to that point. In *Oxford International Primary Maths*, summative assessment opportunities are provided in the Review lesson at the end of each unit in the Student Book, while half-termly summative assessment opportunities are provided through printable resources, available online.

Assessment for learning is an approach brought to prominence by Paul Black and Dylan Wiliam and is based on the notion that students have a full, clear sense of what they are learning, where they have reached in their learning and what they need to do to improve further. It is carried out during lessons and gives teachers continuous data on each student's learning, as well as allowing students to track their own learning, which provides greater motivation. (Black, P., Harrison, C., Lee, C., Marshall, B., and Wiliam, D. 2004. 'Inside the Black Box: Assessment for Learning in the Classroom'. *Phi Delta Kappan*. (86)1, 8–21.)

It is suggested  that there are five key strategies for assessment for learning. These are outlined below with suggestions of how you can do this in your classroom.

1  Being clear about learning objectives and success criteria with the students.

Each activity has at least one learning objective. At the beginning of a lesson, share the activity's learning objective with students. This should be more than simply stating the objective. You should make sure that students understand the objective and how you will measure success. For example, you might say: *I know that you can all count 10 objects* and all count to 10 as a class. Then you point to 20 on a number line and ask: *Does anyone know what this number is?* If a student knows it is 20 praise them, if no-one knows,  tell them it is 20 and say: *By the end of the lesson I will be able to listen to you count to 20.*

2  Planning student discussions that give you evidence of their learning.

Every activity plan in the Teacher's Guide offers the opportunity for small-group or whole-class discussion. There are also examples of probing questions that you can ask to assess students' current understanding. For example, if a group has been counting two sets of objects you can ask: *Were there more or less in the second group? How do you know?*

3  Giving students feedback that helps them move forward.

This allows students to know whether or not they are meeting the success criteria and what they can do next to move their learning on. Developing the example above, if a group has been comparing two sets and understands the concept of 'more' and 'less' you could ask them to make sets that are one more and one less, or even two more and two less.

4   Activating students to act as instructional resources for each other.

Collaborative group work in mixed-attainment groups, as described by Jo Boaler in her research (see under Differentiation earlier), gives students the opportunity to operate both as learners and teachers, with peer learning being highly effective. Not only is understanding of the mathematics enhanced, but students can support each other in assessing their progress.

5   Activating students as owners of their own learning.

The key point here is to listen carefully to the students and adapt your questioning to support individual development and to follow individual interests.

*Questioning is key*

The most skilled mathematics teachers can ask open questions to elicit students' current understandings. Skilful open questioning also allows students to articulate their current understanding carefully and though this process either consolidate their understanding or come to realise where they have made a mistake. The list below offers a series of open questions that can be used whatever mathematics you are teaching.

- *How are these the same/different?*
- *About how many/how long/many more … do you think there will be?*
- *What would happen if …?*
- *How else could you have done that?*
- *Why did you …?*
- *How did you …?*
- *How do you know that is correct?*

If you want students to check their solutions and consolidate their learning it is helpful to ask them to explain how they reached their solution to a friend. Similarly, to support students in reflecting on their learning you might ask the following.

- *What mathematics did you use to solve the problem?*
- *What new mathematics did you learn?*
- *What key words did you use?*
- *What was the most challenging part of the activity?*
- *What did you do when you got stuck?*
- *What other questions could you ask?*
- *Did this remind you of any other areas of mathematics?*

In *Oxford International Primary Maths*, there is an opportunity to ask these reflective questions, and for students to reflect on their learning, at the end of each unit in the Review lesson of the Practice Book.

## Word problems

Word problems are useful as an assessment of children's understanding of the correct mathematics to use in any given situation. In *Oxford International Primary Maths* word problems are included throughout the units and on every Student Book Review page as part of the end-of-unit assessment. Many teachers find teaching word problems a challenge. This area is particularly challenging for students with a limited English vocabulary as word problems are tightly bound to linguistic ability. We have to decode and understand what the problem is asking us to do before we can begin to apply our mathematical knowledge. Some teachers have found the following acronym helpful when working with students on solving word problems.

**R:** Read the problem carefully.

**U:** Understand what the problem is asking you to do.

**C:** Choose the mathematics or arithmetical operations that you need to use to solve the problem.

**S:** Solve the problem.

**A:** Answer the problem.

**C:** Check that the answer is accurate and reasonable.

It is often helpful for students to underline key facts and write down the operations they are going to use before they solve the problem. For example:

> Tony rode his bicycle 7 miles to school with his friend. On his way home he took a short cut which was only 5 miles. How far did he cycle altogether?
>
> *This will be an addition calculation.*

It is a useful activity for students to annotate word problems and write down the operation(s) they will use without carrying out the calculation as this focuses on the skill of understanding the problem and choosing the operations appropriately.

Another activity that helps students to become skilled at solving word problems is asking them to write their own word problems based on a picture or a set of objects. Here is an example.

- How many black cubes are there? (3)
- Two friends took three cubes each. How many were left? (2)
- If I take out the black cubes, how many are left? (5)
- If I share the cubes equally between two people, how many do they each get? (4)

## Wellbeing and *Oxford International Primary Maths*

It is thought that students learn more and feel more connected to their learning when they are active in their lessons. *Oxford International Primary Maths* has active learning at its heart. Most lessons start with a whole-class session that usually includes a range of physical or active activites. You will see this signified by a 'star-jump' icon in the Teacher's Guide.

Many adults and children have felt anxious about their learning of mathematics at some stage. This anxiety is reduced by working collaboratively in all-attainment groups. There is also a reflective session at the end of each lesson and the formative assessment activity in the Practice Book asks students to reflect on their learning across the unit.

Wellbeing is also supported by effective questioning to support and stretch students and by planning group work carefully. These areas have already been discussed above.

## Language support

### The challenges

Ministries of Education at both local and national level are increasingly adopting the policy of English Medium Instruction (EMI), for either one or two subjects or across the whole curriculum. The rationale for doing so varies according to the local context, but improving the levels of achievement in English is an important factor.

In international schools an additional reason is likely to be that students do not share a mother tongue with each other or perhaps the teacher. English is, therefore, chosen as the medium for instruction so that all students are in the same position and to provide the opportunity to develop proficiency in an international language.

This does not mean that the mathematics teacher is now being asked to replace the English teacher, or to have the same skills or knowledge of English (though in many primary schools one teacher may indeed teach both). What it does mean, however, is that mathematics teachers have to view their role differently: they have to become much more language aware. It is this recognition of the need to ensure that the delivery of the content is not negatively impacted by the use of the second language that informs the planning and methodology of EMI.

This raises significant challenges, including:

- the teacher's knowledge of English
- students' level of English (which may vary considerably in international schools)
- resources that provide appropriate language support
- assessment tools which ensure that it is the content and not the language that is being tested
- differentiation that acknowledges different levels of proficiency in both language and content.

### Meeting the challenges positively

Perhaps lack of confidence in their own English proficiency is one of the most common concerns among teachers. However, while it is a factor, success in EMI is not necessarily linked to teachers' proficiency in English. Teachers who have English as their mother tongue may well lack the sensitivity to, or awareness of, the language that a non-native speaker has acquired through learning and studying the second language. Developing this awareness and demonstrating it in both materials and method is the key to effective EMI.

### Classroom language/Teacher Talk

Often non-native-speaker teachers are more concerned about their ability to run and manage the whole class in English than they are about the teaching of the mathematics concepts, as the resources or textbook should help them with the latter. However, this use of English in the class is very important as it provides exposure to the second language, which plays a valuable role in language acquisition. It is also true that the Teacher Talk for purposes such as checking attendance and collecting homework does not have to be totally accurate or accessible to students. When teaching the mathematics concepts, however, it is essential that the Teacher Talk is comprehensible. Some basic strategies to ensure this include:

- simplify your language
- use short, simple sentences and project your voice
- paraphrase (say in a different way) as necessary
- use visuals, write or draw on the board, gestures and body language to clarify meaning
- repeat as necessary
- plan before the lesson
- prepare clear, simple instructions and check understanding.

### Creating a language-rich environment

Primary teachers often excel at providing a colourful and engaging physical environment for students. In the EMI classroom, this becomes even more important. Posters, 'word walls', lists of key structures, students' work, English signs and notices all provide a backdrop that provides the opportunity for language exposure and language acquisition.

### Planning

When planning, look carefully at each stage of the unit and identify the language demands. This means thinking about what language students will need to understand or produce, and deciding how best to scaffold the learning to ensure that language does not become an obstacle to understanding the concept. This involves providing language support and goes beyond the familiar strategy of identifying key vocabulary.

### Support for listening and reading

Listening and reading are receptive skills, requiring understanding rather than production of language. If you are asking students to listen to or read texts in English, ask yourself the following questions when you are planning the unit.

- Do I need to teach any vocabulary before they listen/read?
- How can I prepare them for the content of the text so that they are not listening 'cold'?

- Can I provide visual support to help them understand the key content?
- How many times should I ask them to read/listen?
- What simple question can I set before they listen/read for the first time to focus their attention?
- How can I check more detailed understanding of the text? Can I use a graphic organiser (e.g. tables, charts and diagrams) or gap-fill task to reduce the language demands?
- Do I need to differentiate the task for those students who find reading/listening difficult?
- Could I make the tasks interactive (e.g. jigsaw reading, when students access different information before coming together then share information)?
- How am I going to check their answers and give feedback?

## Support for speaking and writing

Speaking and writing are productive skills because students doing these need to produce language. They are different from the receptive skills of listening and reading where students receive language from other sources. These skills may require more input from the teacher.

When you plan to use a task that requires students to *produce* English (speak or write), you need to think about how to help them do this.

This means that you have to think in detail about what language the task requires (Language Demands, LD) and what strategies you will use to help them use English to perform the task (Language Support, LS).

You need to ask yourself the following questions.

- What *vocabulary* does the task require? (LD)
- Do I need to teach this before they start? How? (LS)
- What *phrases/sentences* will they need? Think about the language for learning mathematics (e.g. predicting and comparing). What structures do they need for these language functions? (LD)
- Will they be able to produce these sentences or should I provide some *scaffolding* [e.g. sentence starters/sentence frames/gapped sentences (see below)]? (LS)

  A square has _____ sides.

  A triangle has _____ sides.

  A quadrilateral has _____ sides.

  A pentagon has _____ sides.

- While I am *monitoring* this task is there any way I can provide further support for their use of English (especially for the less-confident students)? (LS)
- What language will students need to use at the *feedback* stage (e.g. when they present their task)? Do I need to scaffold this? (LD, LS)

## Teaching vocabulary and structures

*Vocabulary*

Learning the key mathematics vocabulary is central to EMI and 'learning' means more than simply understanding the meaning. Knowing a word also involves being able to *pronounce* it accurately and *use* it appropriately. Below is a list of strategies that could be useful.

- Avoid writing the list of vocabulary on the board at the start of the unit and 'explaining' it. The vocabulary should be introduced as and when it arises. Word boxes are provided on each page of the Student Books and Practice Books with the key words for the lesson. This helps students associate the word or phrase with the concept and context.
- Before the lesson, check that you are confident with the pronunciation and spelling of the vocabulary that will be used. Write the vocabulary clearly on the board when you first introduce it in the lesson. If you think students may struggle to pronounce words, decide how best to model the pronunciation.
- Give students a chance to say a word once they have understood it. The most efficient way to do this is through repetition drilling.
- Use visuals whenever possible to reinforce students' understanding of the word.
- Ensure that students are recording the vocabulary systematically in their glossaries at the back of their Student Books, and, if possible, use a word wall that lists the vocabulary under unit or topic headings.
- Remember to use and revise the vocabulary.

*Structures*

In order for students to talk or write about their mathematics, they will need to go beyond vocabulary: they will also need to use those phrases and sentence frames that a particular task requires.

For example, they may need the following expressions in mathematics.

  *X is the same as Y.*

  *The sides are the same length.*

  *The next number in the sequence.*

  *I predict that X will happen.*

  *If X happens, then Y happens.*

  *The next step is …*

You need to build up banks of common mathematics phrases and encourage students to record them. This is an important part of identifying the language demands and providing the necessary support. You do not have to focus on grammar as the language can be taught as phrases rather than specific grammatical structures.

Every unit of the Teacher's Guide begins with useful background information that includes the following.

**The Big idea**: The main mathematical concept covered in the unit is outlined.

**Look out for**: This section focuses on tricky concepts that may need explaining prior to any learning taking place.

**Common misconceptions**: Common errors that students make, or misunderstandings that students have, are identified. This section offers advice on how to deal with these misconceptions.

**Key vocabulary**: This is a list of the key mathematical words used in the unit.

**Coverage in lessons**: The English National Curriculum objectives covered in the unit are listed.

Every lesson in the Student Book and Practice Book has corresponding lesson notes in the Teacher's Guide. These comprehensive lesson notes include the following.

**A mini reproduction**: This shows the relevant pages from the Student Book.

**Global skills**: These are the skills that aim to foster a classroom environment where students develop the skills for success. The skills are: *creative skills* where students are problem solving, investigating or exploring new maths content; *real-world skills* where students are taking part in research, or presenting and interpreting information, or if they are dealing with money and developing their financial literacy; *interpersonal skills* where students are practising their teamwork and communication, often through working in pairs or larger groups; and *self-development skills* where students have the opportunity to reflect on their learning and talk about what went well and what they are still uncertain about.

**The key vocabulary and resources**: Key vocabulary used in the lesson, and the concrete resources required for the activity, are listed.

**Language support**: This includes a range of strategies, including card sorts and card games, word walls, team games to define or explain words, use of similar words to explain meaning and exploration of the origins of words.

The key principles underpinning the language support are listed below.

Words should be introduced and explained carefully.

Words should be explained in context.

Repetition is vital.

Words should be linked to pictures or actions.

Students should develop their own glossaries.

The learning of mathematics vocabulary should be fun.

Language should not be a barrier to effective learning of mathematics.

**Detailed lesson notes**: Comprehensive lesson notes include an Introductory activity and Main activity. These notes refer to the Student Book and Practice Book, where relevant. The notes include probing questions for formative assessment, which are italicised. Icons are used to suggest the groupings that should be used at each point of the activity (whole class, small group, pairs, individual). A separate 'star-jump' icon indicates that the activities give students an opportunity for physical movement (standing up, jumping, moving around) rather than doing activities sitting down.

**Differentiation**: The Teacher's Guide offers strategies for you to *support* those students who may have difficulty accessing the task; to *consolidate* the learning for those students who need a little more practice; and to *extend* the learning for those who need more challenge.

The Teacher's Guide also offers differentiated outcomes. These outcomes are listed in the form of:

**All students**

**Most students**

**Some students**

**Stretch zone**: Each activity in the Student Book and the Practice Book has a Stretch zone question to support deeper learning. The Teacher's Guide provides additional notes on these activities.

**Reflection time**: Suggestions are made on how to bring the class back together to reflect on the learning and share ideas.

**Answers**: Answers to all the Student Book and Practice Book activities are provided.

**Review pages**: The Teacher's Guide has notes on the Review pages of the Student Book (summative assessment), with answers to the assessment questions, and the Practice Book (a formative, reflective review).

**Digital resources**: Where it is appropriate to use digital resources in a lesson, such as sharing the interactive Student eBook page on an interactive whiteboard (IWB), suggestions are embedded in the lesson plan.

**Resources sheets:** These photocopiable resources can be used with some of the main activities. They are referenced in the resources section of the lesson plan and are available on the Oxford Owl website (www.oxfordowl.co.uk).

# Tour of a typical unit



The 'Big question' provides a discussion stimulus about the key idea of the unit.

## 1 Numbers and counting

**?** How do we use numbers?

**In this unit you will:**

- count, read and write numbers to 100
- count in twos, fives and tens
- know and make numbers using objects and pictures
- use words such as equal to, more than, less than (fewer), most, least
- read and write numbers from 1 to 20 in words.

Learning objectives are stated clearly at the beginning of every unit.

### Engage

Which numbers can you see in the classroom?

Which numbers can you see on your way to school?

What is the biggest number you have ever seen?

Further questions allow students to develop communication skills.

6

The Engage spread is bright and colourful, with artwork or photos to spark interest in young students and provide discussion points.

# Student Book Discover and Explore

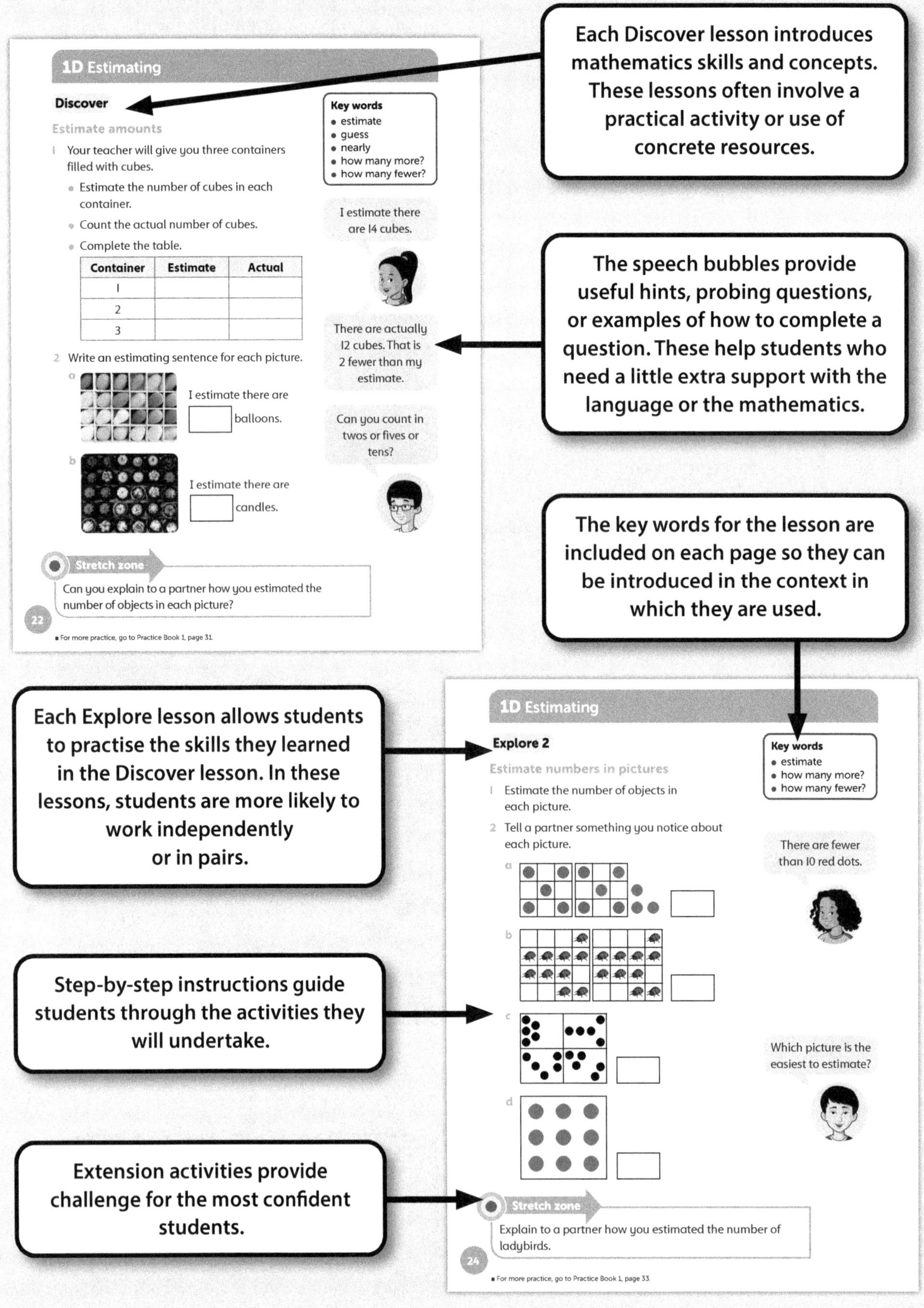



Each Discover lesson introduces mathematics skills and concepts. These lessons often involve a practical activity or use of concrete resources.

The speech bubbles provide useful hints, probing questions, or examples of how to complete a question. These help students who need a little extra support with the language or the mathematics.

The key words for the lesson are included on each page so they can be introduced in the context in which they are used.

Each Explore lesson allows students to practise the skills they learned in the Discover lesson. In these lessons, students are more likely to work independently or in pairs.

Step-by-step instructions guide students through the activities they will undertake.

Extension activities provide challenge for the most confident students.

# 1 Numbers and counting

## Connect

### Make a number poster

Work as a group.

1 Collect some magazines. Talk about which magazines might have numbers in them. What do the numbers tell us?

2 Cut out pictures that have numbers.

3 Make a poster to display in class.

4 Talk in your group about the numbers you have found.

**Stretch zone**

Take photographs of numbers on the way home from school. What job are the numbers doing? Explain your ideas to a partner.

We use numbers to count or to say how many of something there are.



What is the biggest number on your poster?

What is the smallest number on your poster?

## 1 Numbers and counting

### Review

1  Draw the beads and write the numbers in the spaces.

| Beads | Numbers | Words |
|---|---|---|
|  | 5 |  |
|  |  | sixteen |
|  |  |  |
|  |  | three |
|  |  |  |
|  | 12 |  |
|  |  | nineteen |
|  | 1 |  |
|  |  | four |
|  | 14 |  |
|  |  | twenty |

2  Samir has a bracelet with 19 beads. Lina's bracelet has one more bead than Samir's. How many beads are on Lina's bracelet?

Celine's bracelet has 10 more beads than Lina's. How many beads are on Celine's bracelet?

26

A word problem is always included on the Review page.

Students' progress is assessed through the questions and tasks at the end of each unit. In Student Books 2 and 6, these questions reflect the style of the SATs (national Standard Assessment Tests).

# Practice Book Discover and Explore

Practice Book activities can be completed in the school lesson or as homework.

Most of the Discover and Explore lessons in the Student Book have a corresponding page in the Practice Book. These activities provide opportunities for students to consolidate and deepen their learning.

Step-by-step instructions guide students through the activities they will undertake.

**1D Estimating**

Discover     Student Book 1, page 22

- four different sorts of small objects that you can hold in your hands

How many of each object do you think you can hold?

Write your estimate and then take a handful to find out.

Was your estimate more or less than the actual number? How many more or less?

An example is shown in the table.

| Object | Estimate | Actual number | More or less |
|---|---|---|---|
| cherries | 11 | 8 | My estimate was 3 more. |
|  |  |  |  |
|  |  |  |  |
|  |  |  |  |
|  |  |  |  |

**Stretch zone**

Did your estimates get better each time? _______________

If they did get better, can you explain why?

_______________________________________________



31

If students require concrete resources, these are listed in a box at the top of the page. This is particularly useful if students are completing the activities at home.

**1D Estimating**

Explore 2     Student Book 1, page 24

Estimate the number of dots. Do not count them!

Draw a circle around the number that you think is a good estimate.

1    2   5   9

4    6   9   12

2    9   12   14

5    10   13   16

3    3   4   6

6    5   8   10

**Stretch zone**

How many sweets do you think there are in this jar? [ ]

How did you make your estimate?

_______________________________________________

_______________________________________________



33

Extension activities provide challenge for the most confident students.

## 1 Numbers and counting

### Review

 1 Draw a face next to each bubble to show how you feel about your learning.

counting objects

reading and writing numbers

counting in twos, fives and tens

estimating quantities

 2 Tell a partner about one thing you did really well in this unit.

3 Draw or write about things you found easy, challenging or really hard.

What work did you feel confident doing?

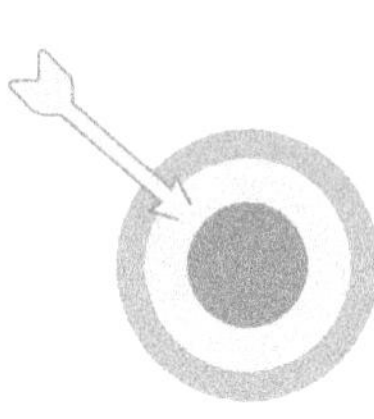

What work was challenging?

Is there any work you might need some extra help with?

34

## The Student Books

The Student Books are write-in textbooks for students to read and use. There are six Student Books: one for each school year at primary school. The Student Books introduce learning through a mixture of practical, discussion and independent activities.

| Student Book | Typical student age range |
| --- | --- |
| Student Book 1 | Age 5–6 |
| Student Book 2 | Age 6–7 |
| Student Book 3 | Age 7–8 |
| Student Book 4 | Age 8–9 |
| Student Book 5 | Age 9–10 |
| Student Book 6 | Age 10–11 |

## The Practice Books

The Practice Books are write-in workbooks for students to read and use. There are six Practice Books: one for each school year at primary school. The Practice Books provide deeper learning opportunities through a range of independent activities, which can be completed in school or at home.

| Practice Book | Typical student age range |
| --- | --- |
| Practice Book 1 | Age 5–6 |
| Practice Book 2 | Age 6–7 |
| Practice Book 3 | Age 7–8 |
| Practice Book 4 | Age 8–9 |
| Practice Book 5 | Age 9–10 |
| Practice Book 6 | Age 10–11 |

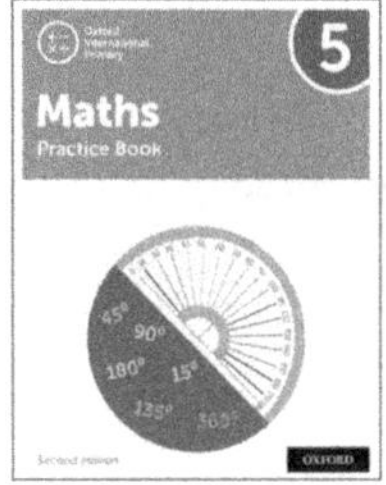

## The Teacher's Guides

There are six Teacher's Guides: one for each school year at primary school. Each Teacher's Guide includes:

- an introduction with advice about delivering mathematics in primary schools using *Oxford International Primary Mathematics*
- a unit overview, giving advice on teaching each unit, including common misconceptions and how to deal with them
- a lesson plan for every lesson in the Student Book and corresponding pages in the Practice Book
- model answers to each question in the Student Book and Practice Book.

## Digital resources
### Interactive eBooks

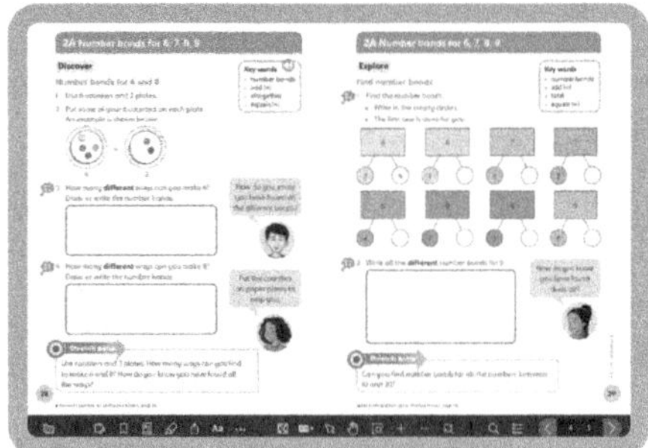

*For the teacher*

Teachers can access the Student Books, Practice Books and Teacher's Guides online in eBook format, on the Oxford Owl website (www.oxfordowl.co.uk).

The enhanced eBooks show the course content on screen, making it easier for teachers to deliver engaging lessons.

*For the students*

Teachers can allocate an eBook version of the Student Books to students for use at home. The Student eBooks include interactive activities, worksheets and audio of all the key vocabulary,

### Assessment resources

The downloadable assessment materials offer you additional opportunities to assess students' progress. The materials include:

- end-of-unit summative assessment
- end-of-year summative assessment.

Every test comes with everything you need to assess and record progress including:

- answers
- mark schemes and guidance on assessment.

### *Oxford Primary Illustrated Maths Dictionary*

The *Oxford Primary Illustrated Maths Dictionary* gives comprehensive coverage of the key maths terminology students use in the course. Entries are in alphabetical order, and each includes a clear and straightforward definition along with a fun and informative colour illustration or diagram to help explain the meaning. The dictionary is suitable for students with English as an Additional Language.

### The curriculum

The Oxford International Curriculum offers a new approach to teaching and learning focused on wellbeing, which places joy at the heart of the curriculum and develops the global skills students need for their future academic, personal and career success.

Through six subjects – English, Maths, Science, Computing, Wellbeing and Global Skills Projects – the Oxford International Curriculum offers a coherent and holistic approach to ensure continuity and progression across every student's educational journey, equipping them with the skills to shape their own future. Through this approach, we can help your students discover the joy of learning and develop the global skills they need to thrive in a changing world.

# 1 Number and place value



## Big idea

The two Big Ideas for this unit are place value for numbers up to a million, and negative numbers.

In the decimal number system, the value of a digit depends on its place, or position, in the number. It is important to know that each place has a value 10 times the value of the place to its right and that this extends indefinitely. Students practise ordering and rounding numbers in order to consolidate their understanding of place value in large numbers. They look at large numbers in contexts, knowing when numbers can be rounded or when they need to be exact.

Clear mental images of the number system help students to understand negative numbers. The use of number lines and empty number lines, as well as real-life examples such as thermometers, enables students to compare and calculate with negative numbers.

## Look out for

- **Students who consider all the digits in a number when** rounding to the nearest 100 000. Give students lots of opportunities to round large numbers by looking at the digit immediately to the right of the value being rounded. For example, rounding 237 428 to the nearest 100 000 does not make use of the 7000 or the 8 ones; only the 30 000 is used to determine whether to round to 200 000 or 300 000.
- **Students who say that to multiply by powers of 10 you 'add zeros'.** This strategy gives the correct answer for the multiplication of integers (for example $257 \times 10 = 2570$). However, this strategy gives incorrect answers for the multiplication of decimals (for example, $24.31 \times 100$ does not equal 24.3100). The activities in this unit show students that, when multiplying or dividing by 10 or 100, the digits change position on the place-value grid.

## Possible misconceptions

- **Students think that, for example, $^-9 > {}^-4$.** Use number lines to help students place negative numbers and see how larger digits mean a smaller number when negatives are used.
- **Students think that you cannot subtract a larger number from a smaller number.** For the 'counting on' and 'counting back' activities in this unit, students need to count back across zero. For example, students may think that $25 - 75$ 'doesn't work', not that $25 - 75 = {}^-50$. It is important to use a number line with negative numbers to support these activities.

## Key vocabulary

- ten thousand, hundred thousand, million
- 4-, 5-, 6-, 7-digit number
- number system
- > greater than, more than, larger than, bigger than
- < less than, fewer than, smaller than
- ≥ greater than or equal to
- ≤ less than or equal to
- power of 10, multiple of 10
- half-way between, estimate
- about the same as, is approximately equal to, just over, just under, round to the nearest
- negative number, positive number, place value, place-value grid, number pairs, known facts, derive
- temperature, below freezing

| Learning objective | E | 1A | 1B | 1C | 1D | 1E | C | R |
|---|---|---|---|---|---|---|---|---|
| Read, write, order and compare numbers up to 10 000 000 and determine the value of each digit. | ✓ | ✓ | ✓ | | ✓ | ✓ | ✓ | ✓ |
| Round any whole number to a required degree of accuracy. | ✓ | | ✓ | | | | ✓ | ✓ |
| Use negative numbers in context, and calculate intervals across zero. | | | | ✓ | ✓ | | | ✓ |
| Solve number and practical problems that involve all of the above. | ✓ | | ✓ | ✓ | ✓ | ✓ | ✓ | ✓ |

# 1 Number and place value

## Big question

- How can I understand the value of very large numbers? How can I use large numbers and negative numbers?

## Global skills

- **Creative skills:** investigating
- **Interpersonal skills:** teamwork
- **Self-development skills:** reflecting on learning

## Key vocabulary

- ten thousand, hundred thousand, million

## Resources

- metre rules and measuring tapes
- cuboid containers of various sizes
- dried chickpeas or similar
- calculators

## Language support

Listen to how students say large numbers. Encourage them to say the number in full and not to say each digit separately. For example: 876 426 is 'eight hundred and seventy-six thousand, four hundred and twenty-six' not 'eight seven six four two six'. Model this in your discussions with students.

### Introductory activity

*How long would it take you to count to one* **million**? Some students may assume that you can count each number in one second and so then calculate how long one million seconds is in days and hours. To challenge this assumption, ask, *How long does it take you to say a number like 876 426 aloud?* First, model how to say this number aloud: *eight hundred and seventy-six thousand, four hundred and twenty-six.* Agree that, for smaller numbers, one second per number or one second per two numbers might be a good estimate, but larger numbers, such as 876 426, may take more than one second to say. *Do you want to revise your estimate for how long it would take to count to 1 million?* Allow students time to discuss this in pairs and share their feedback.

### Main activity

Look together at page 6 of the Student Book. Display on the IWB, if possible. The class should work in mixed-attainment groups of four to six students. *How many chickpeas would fill the classroom?* After their suggestions, ask, *How could you calculate this?* Give students time in their groups to decide how to solve this problem. You can simplify the problem by using a much smaller container and asking, *How many chickpeas would fill this container?* Then ask, *How many small containers would fill a 1-cubic metre container?* As a class, you can then calculate the size of the classroom in cubic metres to solve the problem.

### Differentiation

**Supporting:** Encourage students to take an active role in the discussions and contribute to the group activity.

**Consolidating:** Ask students to support one another in deciding on appropriate calculations and to explain the strategies they are using.

**Extending:** Ask students to check the accuracy of their answers and to justify them.

### Reflection time

Ask groups to share their solutions with the whole class. Each group should tell the presenting group two things that they like in the solution ('We like …') and one target for future problem-solving activities ('Next time, you could …').

You could set further 'large number' challenges for students to work on. Ask, for example, *How many times would the total number of students in this school fit into the National Stadium?* or *How long would it take to walk from your school to the capital city (or a capital city of a neighbouring country)?*

# 1A Place value



## Specific learning focus

- Know what each digit represents in whole numbers up to ten million.

## Global skills

- **Creative skills:** exploring
- **Interpersonal skills:** communication

## Key vocabulary

- hundred thousand, million, ten million

## Resources

- large digit cards 0–9 and sets for each pair of students
- mini whiteboards and markers

## Language support

Listen to each student speaking the numbers aloud. Ask, for example:

- *How many digits does this number have?*
- *What do you know about a 6-/7-digit number?*

Use the vocabulary of place value: ten thousand, hundred thousand, million, ten million and so on.

 Introductory activity

Ask seven students to come to the front of the classroom. Ask each student to take a digit from the large digit cards you have prepared. Ask these students to make a line facing the rest of the class. They should make this line one student at a time. As each student joins the line, ask the rest of the class: *What number is this?* (or ask individual students).

- One student with the number 7 – the class say 'seven'.
- Two students with the number 3 and the number 7 – the class say 'thirty-seven'.
- Three students with the numbers 5, 3 and 7 – the class say 'five hundred and thirty-seven'.
- Four students with the numbers 6, 5, 3 and 7 – the class say 'six thousand, five hundred and thirty-seven' and so on.

Repeat this as many times as you think is necessary.

 Main activity

Give each pair a set of digit cards 0–9. Ask them to take turns choosing a digit and they both write it down to form a 7-digit number, for example 5 438 261. One student then points to a digit in the number and the other has to say the value of that digit. For example, the 3 has a value of 30 000. They can take turns until all digits have been chosen. Repeat with a new 7-digit number made from the digit cards.

Now ask students to work in pairs to complete the activities on page 7 of the Student Book. While students work, ask them to say the numbers aloud to you so that you can check their pronunciation. Ask questions to check students' understanding, for example, *How did you decide that this was the biggest possible number? Are you sure that you can't find a smaller number than that?*

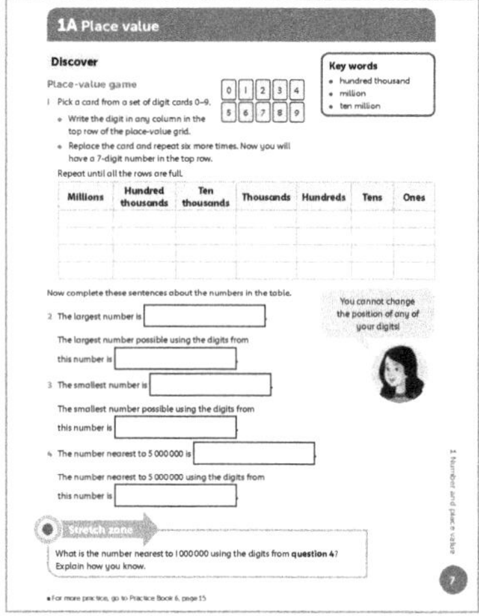

## Differentiation

**Supporting:** Ask students to show you the largest and smallest numbers and explain their reasoning.

**Consolidating:** Ask students to justify their responses to the questions on Student Book page 7.

**Extending:** Challenge students to create 8-digit numbers and to order these numbers smallest to greatest and vice versa.

**Stretch zone:** What is the number nearest to 1 000 000 using the digits from question 4? Explain how you know.

Ask questions to check students' understanding of place value in this number. *Are you sure that you can't swap any of the digits to make the number closer to 1 000 000? Convince me.*

As an additional activity, you could give students some newspapers and magazines and ask them to find examples of large numbers and decimals. Then ask them to write the numbers in words.

## Reflection time

Give out mini whiteboards and markers. Read out a series of 5-, 6- and 7-digit numbers and ask students to write the numbers on their whiteboards. Give students five seconds to write the numbers. Then say, *Show me your numbers.* This helps you to assess individual understanding. Include some numbers with zeros as placeholders, for example 'three hundred and two thousand, one hundred and seven' (302 107).

**Practice Book:** Students complete Practice Book page 15. They can do this directly after the Main activity, as homework, or as the focus of a separate mathematics session to help students consolidate their learning and build fluency.

Students write large numbers (all about the year 2019) in a place-value grid and say each number aloud. Check students' written work, making sure that numbers that contained zeros are written correctly. *If that zero wasn't there, what would that number be? How do you know?*

**Practice Book page 15**

|  | Millions | Hundred thousands | Ten thousands | Thousands | Hundreds | Tens | Ones |
|---|---|---|---|---|---|---|---|
| (Provided) | 5 | 1 | 9 | 5 | 5 | 3 | 4 |
| 1 | 1 | 5 | 5 | 7 | 6 | 9 | 2 |
| 2 | 5 | 5 | 9 | 5 | 7 | 5 | 0 |
| 3 | 5 | 0 | 1 | 9 | 2 | 3 | 1 |
| 4 |  | 3 | 4 | 7 | 2 | 2 | 2 |
| 5 |  |  | 4 | 0 | 0 | 7 | 5 |
| 6 | 7 | 1 | 9 | 9 | 7 | 2 | 9 |

Stretch zone: Check that students have written a suitable fact in millions.

| Differentiated outcomes | |
|---|---|
| **All students** | should recognise and order numbers up to seven digits with support. |
| **Most students** | will recognise and order numbers closest to a given number of millions. |
| **Some students** | may extend to beyond seven digits and be able to explain their strategy. |

## Answers

### Student Book page 7

Answers will vary because students use digit cards to make their own numbers. Check that their answers to the questions are correct for the numbers they have made. Note who has a good understanding of the place value of large numbers and the size of numbers, and who requires more practice in working with these numbers.

# 1A Place value

## Explore  Student Book page 8 • Practice Book page 16

### Specific learning focus

- Multiply and divide any whole number from 1 to 10 000 by 10, 100 or 1000 and explain the effect.

### Global skills

- **Creative skills:** investigating

### Key vocabulary

- place-value grid, power of 10, multiple of 10, tens of thousands, hundreds of thousands

### Resources

- large digit cards 1–9 and sets for each pair of students
- calculators
- mini whiteboards and markers

## Language support

Encourage students to explain their strategies and use the vocabulary of place value: ten thousand, hundred thousand, million. Ask, for example:

- *How many digits does this number have?*
- *What do you know about a 6-digit number?*

 Introductory activity

Draw six boxes on the board. Ask students to copy these six boxes onto their whiteboards.

Choose six digit cards, one at a time, from the large set you have prepared. After each card is chosen, ask students to write the digit in one of their boxes so that, eventually, all six digits have been written in a box. *Make the largest number possible with these digits. Now make the smallest number possible with these digits.* Repeat this with a different set of six digits.

Choose six digits from the large set of cards and make a 6-digit number. Ask students to count in thousands from the number you have made until they cross the ten thousands boundary into the next thousand. *Which digits change each time, and which stay the same?* Do this for counting in **tens of thousands** and **hundreds of thousands**, each time asking students to notice what is happening to each digit in the number. Repeat this process several times and then increase the number of digits so you are counting using 7-digit numbers.

Ask students to work in pairs on the activities on page 8 of the Student Book. Tell them to explain to one another what they notice ('When you multiply by 10/100/1000 …').

Encourage students to explain what they notice very carefully. If any students say to you that 'When you multiply by 10, you add a zero', ask them to multiply 123.4 by 10, as this will convince them that this is not always true: 123.4 × 10 = 1234 and not 123.40.

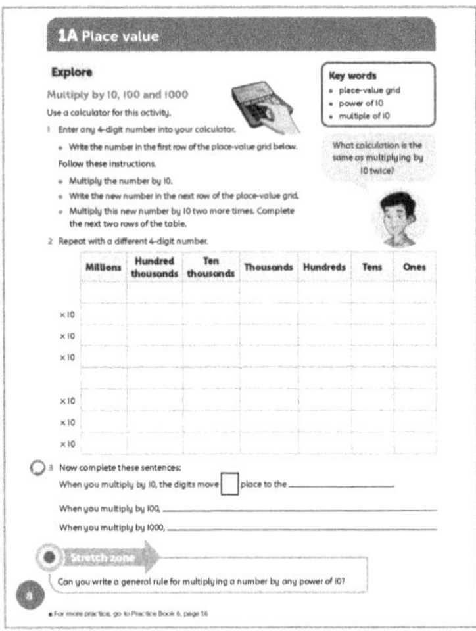

## Differentiation

**Supporting:** Ask students to describe the patterns they notice.

**Consolidating:** Ask students for a general rule for multiplying and dividing by 10, 100 and 1000.

**Extending:** Ask students for a general rule for multiplying and dividing by **powers of 10**.

**Stretch zone:** *Can you write a general rule for multiplying a number by any power of 10?*

Students should recognise the place value aspect of moving digits to the left by the same number of places as there are zeros on the power of 10. For example, for multiplying by 10, the digits move one place to the left; for multiplying by 100, they move two places to the left and so on. Students may also recognise that the reverse is true when dividing by powers of 10, with digits moving to the right the same number of places as there are zeros.



Select one or two students who have a good understanding of place value. Choose students who explained their ideas well in their pairs. Ask them to share their explanation with the whole class. Ask the class, *Can you give me an example that shows this statement is incorrect?*

'When you multiply by 100, you add two zeros.'

**Practice Book:** Students complete Practice Book page 16. They can do this directly after the Main activity, as homework, or as the focus of a separate mathematics session to help students consolidate their learning and build fluency.

This activity will help to consolidate the vocabulary of place value that students have already learned. Encourage students to write a variety of facts, not just to reuse the same ones for each number.

| Differentiated outcomes | |
| --- | --- |
| **All students** | should carry out the calculations using a calculator and notice the patterns. |
| **Most students** | will understand how place value effects the calculations. |
| **Some students** | will generalise beyond multiplying and dividing by 10, 100 and 1000. |

## Answers

### 1, 2 Student Book page 8

Students choose the numbers they will use. Check that their multiplications are correct and that they have written the digits into the correct places in the table.

**3** When you multiply by 10, the digits move one place to the left.

When you multiply by 100, the digits move two places to the left.

When you multiply by 1000, the digits move three places to the left.

### Practice Book page 16

Students will have different facts about each number, similar to the examples given. Check that their facts are correct and varied.

Stretch zone:

**1** 4 000 000

**2** 7 700 000

**3** 600 000

**4** 8 900 000

# 1B Rounding

## Discover
Student Book page 9 • Practice Book page 17

### Specific learning focus
- Round whole numbers to the nearest 10, 100 or 1000.

### Global skills
- **Creative skills:** exploring

### Key vocabulary
- round to the nearest, 4-digit number

### Resources
- large digit cards 0–9 and sets for each pair of students
- mini whiteboards and markers

### Language support

Model key vocabulary. Ask, for example:
- *What is that number approximately equal to?*
- *Which digit do you need to look at?*
- *What is that number rounded up/down to the nearest 10?*
- *What is that number rounded up/down to the nearest 100?*
- *What is that number rounded up/down to the nearest 1000?*

### Introductory activity

Ask four students to come to the front of the class. Ask each of these students to pick one of the large digit cards 0–9 and ask them to stand in a line, facing the class. Ask the rest of the class, in pairs, to work out how many possible arrangements of these four digits there are. As a class, rearrange the four students to form each number in order, smallest to largest. Encourage students to work systematically to check that they have all possible arrangements of the numbers.

### Main activity

Choose one of the 4-digit numbers from the Introductory activity and model, using an empty number line, how to round the number to the nearest 10. In the following example 7254 is used. *Which digit do we need to look at to* **round this to the nearest** *10?* Agree that it is the ones digits that determines whether to round up or down. Use the empty number line to show that 7254 comes between 7250 and 7260. Draw it on the number line. *Is 7254 nearer to 7250 or 7260?*

Repeat, rounding the same number to the nearest 100 and 1000, first agreeing which digit to look at and then modelling on a number line where it will come. For example, for the nearest 100, write 7254 between 7200 and 7300, just slightly closer to 7300; for the nearest 1000, write 7254 between 7000 and 8000 but much closer to 7000 on the line.

Repeat this process several times for different 4-digit numbers.

Look together at page 9 of the Student Book. Display on the IWB, if possible. Ask students to work in pairs and to take it in turns to create a number and then to round it. Their partner should check their answers. Look together at the worked example to show how four digits can be arranged and then how each number can be rounded. Ask individual students, *What is the 'rule' for rounding to the nearest 10, 100 and 1000?* They may spot the third speech bubble on page 9. Note which students have the clearest descriptions of their rules. These students can support you in Reflection time.

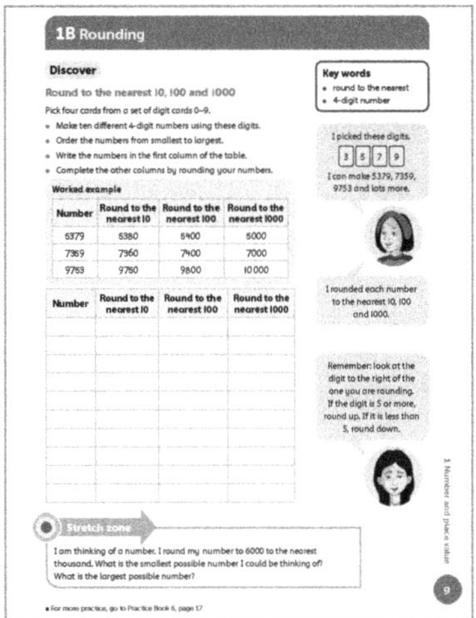

### Differentiation

**Supporting:** Model how to use a number line to support rounding.

**Consolidating:** Ask students to show you how they use a number line to support them.

**Extending:** Ask students to explain the rule for rounding.

**Stretch zone:** *I am thinking of a number. I round my number to 6000 to the nearest thousand. What is the smallest possible number I could be thinking of? What is the largest possible number?*

Check that students are using the agreed class rules to work out what the smallest and largest numbers are. The smallest number is 5500 and the largest number is 6499.

As a follow-up activity, ask students to find examples of large numbers in newspapers or magazines and round them to the nearest thousand. Alternatively, if the numbers are already rounded, ask, *What is the largest and smallest possible number before they were rounded?*

**Reflection time**

Ask two students to work together to explain to the rest of the class what the agreed rule is for rounding numbers to the nearest 10, 100 and 1000. Then use the large digit cards to generate a 4-digit number and write this on the board. Ask all students to round this number to the nearest 10. They should write their responses on their whiteboards. Wait five seconds and ask all students to show you their whiteboards at the same time. Repeat, asking students to round to the nearest 100 and then 1000. Check that everyone agrees on the correct number each time, and ask individual students to explain how they know they are correct.

**Practice Book:** Students complete Practice Book page 17. They can do this directly after the Main activity, as homework, or as the focus of a separate mathematics session to help students consolidate their learning and build fluency.

Students use an empty number line to round a 4-digit number to the nearest 1000. They must first label the number line with the start and end numbers and then mark the 4-digit number in the correct position on the number line, before rounding it.

| Differentiated outcomes | |
| --- | --- |
| **All students** | should round numbers to the nearest 10, 100 and 1000 with support. |
| **Most students** | will round numbers to the nearest 10, 100 and 1000, using a number line. |
| **Some students** | may clearly explain the rule for rounding and apply it to rounding any number. |

## Answers

### Student Book page 9

Students use digit cards to make the numbers they round. Check that the numbers have been rounded correctly to the nearest 10, 100 and 1000. While students are working, ask them to explain the rules they use when rounding numbers.

### Practice Book page 17

Students make their own numbers. Check that students have written their numbers in appropriate places on the number lines and that they have rounded the numbers correctly.

Stretch zone: The largest possible number is 294 499, the smallest possible number is 293 500.

# 1B Rounding

**Explore** Student Book pages 10–12 • Practice Book page 18

## Specific learning focus

- Round whole numbers to the nearest 10.
- Estimate where 4-digit numbers lie on an empty line between nearest thousands.

## Global skills

- **Creative skills:** investigating

## Key vocabulary

- round to the nearest, 4-digit number

## Resources

- large digit cards 1–20

## Language support

Focus on the language of place value by using the terms 'thousand', 'ten thousand' and '4-digit number'. Ask, for example:

- *What are you rounding to?*
- *What is that number to the nearest 10?*
- *What value is in the middle of the empty number line?*

 **Introductory activity**

Use the wall at the front of the classroom. Stick a 0 digit card onto the left-hand side of the wall (as students look at it) and stick a 20 on the right-hand side. Explain that these numbers form two ends of what represents a number line. Ask four students to come to the front of the class. Each of these students chooses a number card and then stands, at the front of the classroom, in ascending order between the 0 and the 20 so that they are in the correct place on this number line.

Now ask these four students to select another digit card each between 0 and 9 and to use these four digits to form one 4-digit number. Explain that students are going to place this 4-digit number on a number line between its two closest thousands. *Which thousand will be at the lower end of the number line? Which thousand will be at the upper end?* For example, for the number 5378 the lower end is 5000 and the upper end is 6000. Write a label, and stick it to the wall, for 5000 at one side at the front of the classroom and for 6000 at the other side. Now ask the class to place the 4-digit number as accurately as they can on this number line.

Repeat this as many times as you think necessary.

## Main activity

Look together at pages 10–12 of the Student Book. Display on the IWB, if possible. For the first part of the activity, on page 10, ask individual students: *What are all the possible numbers with your four digits? How do you know you have found all the possibilities?*

For question 5 on page 11, go through the worked example to show students how to label a number line, depending on the numbers involved. Ask students to complete all the activities on pages 10–12 of the Student Book. You may choose to group less-confident students together so that you can support them in using empty number lines.

Ask groups to write down two facts about the English Premier Football League attendances based on the table on page 12. Keep hold of these facts to share in Reflection time as a whole class.

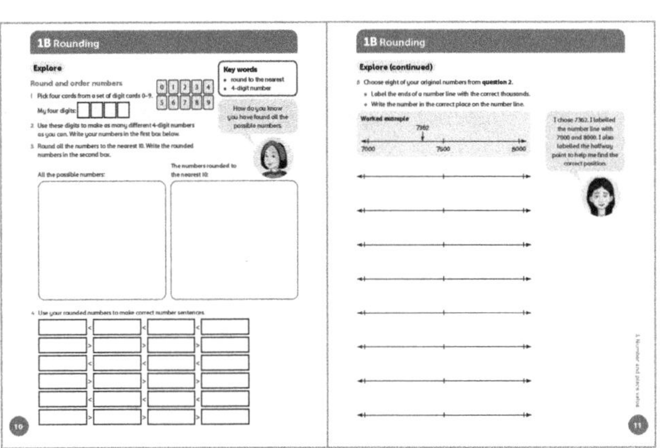

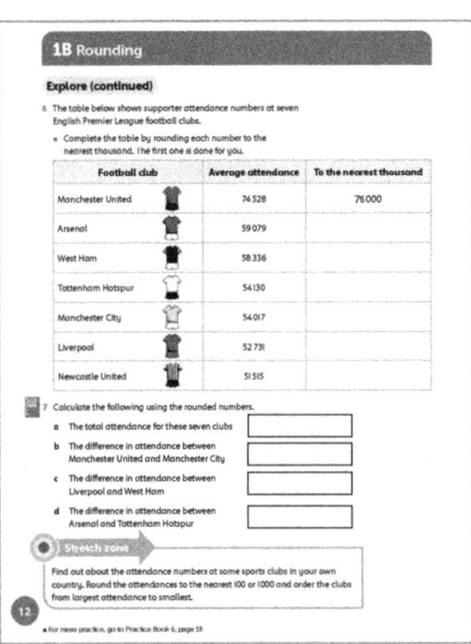

## Differentiation

**Supporting:** Model how to use a number line to support rounding.

**Consolidating:** Ask students to show you how they use a number line to support them.

**Extending:** Ask students to explain the rules for rounding.

**Stretch zone:** *Find out about the attendance numbers at some sports clubs in your own country. Round the attendances to the nearest 100 or 1000 and order the clubs from largest attendance to smallest.*

Ensure that students round the numbers correctly before ordering the clubs' attendances in descending order.

## Reflection time

Ask, *How did you know how to order your 4-digit numbers? How did you work out where to place the number on the empty number line?* It is important to share a range of strategies with the class.

Share students' facts about the English Premier League attendance numbers. *Can anyone think of a new word problem based on the rounded numbers?*

**Practice Book:** Students complete Practice Book page 18. They can do this directly after the Main activity, as homework, or as the focus of a separate mathematics session to help students consolidate their learning and build fluency.

Students generate ten 6-digit numbers and then round each number to the nearest 10 000 and nearest 100 000. They then find the difference between the two rounded numbers. *In real-life situations, can you give me an example of when you might round numbers to the nearest 10 000? What about the nearest 100 000?*

| Differentiated outcomes | |
|---|---|
| **All students** | should round numbers to the nearest 10, 100 and 1000 with support. |
| **Most students** | will round numbers to the nearest 10, 100 and 1000, using a number line. |
| **Some students** | may clearly explain the rule for rounding and apply it to rounding any number. |

## Answers

### Student Book pages 10–12

**1–4** Answers will vary according to the numbers that students make from their digit cards. While students are working, ask them to explain how they decided how to round their numbers.

Check that the rounded numbers are correctly ordered to make number sentences.

**Unit 1 Number and place value**  9

**5** Check that the numbers are correctly marked and labelled on the number lines.

**6** 

| | |
|---|---|
| Arsenal | 59 000 |
| West Ham | 58 000 |
| Tottenham Hotspur | 54 000 |
| Manchester City | 54 000 |
| Liverpool | 53 000 |
| Newcastle United | 52 000 |

**7 a** 405 000

  **b** 21 000

  **c** 5000

  **d** 5000

**Practice Book page 18**

**1** 465 978, 496 587, 856 497, 794 658, 945 678, 974 586, 579 684, 596 487, 849 567 for example.

**2** Numbers in the table are based on the examples in question 1.

| Number | Number rounded to the nearest 10 000 | Number rounded to the nearest 100 000 | Difference |
|---|---|---|---|
| 465 978 | 470 000 | 500 000 | 30 000 |
| 496 587 | 500 000 | 500 000 | 0 |
| 856 497 | 860 000 | 900 000 | 40 000 |
| 794 658 | 790 000 | 800 000 | 10 000 |
| 945 678 | 950 000 | 1 000 000 | 50 000 |
| 974 586 | 970 000 | 1 000 000 | 30 000 |
| 579 684 | 580 000 | 600 000 | 20 000 |
| 596 487 | 600 000 | 600 000 | 0 |
| 849 567 | 850 000 | 900 000 | 50 000 |

Stretch zone: Answers will vary.

# 1C Using negative numbers

## Discover
Student Book pages 13–14 • Practice Book page 19

### Specific learning focus
- Find the difference between a positive and negative integer, and between two negative integers in a context such as temperature or on a number line.

### Global skills
- **Creative skills:** problem solving
- **Real-world skills:** interpreting information

### Key vocabulary
- temperature, below freezing, negative number

### Resources
- A4 cards labelled $^-5$, $^-6$, 0, $^+9$, $^+14$
- mini whiteboards and markers

### Language support

Use the correct vocabulary for numbers in general use, for example, 'positive 9', 'negative 15'. Ask, for example, *Is the number greater than or less than zero?*

Link this to vocabulary such as 'above/below zero (freezing)' for temperature.

## Introductory activity

Ask five students to come to the front of the class. Ask them to pick one card each from the cards you have prepared. Ask students to stand in a line so that they create a reasonably accurate number line. Then ask four other students to come to the front one at a time. Ask each student to write a number on their whiteboard and to stand in the correct position on the number line between two of the students already in the line.

## Main activity

Write the following table on the board:

| Place | London | Moscow | Oslo | Bangkok | Dubai |
|---|---|---|---|---|---|
| Average temperature in May | 17 °C | $^-4$ °C | $^-8$ °C | 35 °C | 42 °C |

Ask pairs to order the temperatures from coldest to warmest. When you take feedback, model the correct way to say the temperatures (for example either 'twenty degrees below freezing' or 'minus twenty degrees').

Ask a student to choose two cities from the table. *Which is the warmest place of the two and by how many degrees?* Students can use the scales on the thermometers or empty number lines to help them work out the temperature differences. Choose three other students to choose different pairs of cities and calculate those cities' temperature differences.

Ask students to complete the activity on pages 13–14 of the Student Book in pairs so that they can discuss and check one another's answers. If you have access to the internet, students can research temperatures in cities around the world. If this is not possible, use atlases that contain this information.

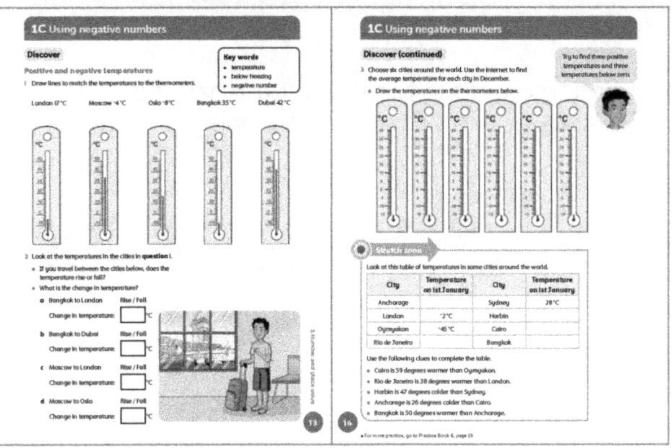

<table>
<tr><td colspan="2">Differentiated outcomes</td></tr>
<tr><td>All students</td><td>should read positive and negative temperatures accurately with support.</td></tr>
<tr><td>Most students</td><td>will find differences between positive and negative temperatures.</td></tr>
<tr><td>Some students</td><td>may find differences between positive and negative temperatures and write their own questions based on temperatures.</td></tr>
</table>

## Differentiation

**Supporting:** Ask students to focus on question 1. Use the thermometers to support students in calculating differences by counting on or counting back.

**Consolidating:** Encourage students to explain their methods for finding differences.

**Extending:** Challenge students to create more complex problems based on temperatures.

**Stretch zone:** *Look at this table of temperatures in some cities around the world.*

The completed table is as follows:

| City | Temperature on 1 January | City | Temperature on 1 January |
|---|---|---|---|
| Anchorage | ⁻12 °C | Sydney | 28 °C |
| London | ⁻2 °C | Harbin | ⁻19 °C |
| Oymyakon | ⁻45 °C | Cairo | 14 °C |
| Rio de Janeiro | 36 °C | Bangkok | 38 °C |

Students can use the information sheets or information from the internet to create their own table. They can then write clues and problems, like those in this Stretch zone, for their partner to solve.

 **Reflection time**

Copy the completed table from the Stretch zone onto the board. Ask students to work in pairs to write down three facts from the table. For example: 'It is 43 degrees warmer in London than Oymyakon.'

**Practice Book:** Students complete Practice Book page 19. They can do this directly after the Main activity, as homework, or as the focus of a separate mathematics session to help students consolidate their learning and build fluency.

Students need to interpret information from a table and then write positive and negative temperatures on thermometers (a vertical number line). *How is a thermometer the same and how is it different from a standard number line?*

## Answers

### Student Book pages 13–14

**1 a**

London 17 °C → thermometer 3

Moscow ⁻4 °C → thermometer 1

Oslo ⁻8 °C → thermometer 4

Bangkok 35 °C → thermometer 2

Dubai 42 °C → thermometer 5

**2 a** Falls by 18 degrees

**b** Rises by 7 degrees

**c** Rises by 21 degrees

**d** Falls by 4 degrees

**3** Students choose their own places and temperatures. Check they have labelled the thermometers correctly.

### Practice Book page 19

Check that students have marked the correct temperature on each thermometer.

Moscow: warmest = 21 °C, coldest = ⁻8 °C

Reykjavik: warmest = 9 °C, coldest = ⁻5 °C

North Pole: warmest = 13 °C, coldest = ⁻28 °C

Stretch zone: Students choose their own places for the table. Check that they have calculated the differences correctly.

# 1C Using negative numbers



## Specific learning focus

- Find the difference between a positive and negative integer, and between two negative integers, in a context such as temperature or on a number line.

## Global skills

- **Creative skills:** investigating
- **Real-world skills:** interpreting information

## Key vocabulary

- positive number, negative number

## Resources

- large A4 cards labelled $^-10$, $^-4$, $0$, $^+3$, $^+8$
- mini whiteboards and markers

## Language support

Use the correct vocabulary for numbers in general use, for example, 'positive 3', 'negative 12'. Ask, for example, *Is the number greater than or less than zero?*

Introduce the phrase 'directed numbers'. Link this to directions on the number line: for example, movement to the right is positive and to the left is negative.

 **Introductory activity**

Draw a number line from $^-10$ to $+10$ on the board. Ask five students to come to the front of the class. Tell them to pick one card each. Ask each student to place their number on the number line on the board.

*What is five more than negative four?*

Write $^-4 + (^+5) =$

*What is seven less than positive three?*

Write $^+3 - (^+7) =$

Remind students: *Move to the right on the number line to add and move to the left on the number line to subtract.*

Write $^-4 - (^-5) =$

Subtracting moves to the left but the negative sign reverses the direction so $^-4 - (^-5) = ^+1$ and we move to the right.

Write $^+4 - (^-7) =$

Subtracting moves to the left but the negative sign reverses the direction so $^+4 - (^-7) = ^+11$ and we move to the right.

 **Main activity**

Write the following sequences of calculations on the board:

| | |
|---|---|
| $5 - 0 =$ | $^-5 + 0 =$ |
| $5 - 1 =$ | $^-5 + 1 =$ |
| $5 - 2 =$ | $^-5 + 2 =$ |
| $5 - 3 =$ | $^-5 + 3 =$ |
| $5 - 4 =$ | $^-5 + 4 =$ |
| $5 - 5 =$ | $^-5 + 5 =$ |
| $5 - 6 =$ | $^-5 + 6 =$ |
| $5 - 7 =$ | $^-5 + 7 =$ |
| $5 - 8 =$ | $^-5 + 8 =$ |
| $5 - 9 =$ | $^-5 + 9 =$ |
| $5 - 10 =$ | $^-5 + 10 =$ |

Ask students to calculate the answers to these in sequence and discuss, with a partner, what they notice about the answers.

Look together at page 15 of the Student Book. Display on the IWB, if possible. Read through each of the statements as a class. Students work in mixed-attainment groups of three or four. Before completing the Student Book page, ask groups to make a poster, which should include: the Student Book statements, whether they are always, sometimes or never true, and some examples to show that their conclusion is correct. Then ask students to copy the answers into page 15 of the Student Book individually.

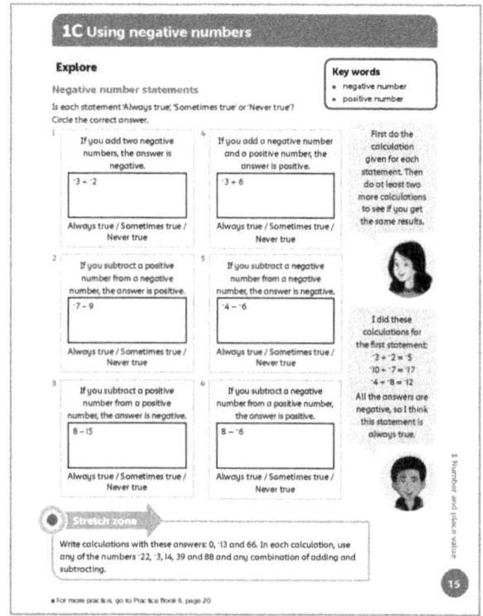

## Differentiation

**Supporting:** Model for students how to use the thermometer images as number lines.

**Consolidating:** Encourage students to explain how they calculated the changes in temperature.

**Extending:** Challenge students to make their own problems – the problems should use any of the positive and negative temperatures listed in the lesson.

**Stretch zone:** *Write calculations with these answers: 0, ⁻13 and 66. In each calculation, use any of the numbers ⁻22, ⁻3, 14, 39 and 88 and any combination of adding and subtracting.*

Some examples are: 88 − 22 = 66 (or 88 + ⁻22 = 66) or 39 − 14 + ⁻22 + ⁻3 = 0

Students can make up similar problems for a partner.

 **Reflection time**

Ask a group to present one of the statements that they know is always true.

Ask a group to present one of the statements that they know is sometimes true.

Ask a group to present one of the statements that they know is never true.

Encourage group discussion about why the general statements are true.

**Practice Book:** Students complete Practice Book page 20. They can do this directly after the Main activity, as homework, or as the focus of a separate mathematics session to help students consolidate their learning and build fluency.

Students need to interpret information from a table and then find the difference between positive and negative temperatures. Remind students that they can use a number line (horizontal or vertical) to help them find the differences.

| Differentiated outcomes | |
| --- | --- |
| **All students** | should find some examples of positive and negative additions and subtractions, with support from their peers. |
| **Most students** | will help to draw conclusions about positive and negative statements and find some examples to support their conclusions. |
| **Some students** | will help to draw conclusions about positive and negative statements, find some examples to support their conclusions and support their peers when working as a group. |

## Answers

### Student Book page 15

| | | | |
| --- | --- | --- | --- |
| **1** Always true | | **4** Sometimes true | |
| **2** Never true | | **5** Sometimes true | |
| **3** Sometimes true | | **6** Always true | |

### Practice Book page 20

| | | |
| --- | --- | --- |
| **1** 29 °C | **4** 17 °C | **7** 37 °C |
| **2** 14 °C | **5** 49 °C | **8** 21 °C |
| **3** 41 °C | **6** 26 °C | **9** 18 °C |

Stretch zone: Students' answers will vary, depending on where they live.

# 1D Comparing numbers

## Discover   Student Book page 16 • Practice Book page 21

### Specific learning focus

- Order and compare positive numbers to one million, and negative integers to an appropriate level.

### Global skills

- **Creative skills:** exploring

### Key vocabulary

- 7-digit number, hundred thousand, million, negative number

### Resources

- large digit cards 1–9 and sets for each pair of students
- mini whiteboards and markers

### Language support

Work with students who are less confident in English and model the pronunciation of the numbers in words. Alternatively, help write down the answers for students who can say the numbers but have difficulty in writing them down in English. When students generate negative numbers, remind them that we say 'negative thirteen' for ⁻13 and not 'minus thirteen', unless we are using temperatures.

 **Introductory activity**

Ask six students to come to the front of the class. Each student should pick a digit card and stand in the order in which they came to the front. Ask the rest of the class to turn to a partner and say the number that the students standing at the front have made. Check that all students can say the number aloud. For example, if the students had picked 7, then 2, then 5, then 3, then 8, then 1, the number formed would be 725 381 (seven hundred and twenty-five thousand, three hundred and eighty-one).

Ask students to talk to a partner and decide what is the largest number that can be formed with the digits (875 321 in the case of the digits chosen above). One of the students should rearrange the students at the front so that they form this number. Ask them to explain how they know that this is the largest possible number. (You place the largest digit in the largest place value column.) Repeat for the smallest number. If necessary, repeat the whole activity until understanding is secure.

 **Main activity**

Write on the board: 5 621 849. Ask students how they might say this number. You can ask them why there are gaps between some of the digits, and explain for those who didn't know that the gaps help them to say the number by showing where the millions and thousands end and so they know where to pause. This means that they can read the numbers as five million, six hundred and twenty-one thousand, eight hundred and forty-nine.

Now write the number in figures on a place-value grid. This will also show students the place-value headings, which help in saying numbers.

| M | HTh | TTh | Th | H | T | O |
|---|-----|-----|----|----|----|----|
| 5 | 6 | 2 | 1 | 8 | 4 | 9 |

Now write the following negative numbers on the board: ⁻27, ⁻81, ⁻94, ⁻36, ⁻50. Ask students whether they can say which number is the largest and why. Look for students who say ⁻94 is the largest because they think of 94 as being the largest of the 2-digit numbers.

Draw a number line from ⁻100 to 0 and ask students to indicate where each negative should be. Ask, *Which end of the number line has the larger numbers? So which number is the largest?* (⁻27) *Which is the smallest?* (⁻94).

In pairs, students complete the activity on page 16 of the Student Book. They use digit cards to generate 7-digit numbers and then order them. If they need support, remind them of the strategy for finding the largest number that was used in the Introductory activity. Challenge them to find the largest and smallest possible numbers with the digits they generate.

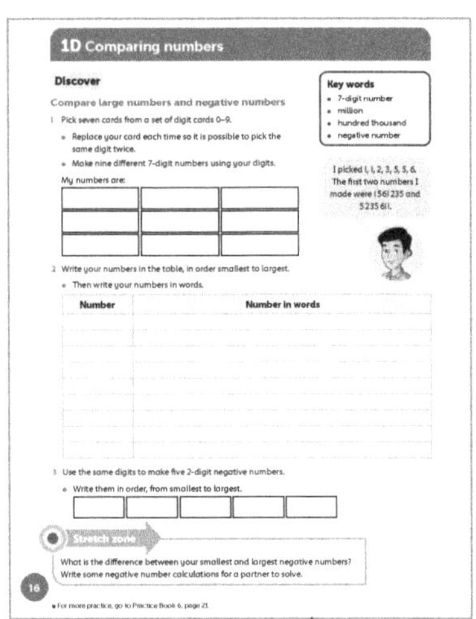

### Differentiation

**Supporting:** Help students to explain their reasoning when ordering 7-digit numbers.

**Consolidating:** Ask students to explain their reasoning when ordering 7-digit numbers and support them when ordering negative numbers.

**Extending:** Ask students to explain their reasoning when ordering 7-digit numbers and negative numbers.

**Stretch zone:** *What is the difference between your smallest and largest negative numbers? Write some negative number calculations for a partner to solve.*

Ask, *Can you write a real-life word problem using your negative numbers?* (You could give students some appropriate contexts such as negative temperatures or depth of sea below sea level.)

Students can follow up this activity by finding examples, in magazines or on the internet, of large numbers and negative numbers in everyday life, and then ordering them to make comparisons.

## Reflection time

Use digit cards to generate five negative numbers of your own and write them on the board. Ask students to order these from smallest to largest on their mini whiteboards. They should work in pairs to support discussion. Draw a 0 to ⁻20 number line on the board at the front of the class. One of the students should mark the numbers you have generated on this number line. Ask students what they notice about ordering negative numbers (the further away from zero the smaller the number – this is the inverse of positive numbers).

**Practice Book:** Students complete Practice Book page 21. They can do this directly after the Main activity, as homework, or as the focus of a separate mathematics session to help students consolidate their learning and build fluency.

Students order various basketball players' large salaries from smallest to largest amounts. They then round the number to the nearest 100 000 and also write the number as a decimal, for example $40 200 000 would be written as $40.2 million.

| Differentiated outcomes | |
|---|---|
| **All students** | should order 7-digit numbers and write them in words with support. |
| **Most students** | will order negative numbers and order and write 7-digit numbers. |
| **Some students** | may support other students in writing large numbers and ordering negative numbers. |

## Answers

### Student Book page 16

Answers will vary according to the numbers that students have made using the digit cards. Check that the numbers have been written correctly in order from smallest to largest and that they have been correctly written in words.

## Practice Book page 21

| | Player | Salary | Rounded to the nearest 100 000 | Written as '$xx.x million' |
|---|---|---|---|---|
| (Provided) | Stephen Curry | $40 231 758 | $40 200 000 | $40.2 million |
| 1 | LeBron James | $37 436 858 | $37 400 000 | $37.4 million |
| 2 | Mike Conley | $32 511 624 | $32 500 000 | $32.5 million |
| 3 | Damian Lillard | $29 802 321 | $29 800 000 | $29.8 million |
| 4 | Kevin Love | $28 942 830 | $28 900 000 | $28.9 million |
| 5 | DeMar DeRozan | $27 739 975 | $27 700 000 | $27.7 million |
| 6 | CJ McCollum | $27 556 959 | $27 600 000 | $27.6 million |
| 7 | Andre Drummond | $27 093 019 | $27 100 000 | $27.1 million |

Stretch zone: Check the calculation based on the research of a teacher's salary.

# 1D Comparing numbers

## Explore    Student Book page 17 • Practice Book page 22

### Specific learning focus

- Order and compare positive numbers to 1 million, and negative integers to an appropriate level.

### Global skills

- **Creative skills:** problem solving
- **Real-world skills:** interpreting information

### Key vocabulary

- 7-digit number, hundred thousand, million, negative number

### Resources

- set of temperature cards: ⁻68 °C, ⁻89 °C, ⁻40 °C, ⁻62 °C and ⁻7 °C and set of location cards containing the following information:
  - Oymyakon, Russia
  - Vostok Station, Antarctica
  - International Falls, Minnesota
  - Winnipeg, Canada
  - Espoo, Finland
- world map
- sticky notes
- mini whiteboards and markers

### Language support

Ask students to work in pairs to encourage discussion. If you can pair up experienced English speakers with those who are less experienced, this will support language development. Ask students to say the numbers they are forming aloud – model the pronunciation if necessary.

 Introductory activity

Ask five students to come to the front of the class. They each pick one of the location cards, find the place (or the country) on the map and put a sticky note on the map next to it, writing the name of the place on the sticky note. (You may need to help them in finding the places.) When all five places have been marked on the map, ask these students to remain at the front of the class and ask five more students to come and select a temperature card. Explain that these are the minimum temperatures recorded at the five locations marked on the map. The second set of students should pair up with the student who they think is holding the matching card so that all location cards are matched with a temperature card.

Pairs should then arrange themselves in order, showing the locations with the temperatures from coldest to warmest temperature. Once all the students are in a line, you can tell them the correct order – Vostok Station, Antarctica: ⁻89 °C, Oymyakon, Russia: ⁻68 °C, Winnipeg, Canada: ⁻62 °C, International Falls, Minnesota: ⁻40 °C, Espoo, Finland: ⁻7 °C. Ask the students with the temperature cards to move so they are showing the class the correct order, from coldest to warmest. As a class, read out the temperatures. Remember to say 'minus seven degrees Celsius' and so on.

 Main activity

Look together at page 17 of the Student Book. Display on the IWB, if possible. Ask students to look at the map and populations. Ask them, in pairs, to read out the populations of the places, making sure that they pronounce the numbers correctly. It will help them to say the numbers aloud to a partner. This helps them speak out their thinking and also allows a partner to check that they are correct.

*Who can tell me a place with a population closest to 500 000?* When they have had time to look at the populations, ask a student to give their answer and explain how they know. They should be encouraged to describe how they rounded the populations. Now ask for a population closest to 2 500 000 and again allow students to round and explain.

Students complete the activities on page 17, working in pairs and checking each other's answers.

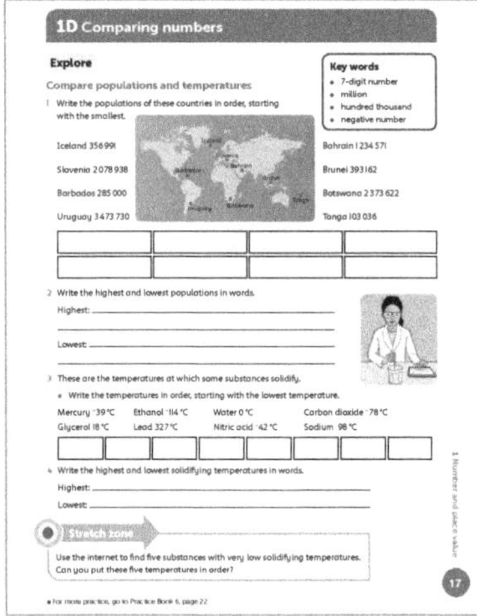

### Differentiation

**Supporting:** Help students to explain their reasoning when ordering the populations.

**Consolidating:** Ask students to explain their reasoning when ordering the populations and support them when ordering negative numbers.

**Extending:** Ask students to explain their reasoning when ordering the populations and negative numbers.

**Stretch zone:** *Use the internet to find five substances with very low solidifying temperatures. Can you put these five temperatures in order?*

Students should order the substances from lowest solidifying temperature to highest. They may then like to research other scientific data and use this to make comparisons.

 **Reflection time**

Discuss answers to the questions in the Student Book and then ask students to discuss with a partner other data that uses either very large numbers or negative numbers. They can then share their ideas. If there is time, do some class research, using the internet, to find data that students show an interest in.

**Practice Book:** Students complete Practice Book page 22. They can do this directly after the Main activity, as homework, or as the focus of a separate mathematics session to help students consolidate their learning and build fluency.

Students order positive and negative temperatures from coldest to warmest. Point out the hint box, which tells them to write just the letter, not the whole place name, in the table. They then answer questions about temperature differences.

| Differentiated outcomes | |
| --- | --- |
| **All students** | should order up to 7-digit numbers and negative numbers with support. |
| **Most students** | will order up to 7-digit and negative numbers. |
| **Some students** | may order up to 7-digit numbers and negative numbers and explain their reasoning. |

## Answers

### Student Book page 17

**1** 103 036, 285 000, 356 991, 393 162, 1 234 571, 2 078 938, 2 373 622, 3 473 730

**2** Highest: three million, four hundred and seventy-three thousand, seven hundred and thirty.

Lowest: one hundred and three thousand and thirty-six.

**3** $^-$114 °C, $^-$78 °C, $^-$42 °C, $^-$39 °C, 0 °C, 18 °C, 98 °C, 327 °C

**4** Highest: three hundred and twenty-seven degrees Celsius.

Lowest: minus one hundred and fourteen degrees Celsius.

### Practice Book page 22

**1**

| Place | Temperature |
| --- | --- |
| H | $^-$48 °C |
| F | $^-$30 °C |
| D | $^-$29 °C |
| C | $^-$26 °C |
| A | $^-$21 °C |
| G | 23 °C |
| B | 26 °C |
| E | 29 °C |

**2** 44 °C   **3** 47 °C   **4** 52 °C   **5** 59 °C

Stretch zone: Check results in comparison to local average low January temperature.

# 1E Using place-value facts

**Discover** Student Book page 18 • Practice Book page 23

## Specific learning focus

- Derive quickly pairs of whole numbers totalling 100, and pairs of one- and two-place decimals totalling 1 and 10.

## Global skills

- **Creative skills:** investigating
- **Self-development skills:** reflecting on learning

## Key vocabulary

- number pairs, known facts, derive

## Resources

- counting sticks
- digit cards 0–9, enough for one set for each pair
- mini whiteboards and markers



## Introductory activity

Use your counting stick to count with the whole class in hundreds starting at 0. Point to 200 and ask what would need to be added to make 1000. Repeat with four other numbers. Repeat the activity, this time counting in tens, stopping at a number, for example 30, and asking what would need to be added to make 100. Finally, repeat the counting again, this time counting in tenths and asking what number needs to be added to, for example, 0.5 to make 1.

## Main activity

Ask one student to come to the front and pick two digit cards. Ask them to make two different 2-digit numbers with the cards and write them on the board. Use the counting stick to help students find the 2-digit numbers that make pairs to 100. For example, if the cards were 4 and 7, you would make 47 and 74. The pairs to 100 would be 47 + 53 and 74 + 26.

Now make the numbers into 4.7 and 7.4 and again ask students to find number pairs to 10 (4.7 + 5.3 and 7.4 + 2.6). Finally, write 0.47 and 0.74 and ask for **number pairs** to 1 (0.47 + 0.53 and 0.74 + 0.26)

Students complete the activity on page 18 of the Student Book, referring to the speech bubbles for examples of what they need to do, and using a counting stick or number line if necessary. Students take it in turns to pick two cards and make a two-digit number. Working as a pair means that they can check one another's answers and support one another in developing the appropriate vocabulary. As students work on the activity, ask them why they think moving the digits one or two places to the right results in totals to 10 and 1. (We are dividing the two numbers by 10 or by 100, so we will also divide the total by 10 and 100.)

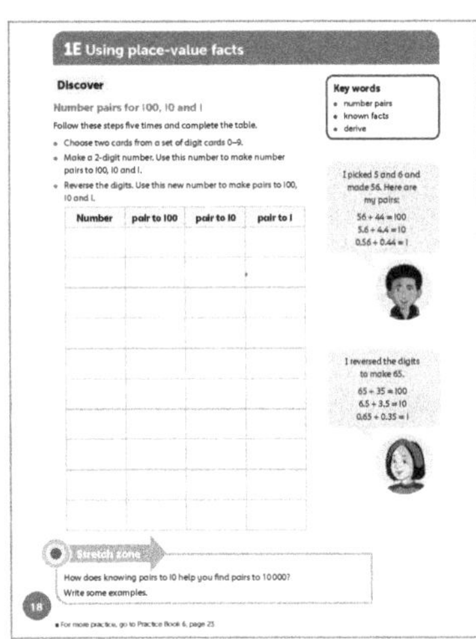

## Differentiation

**Supporting:** Help students to find pairs to 1 using their knowledge of pairs to 10.

**Consolidating:** Ask students why their knowledge of pairs to 10 can help find pairs to 1.

**Extending:** Challenge students to find pairs to 10 using 3-digit numbers with 1 decimal place, for example 2.83 + 7.17 = 10.

**Stretch zone:** *How does knowing pairs to 10 help you find pairs to 10 000? Write some examples.*

Students may write, for example, 4 + 6 helps us to know that 4000 + 6000 = 10 000.

If there is time, develop the idea in the main lesson using four digits, for example: 1321 + 8679 = 10 000, 13.21 + 86.79 = 100, 1.321 + 8.679 = 10

##  Reflection time

Select two digit cards and write the 2-digit number on the board. Working in pairs, students should write the three sums: to 100, to 10 and to 1. You can use this activity to make sure that all students have understood the process. Ask one of the students who explained the reasoning well to you in the Main activity to give the explanation to the whole class.

**Practice Book:** Students complete Practice Book page 23. They can do this directly after the Main activity, as homework, or as the focus of a separate mathematics session to help students consolidate their learning and build fluency.

Students continue their work from the Main activity, this time with given examples of 2-digit numbers. They use these to write the pairs to 1, 10 and 100, drawing on **known facts** each time.

| Differentiated outcomes | |
|---|---|
| **All students** | should find pairs to 1 using their knowledge of pairs to 10. |
| **Most students** | will understand the role that place value plays in finding pairs to 1 using their knowledge of pairs to 10. |
| **Some students** | may find pairs to 1 without needing to refer to pairs to 10. |

## Answers

### Student Book page 18

Students make their own numbers. Check that the addition pairs are correct for each number.

| Number | Addition pair to 100 | Addition pair to 10 | Addition pair to 1 |
|---|---|---|---|
| 38 | $38 + 62 = 100$ | $3.8 + 6.2 = 10$ | $0.38 + 0.62 = 1$ |
| 85 | $85 + 15 = 100$ | $8.5 + 1.5 = 10$ | $0.85 + 0.15 = 1$ |
| 41 | $41 + 59 = 100$ | $4.1 + 5.9 = 10$ | $0.41 + 0.59 = 1$ |
| 16 | $16 + 84 = 100$ | $1.6 + 8.4 = 10$ | $0.16 + 0.84 = 1$ |
| 73 | $73 + 27 = 100$ | $7.3 + 2.7 = 10$ | $0.73 + 0.27 = 1$ |
| 24 | $24 + 76 = 100$ | $2.4 + 7.6 = 10$ | $0.24 + 0.76 = 1$ |
| 11 | $11 + 89 = 100$ | $1.1 + 8.9 = 10$ | $0.11 + 0.89 = 1$ |
| 89 | $89 + 11 = 100$ | $8.9 + 1.1 = 10$ | $0.89 + 0.11 = 1$ |
| 54 | $54 + 46 = 100$ | $5.4 + 4.6 = 10$ | $0.54 + 0.46 = 1$ |
| 67 | $67 + 33 = 100$ | $6.7 + 3.3 = 10$ | $0.67 + 0.33 = 1$ |
| 14 | $14 + 86 = 100$ | $1.4 + 8.6 = 10$ | $0.14 + 0.86 = 1$ |
| 51 | $51 + 49 = 100$ | $5.1 + 4.9 = 10$ | $0.51 + 0.49 = 1$ |
| 40 | $40 + 60 = 100$ | $4.0 + 6.0 = 10$ | $0.4 + 0.6 = 1$ |
| 5 | $5 + 95 = 100$ | $0.5 + 9.5 = 10$ | $0.05 + 0.95 = 1$ |
| 92 | $92 + 8 = 100$ | $9.2 + 0.8 = 10$ | $0.92 + 0.08 = 1$ |

Stretch zone: Divide each number in the pair by 100 by moving the digits two places to the right.

# 1E Using place-value facts

## Explore  Student Book page 19 • Practice Book page 24

### Specific learning focus

- Derive quickly pairs of whole numbers totalling 100, and pairs of one- and two-place decimals totalling 1 and 10.

### Global skills

- **Creative skills:** exploring
- **Self-development skills:** reflecting on learning

### Key vocabulary

- number pair, known facts, derive

### Resources

- mini whiteboards and markers

### Language support

Encourage students to discuss their answers with you, focusing on the students you noticed were struggling to explain their thinking in the previous lesson. Continue to model the key language:

- *What is the sum of those two numbers?*
- *What is the addition pair to 100, 10, 1?*

 Introductory activity

Write 10 on the board. Give each pair two minutes to write as many addition pairs to 10 as they can. The numbers they use must include one decimal place. Choose a pair of students and take one of their answers. Write it on the board. Ask students to tell you the corresponding addition pairs to 100 and to 1. Repeat several times.

 Main activity

Draw students' attention to the spider diagram on page 19 of the Student Book. Look together at the example already completed. Ask students to describe the connection between the three number sentences. They should be able to say something about how the numbers have been divided by 10 from one line to the next. Ask a student to choose two digits and then, as a class, complete one more box on the diagram. For example, if they choose 3 and 1, then write together $31 + 69 = 100$, $3.1 + 6.9 = 10$ and $0.31 + 0.69 = 1$.

Students then work as individuals to complete the activities on page 19. While they are working, observe any students you were concerned about in the previous activity so that you can check their understanding and support them if necessary, for example by using coins and making pence up to a pound or cents up to a dollar.

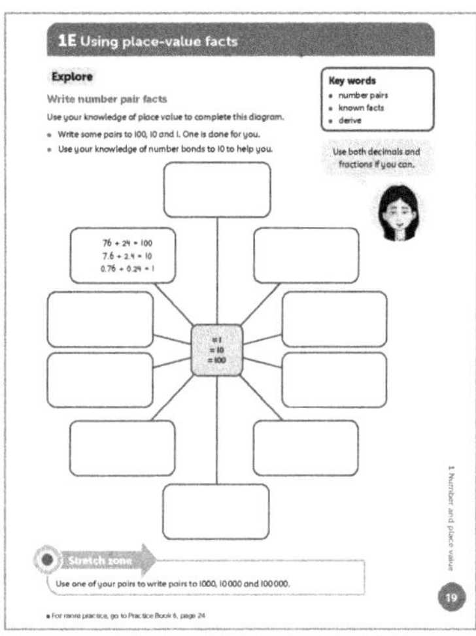

## Differentiation

**Supporting:** Help students find pairs to 1 using their knowledge of pairs to 10 and 100, using a number line or a place-value grid and counters if necessary.

**Consolidating:** Ask students why their knowledge of pairs to 10 and 100 can help them find pairs to 1.

**Extending:** Challenge students to find pairs to 1 with numbers with three decimal places.

**Stretch zone:** *Use one of your pairs to write pairs to 1000, 10 000 and 100 000.*

Students might use $76 + 24 = 100$, for example, to make the following pairs:

$760 + 240 = 1000$, $7600 + 2400 = 10\,000$, and $76\,000 + 24\,000 + 100\,000$.

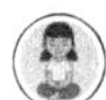 **Reflection time**

As in the previous activity, use this time as an assessment opportunity. Ask different students for a pair that they used to total 100. Working individually, they should then write the corresponding pairs to 10 and to 1.

**Practice Book:** Students complete Practice Book page 24. They can do this directly after the Main activity, as homework, or as the focus of a separate mathematics session to help students consolidate their learning and build fluency.

Students are given the total that they need to make, which is always a power of 10. They write six different pairs of numbers that make that total. The totals they need to make range from 0.01 to 10 000.

| Differentiated outcomes | |
|---|---|
| **All students** | should find pairs to 1 using their knowledge of pairs to 10 and 100. |
| **Most students** | will understand the role that place value plays in finding pairs to 1 using their knowledge of pairs to 10 and 100. |
| **Some students** | may find pairs to 1 without needing to refer to pairs to 10 and 100. |

## Answers

### Student Book page 19

Students choose their own numbers for the spider diagrams. Check that the number sentences written are correct and that the numbers total 100, 10 or 1.

### Practice Book page 24

Students write their own pairs of numbers that total given numbers. Check that the addition pairs are correct.

Stretch zone: Students should use $34 + 66 = 100$ to derive $0.00034 + 0.00066 = 0.001$, then **derive** $0.001 - 0.00034 = 0.00066$.

# 1 Number and place value

## Connect  Student Book page 20

### Big idea

I can use place value to understand large numbers. I can order large numbers and negative numbers based on the value of each digit. I can round numbers when an approximate number is needed.

### Global skills

- **Creative skills:** problem solving
- **Interpersonal skills:** teamwork
- **Self-development skills:** reflecting on learning

### Key vocabulary

- double, thousand, ten thousand, hundred thousand, million

### Resources

- mini whiteboards and markers
- large sheets of paper

### Language support

Encourage students to work mentally where possible, saying aloud their calculations as they double repeatedly. As the numbers get very large, help students to approximate by asking, for example:

- *What is 1024 to the nearest 1000?*
- *What is 256 000 to the nearest 10 000?*

 **Introductory activity**

Ask a student to choose a single digit, say 3. As a class, start from 3 and keep **doubling** as far as they can mentally: 3, 6, 12, 24, 48, 96, 192. At this point, suggest they round to 200 and continue: 200, 400, 800, 1600, 3200, 6400, 12 800. Maybe round again to 13 000 and continue: 13 000, 26 000, 52 000, 104 000, …

Repeat with a new single digit (but not 6 as this would match the doubling from 3).

 **Main activity**

Look together at page 20 of the Student Book. Display on the IWB, if possible. Introduce students to the chessboard and rice problem. Read through the story and look at the photograph, then ask students to discuss in pairs how much rice they think will be needed and offer approximate answers, perhaps 1000 grains, 1 000 000 grains or even 1000 sacks of rice.

Explain that students will be working in small groups to calculate and estimate the answers to the questions on page 20. They can use the large sheets of paper to write their calculations once they become too large to do mentally or with a calculator.

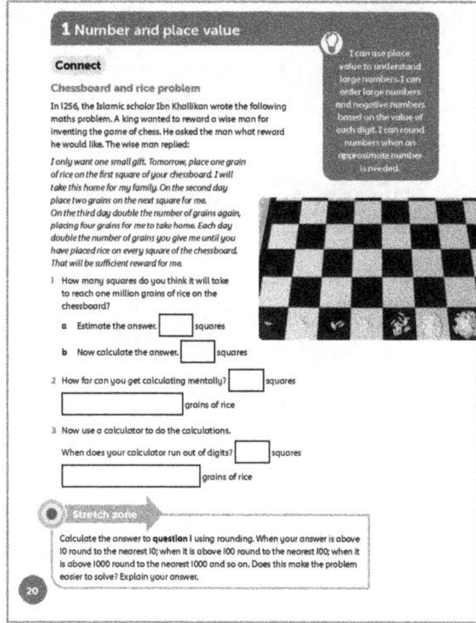

**Stretch zone:** *Calculate the answer to **question** 1 using rounding. When your answer is above 10 round to the nearest 10; when it is above 100 round to the nearest 100; when it is above 1000 round to the nearest 1000 and so on. Does this make the problem easier to solve? Explain your answer.*

Using this method students should get: 1, 2, 4, 8, 16 (round to 20), 40, 80, 160 (round to 200), 400, 800, 1600 (round to 2000), 4000, 8000, 16 000 (round to 20 000), 40 000, 80 000, 160 000 (round to 200 000) and so on.

 **Reflection time**

Ask groups to report back on their estimating and to give their reasoning for the numbers of grains of rice at each stage. Ask students to compare and comment on each other's estimations and calculations.

You can share the following step-by-step total of the number of grains of rice on the chessboard if students are interested.

By the end of row 1, the king would have needed 128 grains of rice. He then moved onto the second row: 256 grains on the ninth square, 512 on the tenth square, then 1024, then 2048, doubling each time until he needed to put 32 768 grains of rice on the last square of the second row. By the end of the third row, the king would have needed 8.4 million grains of rice. By the end of the fourth row, 2.1 billion grains were needed. By the time the king had filled the board he would have needed 18 446 744 073 709 551 615 grains of rice!

| Differentiated outcomes | |
|---|---|
| **All students** | should reach an estimate for the grains of rice on the first row of the chessboard with support. |
| **Most students** | will reach an estimate for how many squares will be covered by 1 million grains of rice. |
| **Some students** | may estimate the total grains of rice for the chessboard. |

### Answers

**Student Book page 20**

Answers will vary as students will estimate differently and reach their own mental limit for calculations.

# 1 Number and place value



## Global skills

- **Self-development skills:** reflecting on learning

## Student Book

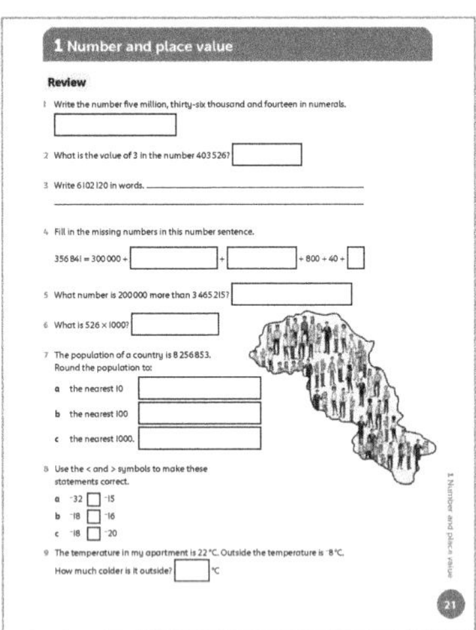

With young students, assessment activities are most effective when carried out as an everyday classroom activity. Students should have digit cards, number lines and place-value cards available to support them as they complete the Student Book Review page. Watch as students round and order very large numbers. Listen as they say the value of each digit and say aloud the whole numbers. Listen to check that they understand how the place of a digit affects its value. If there are errors, help them partition so they can understand the value of different digits.

Students should be able to use known facts to work out new ones by multiplying or dividing by 10, 100 or 1000. It may help to model how to place numbers on number lines, using the most significant digits, and to model the partitioning of large numbers using place-value cards.

## Answers

### Student Book page 21

**1**  5 036 014

**2**  3000

**3**  Six million, one hundred and two thousand, one hundred and twenty.

**4**  $356\,841 = 300\,000 + 50\,000 + 6000 + 800 + 40 + 1$

**5**  3 665 215

**6**  526 000

**7**  **a**  8 256 850     **b**  8 256 900     **c**  8 257 000

**8**  **a**  $^-32 < {}^-15$     **b**  $^-18 < {}^-16$     **c**  $^-18 > {}^-20$

**9**  30 °C

## Practice Book

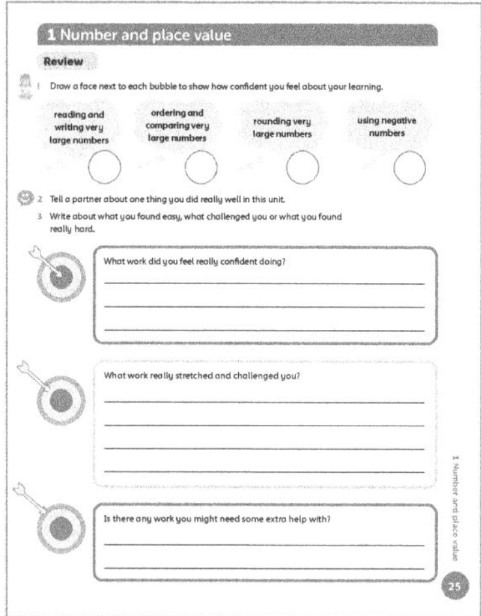

It is appropriate to complete this Practice Book Review page as a whole-class discussion. You may choose to keep a record of the class discussion or a copy of the Review page for your own records. The Review provides an opportunity for students to reflect on their learning from the unit, to discuss any areas of mathematics that they feel went particularly well, and any areas that they feel less confident about. Ensure that all students have a copy of the Student Book as a reminder of the areas of mathematics that they have worked on in this unit.

Allow students plenty of time for discussion before asking them to complete the Practice Book page individually, and then, if appropriate, to share their responses with the rest of the class. If students complete this self-assessment at home, encourage them to discuss this with adults.

Make a note of areas that students still feel unsure about. As number is at the heart of many of the other units, you can revisit this regularly. You can also build number into everyday practice, for example by looking for examples of very large numbers in the media.

## Additional material

There are additional end-of-unit assessments available on the *Oxford Owl for School* website.

# 2 Addition, subtraction, multiplication and division

## Overview

### Big idea

This unit develops both mental strategies and written methods. Mental calculation has a central place in the teaching and learning of mathematics. Students develop their sense of number by developing effective and efficient mental methods. This gives them confidence and ability to manipulate numbers and to draw on facts that they already know to develop new facts.

In order to calculate mentally, students must have clear mental images of the number system. It is important to continue to use images of the number system such as empty number lines for addition and subtraction and arrays for multiplication and division. Students have become used to these images in earlier years.

In this unit, students build on the habit of looking at questions to decide what method is appropriate to use and, in particular, whether to use a mental or a written method. Students learn to add and subtract large numbers, including using written algorithms involving carrying figures and decomposition, and applying strategies to problem-solving situations, including using estimation and near multiples.

There is further development of written methods for multiplication and division. Students have worked with these operations in previous years, using methods such as partitioning. In this unit, the grid method for multiplication is revisited and the use of the column method is consolidated. It is important that students have a conceptual understanding of the column method so that they can use it efficiently and accurately. In division, students are encouraged to present remainders as fractions and decimals and to decide whether to round their answer up or down depending on the context. Strategies that students know from earlier stages are extended in this unit, including short and long division, and there is further consolidation of factors, multiples and prime numbers.

### Look out for

- **Students who use the formal written methods (algorithms) they know for all calculations, even when they are not appropriate**. Help students to build a repertoire of mental and written strategies. Encourage students to look at the numbers in a calculation and decide how to approach it.
- **Students who do not apply the correct operation to word problems**. Give students experience with a range of contexts so that they know the key words to look out for and are aware of when to add, subtract, multiply or divide.

### Possible misconceptions

- **When estimating an answer, students use the wrong digits to estimate and round numbers**. Help them by modelling the use of leading digits and suitable place value for the size of the numbers.
- **Students may confuse the terms 'factor' and 'multiple'.** Help students to focus on the idea of a multiple of a number being 'lots of' that number, so multiples of 3, for example, are 3, 6, 9, 12 and so on. Factors are the numbers that divide exactly into a number, so factors of 12 include 3 and 6, but also 1, 2, 4 and 12.

### Key vocabulary

- addition, subtraction, mental strategy, bonds to 10, sum, total, difference
- multiplication, long multiplication, division, divisibility rules, multiples of 10
- known facts, derive, estimate, smallest, largest, place value, written strategy
- column method, grid method, written method, recombine
- how much bigger/smaller?, how much is left?
- partition, product, dividend, divisor, quotient, remainder
- prime number, composite number, factor, multiple, prime factor, common multiple, brackets, order of operations, BIDMAS
- commutative law, associative law, distributive law

| Learning objective | E | 2A | 2B | 2C | 2D | 2E | 2F | 2G | 2H | 2I | 2J | 2K | 2L | C | R |
|---|---|---|---|---|---|---|---|---|---|---|---|---|---|---|---|
| Multiply multi-digit numbers up to four digits by a 2-digit whole number using the formal written method of long multiplication. | ✓ | | | | | | ✓ | | | | | | | ✓ | ✓ |
| Divide numbers up to four digits by a 2-digit whole number using the formal written method of long division, and interpret remainders as whole number remainders, fractions, or by rounding, as appropriate for the context. | | | | | | | | ✓ | | ✓ | | | | | ✓ |
| Divide numbers up to four digits by a 2-digit number using the formal written method of short division where appropriate, interpreting remainders according to the context. | | | | | | | | | ✓ | ✓ | | | | ✓ | ✓ |
| Perform mental calculations, including with mixed operations and large numbers. | ✓ | ✓ | ✓ | ✓ | ✓ | ✓ | | ✓ | ✓ | | | | | ✓ | ✓ |
| Identify common factors, common multiples and prime numbers. | | | ✓ | | | | | | | | ✓ | | | | |
| Use knowledge of the order of operations to carry out calculations involving the four operations. | | | | | | | | | | | | ✓ | ✓ | | |
| Solve addition and subtraction multi-step problems in contexts, deciding which operations and methods to use and why. | | | ✓ | | ✓ | ✓ | ✓ | | | | | | | | ✓ |
| Solve problems involving addition, subtraction, multiplication and division. | | | | | | | | ✓ | ✓ | ✓ | | | | | ✓ |
| Use estimation to check answers to calculations and determine, in the context of a problem, an appropriate degree of accuracy. | | | | | ✓ | | ✓ | | ✓ | ✓ | ✓ | | | | |

# 2 Addition, subtraction, multiplication and division

## Engage  Student Book page 22

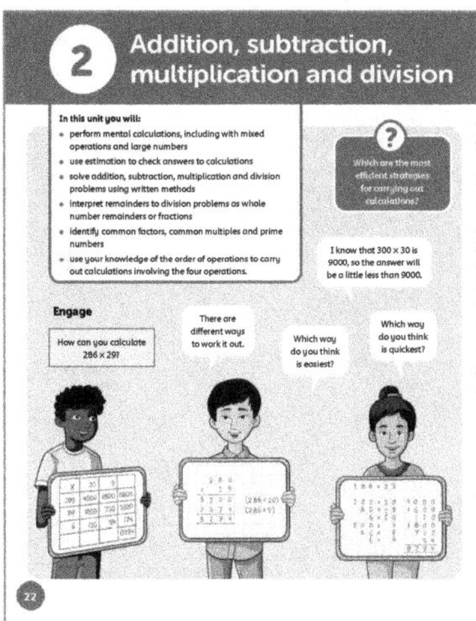

## Big question

- Which are the most efficient strategies for carrying out calculations?

## Global skills

- **Creative skills:** problem solving
- **Interpersonal skills:** communication
- **Self-development skills:** reflecting on learning

## Key vocabulary

- multiplication, grid method, column method, long multiplication, product, estimate

## Resources

- mini whiteboards and markers

## Language support

Each time a student gives you an answer, whether it is correct or not, remember to ask: *How did you work that out? What strategy did you use?* Explaining their thinking helps students to develop mental and written calculation strategies and to develop mathematical language. Support the language that students need to explain their methods, for example:

- 'The answer is …'
- 'I worked it out by …'

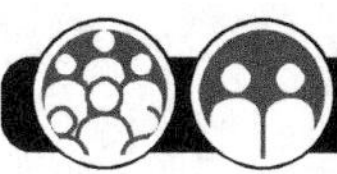

## Introductory activity

Write 28 + 17 on the board. Ask pairs of students to perform this calculation mentally. Then ask students to talk to a partner and explain the strategy they used. To help students with the necessary language, you can write on the board: 'The answer is …' and 'I worked it out by …'. Take feedback from the class until you have all the different strategies.

Repeat this activity with '4 × 14' and 'What is a half of 78?' On each occasion, spend time sharing the strategies until you have heard each different strategy that was used. Ask students to discuss which they think is the most efficient strategy.

## Main activity

Give students an addition or subtraction, for example 483 + 194 or 504 − 376, and discuss different mental and written methods of working it out. Ask students to estimate the answer and then calculate it on their whiteboards before sharing their methods with the class. Decide as a class which methods were most efficient.

Explain that, as with addition and subtraction, different methods can be used to solve multiplication and division calculations and that students should always choose the most efficient method.

Look together at page 22 of the Student Book. Display on the IWB, if possible. This multiplication calculation has been completed using the **grid method**, **column method** and expanded method of multiplication. Ask students to spend some time, in pairs, looking at all three methods, comparing them and seeing how they are the same and different. Ask pairs to discuss each strategy, and to decide which strategy they prefer.

### Differentiation

**Supporting:** Ask students to explain which strategies they understand so that you can assess their current understanding.

**Consolidating:** Ask students to explain the strategy they are using. Ask whether they think another student has a better strategy.

**Extending:** Ask students to use a range of strategies to carry out calculations and choose the most efficient.

## Reflection time

Ask pairs to feed back to the class their choice of efficient method for the calculation. They may have different strategies. It is important to value all the different strategies. Focus the discussion on the effectiveness of the strategies. *Do you think this will always be your preferred strategy? Do you think you should always use a written strategy for large numbers or can we sometimes use a mental strategy?* Create a class poster of different calculation strategies. You can refer to the poster throughout the unit.

# 2A Mental strategies for addition and subtraction



## Specific learning focus

- Recall addition and subtraction facts for numbers to 20 and use them in mental calculations.

## Global skills

- **Creative skills:** problem solving
- **Real-world skills:** interpreting information

## Key vocabulary

- addition, subtraction, mental strategies, bonds to 10

## Resources

- mini whiteboards and markers

## Language support

Support students as they describe the strategies they are using.

*How did you work that out?* Leave thinking time, and offer prompts such as:

- *What did you do next? Then what did you do?*
- *Why did you decide to use that strategy?*

### Introductory activity

Give each pair of students one whiteboard. Write the number 50 on the board. Give pairs two minutes to write as many questions as they can with the answer 50. They can use any operation. After two minutes, take one suggestion from each pair and record it.

*How many ways can I make 15 by adding two whole numbers?* Again, give the pairs two minutes to answer. This time ask one pair to report back. *How do you know whether you have all the possibilities?*

### Main activity

Ask students in pairs to work on the activities on page 23 of the Student Book so they can compare answers and discuss strategies. You may need to give some students support in working mentally. Refer students to the speech bubbles on the Student Book page, which offer some hints about strategies they can use, such as looking for doubles and bonds to 10. Also emphasise the importance of checking an answer by using a second method, especially when adding lots of numbers.

As students work, ask them to share some of their strategies. Ask pairs who successfully complete question 1 (adding the number of students in different classes) to share their **mental strategies** with the rest of the class. This helps all students to understand a range of strategies.

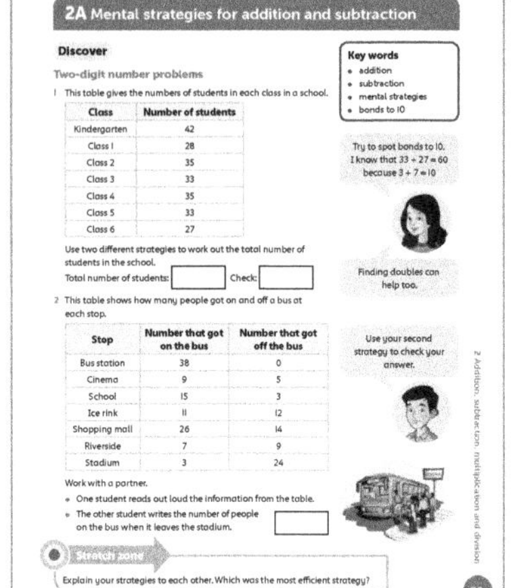

## Differentiation

**Supporting:** Help students find pairs to 10 and doubles when adding mentally.

**Consolidating:** Ask students to explain their strategies for adding and subtracting mentally.

**Extending:** Challenge students to create their own word problems based on the numbers in question 2.

**Stretch zone:** *Explain your strategies to each other. Which was the most efficient strategy?*

Question students about the strategies they used, for example: *Did you work through all the numbers in the order they are written in the table or did you first do all the addition and then all the subtraction? Did you first add all the people who got off the bus and then take away that total number?*

### Reflection time

List the number of students in your school on the board, class by class. Ask students to work in pairs to find the total number of students in the school. Ask one student in each pair to use a written method and the other student to calculate mentally. Do the calculation yourself in advance so that you have good grouping strategies that you can share. You can also show that you can use different efficient mental methods.

**Practice Book:** Students complete Practice Book page 26. They can do this directly after the Main activity, as homework, or as the focus of a separate mathematics session to help students consolidate their learning and build fluency.

Students use mental strategies to solve addition and subtraction problems with 4-digit numbers. They can apply similar strategies to those they have learned in the main lesson. Use similar questioning such as: *What will you do first to work that out? What did you do next? Then what will you do? Why did you decide to use that strategy?*

| Differentiated outcomes | |
|---|---|
| **All students** | should complete the addition and subtraction problems with support. |
| **Most students** | will complete the addition and subtraction problems. |
| **Some students** | may use a varied range of mental strategies to complete addition and subtraction problems. |

## Answers

### Student Book page 23

**1** The total number of students is 233.

| Stop | Number that got on the bus | Number that got off the bus | Number of people on the bus |
|---|---|---|---|
| Bus Station | 38 | 0 | 38 |
| Cinema | 9 | 5 | 42 |
| School | 15 | 3 | 54 |
| Ice rink | 11 | 12 | 53 |
| Shopping mall | 26 | 14 | 65 |
| Riverside | 7 | 9 | 63 |
| Stadium | 3 | 24 | 42 |

### Practice Book page 26

Check that students' strategies are appropriate. While students are working, ask them to explain why they chose the methods they used.

**1** 4450

**2** 2830

**3** 2226

**4** 550

Stretch zone: Yes, estimating suggests that the total is more than 10 000.

## 2A Mental strategies for addition and subtraction

**Explore**   Student Book page 24 • Practice Book page 27

### Specific learning focus

- Use place value and number facts to add or subtract whole numbers.

### Global skills

- **Creative skills:** investigating

### Key vocabulary

- sum, difference, addition, subtraction

### Resources

- cards showing 1, 2, 3, 4, 5, 6, 7
- mini whiteboards and markers

### Language support

Support students in describing the strategies they use. Ask: *How did you work that out?*

Leave thinking time and offer prompts such as:

- *What did you do next?*
- *Why did you decide to use that strategy?*

### Introductory activity

Prepare seven large cards with 1, 2, 3, 4, 5, 6 and 7 on them. Ask a student to come to the front of the class and pick each card at random to create an addition of a 4-digit number and a 3-digit number. For example, if they pick the cards in the order 4, 6, 1, 5, 3, 7, 2, they would write 4615 + 372. Tell students to work in pairs to calculate the answer mentally, write it on their whiteboards and show you.

Now ask students to mentally subtract the two numbers. Tell students to work in pairs to calculate the answer mentally, write it on their whiteboards and show you.

*What is the largest possible total of a 4-digit number and a 3-digit number that can be made using these digits? What is the smallest possible total?*

### Main activity

Use the same digit cards and create two additions using two or more numbers of different lengths and a subtraction using two numbers. For example, you could write on the board: 742 + 15 + 36, 43 561 + 27 and 7241 – 365. In each case, ask students to work in pairs to estimate, then mentally calculate the answers. Students can share their answers and strategies.

Now ask students to work on the activities on page 24 of the Student Book in pairs so they can compare answers and discuss strategies. As students are working on the first question, ask direct questions to encourage extended responses, for example:

- *Can you make an addition that gives the largest total?*
- *What could be the smallest answer?*
- *How do you know that is the largest total?*
- *Convince me that that is the smallest total possible.*

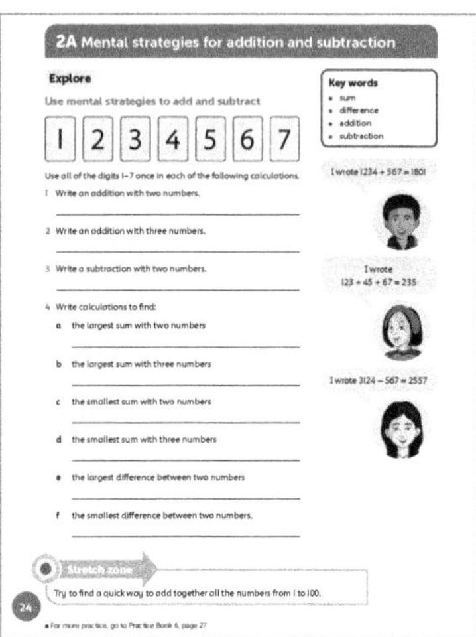

## Differentiation

**Supporting:** Help students to find sums and differences of 4-digit and 3-digit numbers by modelling mental strategies.

**Consolidating:** Ask students to describe their mental strategies.

**Extending:** Ask students how they know that their answers are the largest and smallest possible sums and differences.

**Stretch zone:** *Try to find a quick way to add together all the numbers from 1 to 100.*

Encourage students towards the method used by the famous mathematician Gauss, who solved this quickly by adding 50 pairs of numbers that each made 101 (1 + 100, 2 + 99, 3 + 98, 4 + 97 and so on), giving the answer 5050.

 **Reflection time**

Discuss with the class their responses to the different calculations and ask them to share their strategies for finding the answers. Ask students to check each calculation mentally and to share their alternative strategies with the whole class, in order to check answers.

**Practice Book:** Students complete Practice Book page 27. They can do this directly after the Main activity, as homework, or as the focus of a separate mathematics session to help students consolidate their learning and build fluency.

Students are given the answer to a calculation and they have to write possible questions that give that answer. They can use addition or subtraction. *Will you add just two numbers or will you add more than two numbers to get that answer? Can you write an addition and a subtraction that both give that answer?*

| Differentiated outcomes | |
| --- | --- |
| **All students** | should find simple totals and differences. |
| **Most students** | will find the largest possible totals and differences. |
| **Some students** | may generalise to find the largest totals and differences. |

## Answers

### Student Book page 24

**1–3** Check that students have formed and calculated appropriate addition and subtraction problems.

**4  a**  765 432 + 1 = 765 433

   **b**  76 543 + 2 + 1 = 76 546

   **c**  1346 + 257 = 1603

   **d**  125 + 36 + 47 = 208

   **e**  765 432 − 1 = 765 431

   **f**  1234 − 765 = 469

### Practice Book page 27

Students make up their own calculations. Check that the calculations include large numbers and decimals and have the required answers.

Stretch zone: Check that students have made a calculation with the answer 100 501, for example 50 345 + 50 156.

# 2B Mental strategies for multiplication and division

## Discover 1
**Student Book page 25 · Practice Book page 28**

### Specific learning focus

- Know and apply tests of divisibility by 2, 3, 4, 5, 6, 8, 9 and 10.
- Multiply near multiples of 10 by multiplying by the multiple of 10 and adjusting.

### Global skills

- **Creative skills:** exploring
- **Self-development skills:** reflecting on learning

### Key vocabulary

- multiplication, division, divisibility rules, mental strategy

### Resources

- counters
- prepared cards labelled 15, 55, 56, 60, 72, 80, 93, 120, 145, 208, 447, 480, 520, 576

### Language support

Asking students to write down the **divisibility rules** allows you to follow their thinking and develops their mathematical vocabulary.

Other key vocabulary in this activity is '11 and 9 are near multiples of 10'. Ensure that students know why 11 and 9 are called near multiples.

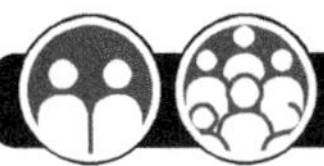

## Introductory activity

Give pairs of students 36 counters. Ask them to arrange the counters in different arrays. They should write down the dimensions of all the arrays they make. Take feedback from the class and agree on all the possible arrays. Remind the class that the dimensions of the arrays are the factors of the number 36.

Write on the board the factors of 36: 1, 2, 3, 4, 6, 9, 12, 18 and 36. Explain that these factors of 36 formed from the array dimensions are the numbers that 36 is exactly divisible by.

## Main activity

Students work in groups on the first activity in the Student Book on page 25. They should discuss and complete the table for divisibility tests. Ask students to explain how they know the rules as they work, and to provide examples and counter examples for each number.

Write on the board the calculations: $15 \times 11 =$ ; $15 \times 9 =$ .

Explain that to multiply by a near multiple of 10, that is, 9 or 11, they can multiply by 10 and then adjust the answer. (For the example above: $15 \times 11 = 15 \times 10 + 15 \times 1 = 165$, $15 \times 9 = 15 \times 10 - 15 \times 1 = 135$.)

Students can now use this strategy to complete question 2 on page 25 of the Student Book.

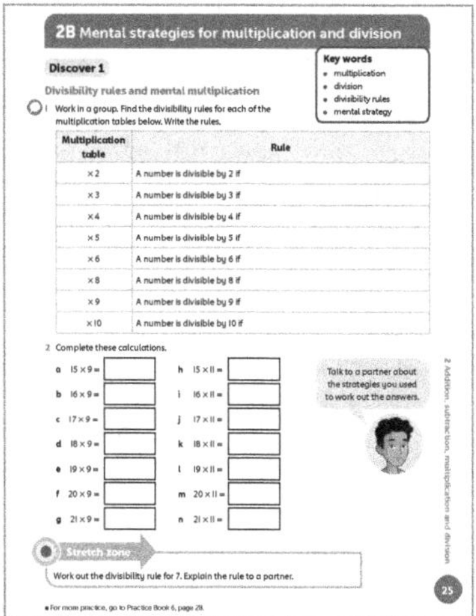

## Differentiation

**Supporting:** Help students to use multiplication tables, and knowledge of factors, to test rules of divisibility.

**Consolidating:** Ask students to describe how they know that a number is divisible by a single digit using the divisibility tests.

**Extending:** Challenge students to extend the 'near 10' multiple strategy to help them multiply by 8 or by 12.

**Stretch zone:** *Work out the divisibility rule for 7. Explain the rule to a partner.*

The divisibility rule for 7 can be explained in two steps.

Step 1: Take the last digit of the number and double it.

Step 2: Subtract this from the remaining digits in the number. If the answer is 0 or a multiple of 7, then the original number is also divisible by 7. If the answer is still a big number, go back to step 1 and repeat.

Example 1: 455

Double 5 = 10

45 – 10 = 35

35 is divisible by 7 so 455 is also divisible by 7.

Example 2: 2961

Double 1 = 2

296 – 2 = 294

Repeat:

Double 4 = 8

29 – 8 = 21

21 is divisible by 7 so 2961 is also divisible by 7.

Students can test the rule on 2- and 3-digit numbers.

Ask each group to explain the divisibility rule they found for each number. Tell the other students to write the rules. Draw a three-way Venn diagram on the board, with the circles labelled:

'Multiples of 3''Multiples of 5''Multiples of 8'

Ask students to come to the front one at a time. Tell each student to choose one of the prepared cards showing 2-digit and 3-digit numbers. Then ask them to place their card in the appropriate place on the Venn diagram. Each time, ask, *Is it a multiple of 3/5/8?*

Ask students to explain their answers.

**Practice Book:** Students complete Practice Book page 28. They can do this directly after the Main activity, as homework, or as the focus of a separate mathematics session to help students consolidate their learning and build fluency.

Students use their knowledge of divisibility rules to write calculations with either no remainder or a given remainder..

| Differentiated outcomes | |
|---|---|
| **All students** | should understand divisibility rules to 10 and mental methods for multiplying near multiples of 10 with support. |
| **Most students** | will understand divisibility rules to 10 and find strategies to multiply near multiples of 10. |
| **Some students** | may know the strategies to multiply near multiples of 10 and the tests of divisibility. |



### Student Book page 25

**1** Divisible by 2 if it ends in 0, 2, 4, 6 or 8

Divisible by 3 if the digits add to a multiple of 3

Divisible by 4 if the last 2 digits are a multiple of 4

Divisible by 5 if it ends in 0 or 5

Divisible by 6 if it is divisible by 2 and 3.

Divisible by 8 if the last 3 digits are a multiple of 8

Divisible by 9 if the digits add to a multiple of 9

Divisible by 10 if it ends in 0

**2**
**a** $15 \times 9 = 135$     **h** $15 \times 11 = 165$
**b** $16 \times 9 = 144$     **i** $16 \times 11 = 176$
**c** $17 \times 9 = 153$     **j** $17 \times 11 = 187$
**d** $18 \times 9 = 162$     **k** $18 \times 11 = 198$
**e** $19 \times 9 = 171$     **l** $19 \times 11 = 209$
**f** $20 \times 9 = 180$     **m** $20 \times 11 = 220$
**g** $21 \times 9 = 189$     **n** $21 \times 11 = 231$

### Practice Book page 28

Students make their own 5-digit numbers. Check that the numbers chosen match the given information and that the answers are correct.

Stretch zone: Students should be able to say that dividing a number by 10, if the number ends in 1, the remainder will be 1, if the number ends in 2, the remainder will be 2, and so on, up to 9. When dividing by 5, if the number ends in 1 or 6, the remainder will be 1, if the number ends in 2 or 7, the remainder will be 2, and so on, up to 4 or 9.

---

# 2B Mental strategies for multiplication and division

## Discover 2
**Student Book page 26 • Practice Book page 29**

### Specific learning focus
- Use known multiplication facts to derive others.

### Global skills
- **Creative skills:** investigating
- **Self-development skills:** reflecting on learning

### Key vocabulary
- multiplication, mental strategy

### Resources
- mini whiteboards and markers



Model phrases and questions to support students with the necessary language, for example:

- *I can multiply by 50 by …*
- *I can multiply by 0.25 by …*
- *What is an easy way to multiply by 40?*
- *An easy way to multiply by 0.8 is to …*

Begin the lesson by practising multiplication and division facts. Call out an answer from a multiplication table. Ask students to write the multiplication facts and corresponding divisions. For example, for 42:

$7 \times 6 = 42, 42 \div 7 = 6, 6 \times 7 = 42, 42 \div 6 = 7$

Write $8 \times 18$ on the board. Ask pairs to discuss how to calculate this. Take feedback. Accept any strategies that give the correct answer. Focus on the idea of multiplying by 20 and adjusting, using the idea that 18 is near to a multiple of 10.

Demonstrate how you can do this for $8 \times 18$ by using these number sentences:

$8 \times 20$

$8 \times 2 \times 10 = 160$

$160 - 16 = 144$

Ask students to explain why we subtract 16 at the end.

Repeat for other numbers such as $12 \times 18$ and $14 \times 19$. Ask students to work these out. Ask them to show you their answers on their whiteboards.

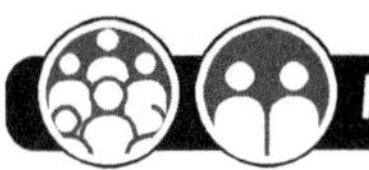 **Main activity**

Ask students to multiply 23 by 2:

$23 \times 2 = 46$

Now ask students to use this fact to calculate the following:

$23 \times 4$     $23 \times 8$     $46 \times 2$     $92 \times 2$     $92 \times 4$

*How could you use the first number facts to help you calculate the others?*

Agree that if one of the numbers in a product doubles, then so will the answer.

Ask pairs to work out the answers to check that these strategies work. Ask pairs to discuss why these strategies work and share their reasoning with the class.

Students then work on the activities on page 26 in the Student Book in pairs. You may choose to work with any students who need extra support to help them come to an understanding of these strategies. Ask questions such as: *What fact can I use to work out $0.25 \times 30$? How will knowing $5 \times 10$ help me work out $0.5 \times 10$? How will knowing $0.25 \times 6$ help me work out $0.25 \times 60$?*

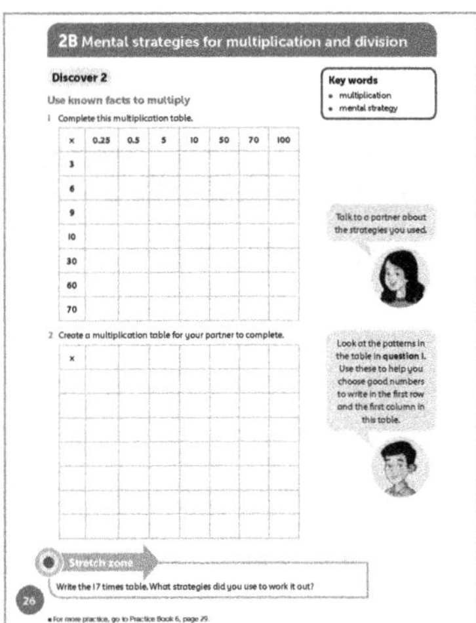

## Differentiation

**Supporting:** Help students to identify strategies that can help them multiply by using known facts.

**Consolidating:** Ask students to explain their thinking carefully when using known facts.

**Extending:** Challenge students to explore alternative strategies for multiplying when solving calculations such as $21 \times 89$.

**Stretch zone:** *Write the 17 times table. What strategies did you use to work it out?*

Encourage students to use variations of known strategies, for example near multiples (17 is near 20) or partitioning ($17 = 10 + 7 = 10 + 5 + 2$).

### Reflection time

Ask students to work on the calculations below by drawing on the strategies they have been using or by using other strategies. They should share the strategies they are using with their partner.

$34 \times 2$     $34 \times 4$     $34 \times 8$     $17 \times 4$     $170 \times 8$

Ask students to share their strategies with the whole class. Ask them to describe the strategy by comparing what changes from one calculation to the next and how this affects the answer.

**Practice Book:** Students complete Practice Book page 29. They can do this directly after the Main activity, as homework, or as the focus of a separate mathematics session to help students consolidate their learning and build fluency.

| Differentiated outcomes | |
|---|---|
| **All students** | should calculate multiplications using known facts, with support. |
| **Most students** | will calculate multiplications using known facts. |
| **Some students** | may use a variety of strategies to multiply using known facts. |

**Student Book page 26**

**1**

| × | 0.25 | 0.5 | 5 | 10 | 50 | 70 | 100 |
|---|---|---|---|---|---|---|---|
| **3** | 0.75 | 1.5 | 15 | 30 | 150 | 210 | 300 |
| **6** | 1.5 | 3 | 30 | 60 | 300 | 420 | 600 |
| **9** | 2.25 | 4.5 | 45 | 90 | 450 | 630 | 900 |
| **10** | 2.5 | 5 | 50 | 100 | 500 | 700 | 1000 |
| **30** | 7.5 | 15 | 150 | 300 | 1500 | 2100 | 3000 |
| **60** | 15 | 30 | 300 | 600 | 3000 | 4200 | 6000 |
| **70** | 17.5 | 35 | 350 | 700 | 3500 | 4900 | 7000 |

**2** Check that the table drawn up for a partner is appropriate and that they have completed it correctly.

**Practice Book page 29**

| × | | 4 | 8 | 16 | 25 | 50 | 150 |
|---|---|---|---|---|---|---|---|
| **1** | 12 | 48 | 96 | 192 | 300 | 600 | 1800 |
| **2** | 13 | 52 | 104 | 208 | 325 | 650 | 1950 |
| **3** | 14 | 56 | 112 | 224 | 350 | 700 | 2100 |
| **4** | 16 | 64 | 128 | 256 | 400 | 800 | 2400 |
| **5** | 18 | 72 | 144 | 288 | 450 | 900 | 2700 |
| **6** | 20 | 80 | 160 | 320 | 500 | 1000 | 3000 |

Check that students have used efficient strategies.

Stretch zone: One strategy could be to find $\frac{3}{4}$ of the number to be multiplied then multiply that by 100, for example: $75 \times 28 = \frac{3}{4} \times 28 \times 100 = 21 \times 100 = 2100$.

---

## 2B Mental strategies for multiplication and division

### Explore 1   Student Book page 27 • Practice Book page 30

### Specific learning focus

- Multiply pairs of multiples of 10.
- Multiply near multiples of 10 by multiplying by the multiple of 10 and adjusting.
- Multiply decimals by multiples of 10.

### Global skills

- **Creative skills:** exploring
- **Self-development skills:** reflecting on learning

### Key vocabulary

- multiples of 10, near multiples of 10, mental strategy

### Resources

- none needed

### Language support

Discussing in pairs helps students to develop their own mathematical vocabulary. Explaining their strategy to the class allows you to assess their understanding.

 **Introductory activity**

Teach this activity soon after 2B Discover. Explain how to multiply by a decimal, so for example to multiply $20 \times 7.2$, multiply the decimal by 10 first, do the calculation, then divide the answer by 10. So, $20 \times 7.2$ would be $20 \times 72 = 1440$ then $1440 \div 10 = 144$.

Ask students to discuss the following with a partner.

- What strategy can you use to work out $10 \times 40$ or $20 \times 7.2$?
- How can you work out $31 \times 30$ mentally?
- How can you calculate $0.3 \times 40$ in your head?

Arrange students into mixed-attainment pairs. The more-confident student should explain the strategy to the less-confident student, who then shares it with the rest of the class.

Model how to multiply **near multiples of 10** as a reminder from 2B Discover. Write on the board the calculation 19 × 20. Ask students to suggest strategies for solving this. For example, they might calculate 20 × 20 then take away 20. What other strategies can they think of?

Can they apply their strategies to calculate 29 × 30? Can they apply them for 31 × 70?

Can they use the strategy of multiplying by 10 first to calculate 0.4 × 60? Can they use this strategy for 0.7 × 50?

Similarly, you may need to point out other strategies, such as doubling the answer when multiplying by 30, to get the answer when you multiply by 60.

Ask students to work in pairs on the activities on page 27 of the Student Book so they can discuss the strategies they are using and check answers. You can use this activity as an assessment of prior learning.

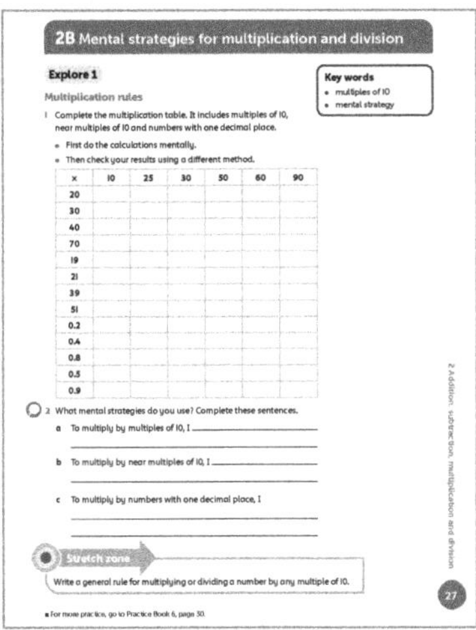

## Differentiation

**Supporting:** Model the strategies for multiplying by near multiples of 10 and by numbers with one decimal place.

**Consolidating:** Ask students to explain their strategies for multiplying by near multiples of 10 and by numbers with one decimal place.

**Extending:** Ask students to explain to a partner and compare their strategies for multiplying by near multiples of 10 and by numbers with one decimal place.

**Stretch zone:** *Write a general rule for multiplying or dividing a number by any multiple of 10.*

Ask students to write short revision notes to explain how to multiply multiples and near multiples of 10. Ask them to include a range of examples including numbers with one decimal place.

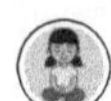 Reflection time

During the Main activity, select students or pairs who can explain their strategies clearly. Ask these students to present to the whole group as the learning review. Other students should share alternative strategies. Ask students to discuss which strategies they think are the most efficient.

**Practice Book:** Students complete Practice Book page 30. They can do this directly after the Main activity, as homework, or as the focus of a separate mathematics session to help students consolidate their learning and build fluency.

Students complete multiplication grids with multiples of 10, near multiples of 10 and decimals. They use their answers in previous grids to help complete the later grids, using known facts.

| Differentiated outcomes | |
|---|---|
| **All students** | should complete a multiplication grid with support. |
| **Most students** | will complete a multiplication grid independently. |
| **Some students** | may support other students by explaining the strategies. |

## Answers

### Student Book page 27

| 1 × | 10 | 25 | 30 | 50 | 60 | 90 |
|---|---|---|---|---|---|---|
| **20** | 200 | 500 | 600 | 1000 | 1200 | 1800 |
| **30** | 300 | 750 | 900 | 1500 | 1800 | 2700 |
| **40** | 400 | 1000 | 1200 | 2000 | 2400 | 3600 |
| **70** | 700 | 1750 | 2100 | 3500 | 4200 | 6300 |
| **19** | 190 | 475 | 570 | 950 | 1140 | 1710 |
| **21** | 210 | 525 | 630 | 1050 | 1260 | 1890 |
| **39** | 390 | 975 | 1170 | 1950 | 2340 | 3510 |
| **51** | 510 | 1275 | 1530 | 2550 | 3060 | 4590 |
| **0.2** | 2 | 5 | 6 | 10 | 12 | 18 |
| **0.4** | 4 | 10 | 12 | 20 | 24 | 36 |
| **0.8** | 8 | 20 | 24 | 40 | 48 | 72 |
| **0.5** | 5 | 12.5 | 15 | 25 | 30 | 45 |
| **0.9** | 9 | 22.5 | 27 | 45 | 54 | 81 |

2 a To multiply by multiples of 10, I move the digits one place to the left, then multiply by how many tens there are.

b To multiply by near multiples of 10, I move the digits one place to the left, multiply by how many tens there are, then adjust by adding or subtracting.

c To multiply by numbers with one decimal place, I multiply the decimal by 10 first to get a whole number, then multiply, then divide by 10 by moving the digits one place to the right.

**1**

| × | **20** | **40** | **30** | **60** | **90** |
|---|---|---|---|---|---|
| **5** | 100 | 200 | 150 | 300 | 450 |
| **60** | 1200 | 2400 | 1800 | 3600 | 5400 |
| **70** | 1400 | 2800 | 2100 | 4200 | 6300 |
| **10** | 200 | 400 | 300 | 600 | 900 |
| **50** | 1000 | 2000 | 1500 | 3000 | 4500 |

**2**

| × | **20** | **40** | **30** | **60** | **90** |
|---|---|---|---|---|---|
| **9** | 180 | 360 | 270 | 540 | 810 |
| **61** | 1220 | 2440 | 1830 | 3660 | 5490 |
| **69** | 1380 | 2760 | 2070 | 4140 | 6210 |
| **11** | 220 | 440 | 330 | 660 | 990 |
| **49** | 980 | 1960 | 1470 | 2940 | 4410 |

**3**

| × | **16** | **25** | **50** | **30** | **60** |
|---|---|---|---|---|---|
| **0.1** | 1.6 | 2.5 | 5 | 3 | 6 |
| **0.4** | 6.4 | 10 | 20 | 12 | 24 |
| **0.8** | 12.8 | 20 | 40 | 24 | 48 |
| **0.2** | 3.2 | 5 | 10 | 6 | 12 |
| **0.5** | 8 | 12.5 | 25 | 15 | 30 |

Stretch zone:

$5000 \times 600 = 3\,000\,000$

$69 \times 800 = 55\,200$

$0.8 \times 900 = 720$

## 2B Mental strategies for multiplication and division

### Explore 2  Student Book page 28 • Practice Book page 31

#### Specific learning focus

- Use known multiplication and division facts to derive new ones.

#### Global skills

- **Creative skills:** exploring
- **Self-development skills:** reflecting on learning

#### Key vocabulary

- known facts, derive

#### Resources

- large sheets of paper

#### Language support

Model for students the language of reasoning when using a known fact to derive new ones, for example:

- I know that $5 \times 3 = 15$, so I can work out $500 \times 3 = 1500$ *because* 500 is 100 times bigger than 5, so the answer *must be* 100 times bigger than 15.
- I know that $24 \div 8 = 3$, so I can work out $2.4 \div 8 = 0.3$ *because* 2.4 is 10 times smaller than 24, so the answer *must be* 10 times smaller than 3.

 **Introductory activity**

Write the following calculation sequences on the board:

| | | | |
|---|---|---|---|
| $7 \times 4 =$ | $9 \times 5 =$ | $36 \div 3 =$ | $72 \div 8 =$ |
| $70 \times 4 =$ | $9 \times 50 =$ | $360 \div 3 =$ | $72 \div 80 =$ |
| $700 \times 4 =$ | $9 \times 500 =$ | $3600 \div 3 =$ | $72 \div 800 =$ |

Ask students to say the answers to the top line of calculations and ask how they know. These should be recalled as known number facts.

Look at the second line and ask a student to choose one calculation and say how it is the same and how it is different from the one above. They should see that one of the numbers has been made 10 times bigger or smaller. Ask what effect this will have on the answer. Repeat for the other calculations in the second line.

Ask students to complete the answers in the third line of calculations using what they know from the ones above.

 **Main activity**

Look together at page 28 of the Student Book. Display on the IWB, if possible. Ask students to look at the Think back text. Discuss together how this shows ways of **deriving** new facts from **known facts**. Ask students to suggest strategies they can use to derive new facts. For example, they might suggest the following.

> 'Multiply one of the numbers by 10 and multiply the answer by 10.'
> 'Divide one of the numbers by 10 and divide the answer by 10.'
> 'Double one of the numbers and double the answer.'

Students should then work in pairs to find derived facts from $8 \times 7 = 56$ using these kinds of strategies.

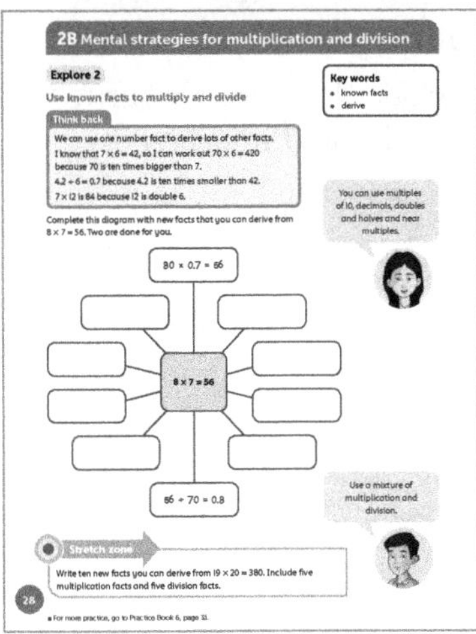

## Differentiation

**Supporting:** Help students identify which parts of a calculation they can change to derive a new fact.

**Consolidating:** Ask students to explain how they derived their new facts.

**Extending:** Challenge students to extend their new facts by multiplying or dividing by larger powers of 10. For example, from $4 \times 3 = 12$, derive $4000 \times 0.03 = 120$ and so on.

**Stretch zone:** *Write ten new facts you can derive from $19 \times 20 = 380$. Include five multiplication facts and five division facts.*

Encourage students to use multiplication and division by numbers other than 2 or powers of 10, for example $19 \times 20 = 380$ so $19 \times 60$ is $380 \times 3$, which is 1140.

 **Reflection time**

Ask pairs to share their diagrams and choose two calculations to explain to the rest of the class. For the first, they should explain their strategy for deriving the new fact from $8 \times 7 = 56$, and for the second, ask the class if they can work out what the strategy was.

**Practice Book:** Students complete Practice Book page 31. They can do this directly after the Main activity, as homework, or as the focus of a separate mathematics session to help students consolidate their learning and build fluency.

Students complete a diagram of related facts for the central fact: $21 \times 8 = 168$. They use near multiples and powers of 10 to complete the diagram.

| Differentiated outcomes | |
| --- | --- |
| **All students** | should derive new facts from known ones, with support. |
| **Most students** | will derive new facts from known ones. |
| **Some students** | may derive new facts from known ones using a wider range of strategies. |

### Answers

**Student Book page 28**

Students choose their own facts derived from $8 \times 7 = 56$ to complete the diagram. Check that the facts are correct.

**Practice Book page 31**

Students choose their own facts derived from $21 \times 8 = 168$ to complete the diagram. Check that the facts are correct.

Stretch zone: Repeat for facts derived from $15 \times 17 = 255$.

# 2C Adding and subtracting near multiples

**Discover**  Student Book page 29 • Practice Book page 32

### Specific learning focus

- Add or subtract a near multiple of 10, 100 or 1000, and adjust.
- Add or subtract near multiples of 1 when adding numbers with one decimal place.

### Global skills

- **Creative skills:** investigating
- **Self-development skills:** reflecting on learning

### Key vocabulary

- addition, subtraction, mental strategy

### Resources

- mini whiteboards and markers

### Language support

It is important that students get used to describing their strategies. Sometimes students will find it easier to do this in their first language. If they find it difficult to describe the strategy in English, ask them to use their first language and support them in translating this into English.

## Introductory activity

Write on the board:

$199 + 345 =$

Students should work in pairs. One of the pair should give the answer to their partner and explain how they know that this is the answer. They can use whiteboards if necessary. Take feedback from one pair. Listen for individuals who can describe their strategy well. Repeat for $3723 - 1998$ and $7.2 + 5.9$. Students should alternate roles.

## Main activity

Write on the board:

$84 + 29 =$

Ask students to work this out using a mental strategy. Choose three students to describe their strategy. Did they all use the same method? Were different methods used? It may be appropriate to model the calculations with a number line or to show students how to adjust the answer by rounding to the nearest 10, 100 or 1000 before calculating.

For example: $84 + 19$ is the same as $(84 + 20) - 1$.

Look together at page 29 of the Student Book. Display on the IWB, if possible. Look together at the speech bubbles and discuss the strategies used. Now ask students, in pairs, to work out the first two calculations in the table. One student carries out the first calculation and the other one checks it, then they swap for the second question. They each describe their strategy before they write it in the table. The best strategy for some of the examples may be counting on. You should also accept any effective mental method. Discuss the most efficient methods with pairs of students.

Students then work in the same pairs to complete the activity on page 29 of the Student Book.

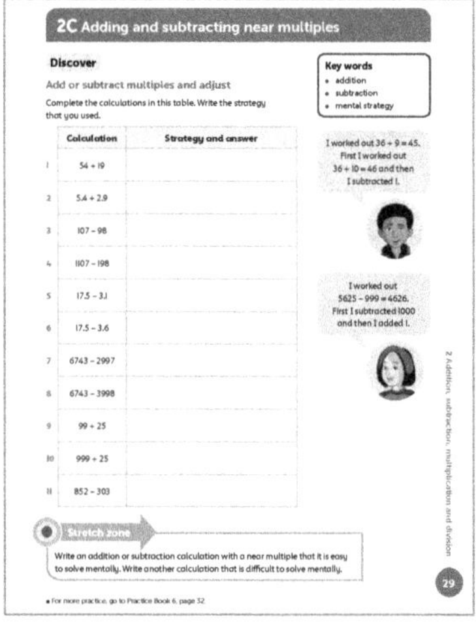

## Differentiation

**Supporting:** Work with a small group to model the strategy of rounding and adjusting.

**Consolidating:** Ask students to carry out the calculations and explain their strategies.

**Extending:** Ask students why they think their strategy is the most efficient.

**Stretch zone:** *Write an addition or subtraction calculation with a near multiple that it is easy to solve mentally. Write another calculation that is difficult to solve mentally.*

Students should write word problems involving near multiples that can be solved by the strategies used in the Main activity. They could use real-life contexts such as money or measurements.

 ## Reflection time

Still working in their pairs, one of each pair should set a calculation for their partner to solve. This should be a calculation similar to those on page 29 of the Student Book. Repeat this four times, with pairs swapping roles each time. Select two examples for the whole class to carry out. For each calculation, make sure that students explain their strategy carefully.

**Practice Book:** Students complete Practice Book page 32. They can do this directly after the Main activity, as homework, or as the focus of a separate mathematics session to help students consolidate their learning and build fluency.

Students write calculations that fit given rules. For example, the calculation must include numbers that are near multiples, or have one decimal place. They are also restricted to using only six given digits in any of the numbers they use.

| Differentiated outcomes | |
|---|---|
| **All students** | should carry out the calculations with support. |
| **Most students** | will carry out the calculations and explain their strategies. |
| **Some students** | may carry out the calculations and use the most efficient strategy. |

## Answers

### Student Book page 29

Students provide their own explanations for the calculations.

**1** $54 + 19 = 73$

**2** $5.4 + 2.9 = 8.3$

**3** $107 - 98 = 9$

**4** $1107 - 198 = 909$

**5** $17.5 - 3.1 = 14.4$

**6** $17.5 - 3.6 = 13.9$

**7** $6743 - 2997 = 3746$

**8** $6743 - 3998 = 2745$

**9** $99 + 25 = 124$

**10** $999 + 25 = 1024$

**11** $852 - 303 = 549$

**Practice Book page 32**

Check that students' calculations match the given rules. While students are working, ask them to explain why they chose the numbers and operations that they used. Answers will vary because students choose their own calculations from the numbers available.

Stretch zone: Check that students have written suitable calculations, for example:

$19\,999 + 5001 = 25\,000$, $18\,997 + 6003 = 25\,000$

---

## 2C Adding and subtracting near multiples

### Explore    Student Book page 30 · Practice Book page 33

### Specific learning focus

- Add or subtract a near whole unit of money, and adjust.

### Global skills

- **Creative skills:** problem solving
- **Real-world skills:** financial literacy

### Key vocabulary

- sum, difference, addition, subtraction

### Resources

- mini whiteboards and markers

### Language support

As with the previous activity, it is important that students get used to describing the strategies they are using. It may be more effective for them to do this in their first language. If they find it difficult to describe the strategy in English, ask them to use their first language and support them in translating this into English. They may also find it easier to use their own currency in money problems.

 **Introductory activity**

Say that a book costs $11.99. Ask students to work out the cost of two books. Give students enough time to work out the answer. They record their answers on their whiteboards and show you, on a count of three. Record the answer on the board for students to refer to. Next, ask students to work out the cost of 10 books, then 20 books, 50 books and 100 books. *Can you use earlier answers to help you work out later ones? For example, If you know the cost of 2 books, how does this help you find the cost of 20 books?*

 **Main activity**

Ask students to refer to page 30 of the Student Book. Ask, *I buy a sports magazine and a puzzle magazine. How much do I spend?*

Give students time to carry out the calculation in pairs. Ask them to hold up their whiteboards so you can see the answer. Select one student to describe the strategy they used. If any students have made an error, ask them to talk through their calculation so that they can find their mistake. Repeat with the question:

*I go to the shops with $23.55 and buy a fashion magazine. How much do I have left?*

Pairs should take it in turns to set one another word problems to solve. They only write the problem down when they have both carried out the calculation and can describe the strategy they have used. As students work on the activity, circulate among them and collect a range of examples to use in Reflection time.

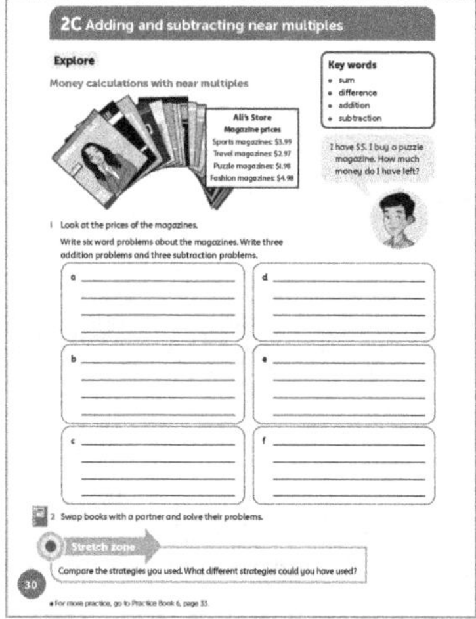

### Differentiation

**Supporting:** Ask students to create questions for you so you can model how to solve them.

**Consolidating:** Ask students to challenge themselves to create harder problems.

**Extending:** Encourage students to create two- and three-step problems.

**Stretch zone:** *Compare the strategies you used. What different strategies could you have used?*

Ensure that students are able to describe the strategies they used and understand when there might be a more efficient strategy.

As a follow-up activity, students could visit shops and set word problems using real prices from the shops. They can set these for one another to solve or use them with family members.

### Reflection time

Ask a student to come to the front and set the class one of their problems. (Choose someone that you identified during the Main activity.) Students should solve this in pairs and tell the student who set the problem how they carried out the calculation. Repeat with a range of problems you have seen.

**Practice Book:** Students complete Practice Book page 33. They can do this directly after the Main activity, as homework, or as the focus of a separate mathematics session to help students consolidate their learning and build fluency.

Students solve problems based on the cost of stationery items in a shop. They use their knowledge of near multiple strategies and apply this to multiplication to solve the problems.

| Differentiated outcomes | |
| --- | --- |
| **All students** | should create simple problems. |
| **Most students** | will create more complex problems. |
| **Some students** | will create two- and three-step problems. |

## Answers

### Student Book page 30

Students make up their own addition and subtraction problems. Check that their answers are correct.

### Practice Book page 33

1  $14.95

2  $24.90

3  $11.97

4  $29.94

5  $29.88

6  $29.90

7  To make customers think things are cheaper.

Stretch zone: 5 notebooks cost $9.98 so Sam could charge $2 per notebook.

## 2D Estimating first in calculations

### Discover  Student Book page 31 • Practice Book page 34

#### Specific learning focus

- Estimate, then calculate, the addition of two three-digit numbers.

#### Global skills

- **Creative skills:** investigating

#### Key vocabulary

estimate, sum, smallest, largest, place value

#### Resources

- mini whiteboards and markers
- large digit cards 4, 5, 6, 7, 8, 9 and + sign

#### Language support

Use the language of estimation to help students work out approximate answers. For example, in 695 + 294, estimate these as 700 + 300 so the answer is approximately 1000.

 Introductory activity

Ask six students to come to the front and choose one of the six number cards 4, 5, 6, 7, 8 or 9. Ask them to stand in a line. Hold the + sign so that you make an addition of two 3-digit numbers (for example 548 + 769). Ask students, working in pairs, to estimate then calculate the answer mentally and write the answer on their whiteboards. Ask, *Which method did you use to calculate the answer mentally?* Discuss how the estimate is a good way to check whether the answer looks approximately correct.

Ask students to move one student in the line to a different position, to make the answer bigger (for example 584 + 769.) Ask pairs to estimate, then calculate this mentally. Discuss the difference between the two answers. Next, ask students to move a different person in the line, this time to make a smaller answer. Again, ask students to estimate and then calculate the answer mentally. Discuss the difference between the two answers.

 Main activity

Refer students to the speech bubbles on page 31 of the Student Book. Discuss how estimating helps them to get as close as possible to the target numbers. For example, with a target of 900, they might realise quickly that the hundreds digits should be 400 and 500.

Ask students whether they can use the numbers 4, 5, 6, 7, 8, 9 once each to make a calculation with the largest possible total. Is there more than one way to do it? What strategies do they use? Ask students to share their answers and explain their strategies. Can anyone make a larger answer?

Students complete the rest of the activities on page 31 in pairs. Encourage pairs and groups to share their answers.

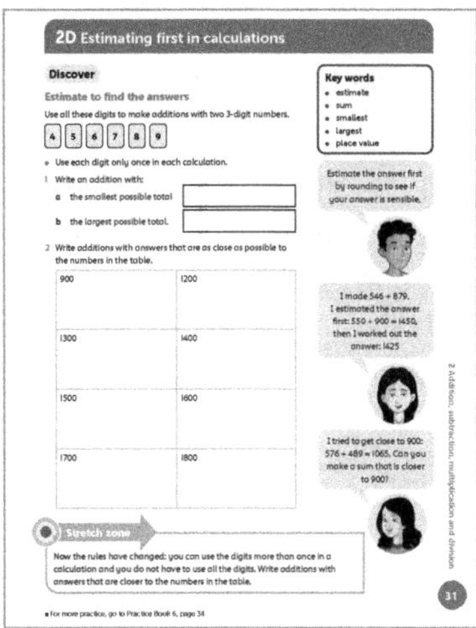

## Differentiation

**Supporting:** Focus on adding two 3-digit numbers, and model estimation.

**Consolidating:** Ask students to explain how they used estimation to get as close as possible to the target number.

**Extending:** Encourage students to create their own calculations, which can be solved easily using estimation and adjusting.

**Stretch zone:** *Now the rules have changed: you can use the digits more than once in a calculation and you do not have to use all the digits. Write additions with answers that are closer to the numbers in the table.*

Encourage students to adjust their estimations to find answers that are now closer to the target numbers.

 Reflection time

For each number in the table on page 31, ask a pair of students to share their estimation and answer. They can explain their strategy to the class and say how they used estimating to help them get close to the target numbers. Then discuss whether the strategy was efficient. See whether other students can suggest alternative strategies for each calculation.

**Practice Book:** Students complete Practice Book page 34. They can do this directly after the Main activity, as homework, or as the focus of a separate mathematics session to help students consolidate their learning and build fluency.

Students estimate and then calculate addition and subtraction calculations. They can use a mental or written method to solve the calculations.

| Differentiated outcomes | |
|---|---|
| **All students** | should estimate the addition of two 3-digit numbers and calculate the answer mentally with support. |
| **Most students** | will estimate the addition of two 3-digit numbers and calculate the answer mentally. |
| **Some students** | may use estimation to spot errors as well as calculate mentally. |

## Answers

### Student Book page 31

**1 a** smallest possible total: 468 + 579

   **b** largest possible total: 864 + 975

**2** 900: 468 + 579 = 1047. They have to use the 4 and 5 as hundreds, then the smallest possible number from the remaining digits.

   1200: 648 + 579 = 1227

   1300: 748 + 569 = 1317

   1400: 748 + 659 = 1407

   1500: 847 + 659 = 1506

   1600: 846 + 759 = 1605

   1700: 946 + 758 = 1704

   1800: 946 + 857 = 1803

### Practice Book page 34

**1** Estimate 180 + 30 = 210; Answer = 203.9

**2** Estimate 120 – 20 = 100; Answer = 101.3

**3** Estimate 210 + 700 = 910; Answer = 908

**4** Estimate 160 + 1000 = 1160; Answer = 1155

**5** Estimate 100 – 90 = 10; Answer = 9.7

**6** Estimate 400 + 150 = 550; Answer = 544

**7** Estimate 970 + 430 = 1400; Answer = 1391

**8** Estimate 480 – 140 = 340; Answer = 338.8

**9** Estimate 100 – 40 = 60; Answer = 54.1

**10** Estimate 60 + 100 = 160; Answer = 155.63

Stretch zone: 44.5 + 54.1 = 98.6, 155.63 – 98.75 = 56.88

# 2D Estimating first in calculations

## Explore
Student Book page 32 · Practice Book page 35

### Specific learning focus

- Estimate, then calculate, the addition or subtraction of two 3-digit numbers, including tenths.

### Global skills

- **Creative skills:** problem solving

### Key vocabulary

- written method, column method, recombine

### Resources

- mini whiteboards and markers

### Language support

Model the language of exchange and recombining when adding or subtracting, for example as follows.

- Exchange a ten for 10 ones.
- Exchange a hundred for 10 tens.
- Recombine 12 ones to make 1 ten and 2 ones.

##  Introductory activity

Write these calculations on the board:

$887 + 296 =$ $694 - 298 =$

Ask students to work with a partner and estimate the answers to each calculation. For each calculation, compare estimates from different pairs and check that the estimates are the same or approximately equal. If not, ask a few pairs to explain how they estimated and agree on a suitable strategy for each calculation, for example: $900 + 300 = 1200$ and $700 - 300 = 400$. Now ask students to write the calculations on their whiteboards and work out each one using a column method. Remind them about exchanging and recombining when necessary.

Share the answers and methods. Check that the exchanging has been completed correctly. Look together at the subtraction calculation. Ask whether it could have been done mentally. Check back against the estimates to see whether the answer looks approximately correct. Some students may suggest using a number line for this subtraction. You might suggest that by adding 2 to each number, the difference remains the same but the calculation is then easier: $696 - 300 = 396$.

##  Main activity

Write on the board:

```
  485
+ 778
-----
 1153
```

Give students a moment to check the column addition. Ask, *Is the answer correct? How do you know?* Students may say they estimated it as $500 + 800 = 1300$ and this is a long way from the 1153 answer.

Ask students to see whether they can see what is wrong with the calculation. They may be able to say that the exchange and **recombine** method has not been used properly. In the ones column, $5 + 8$ makes 13, and although the 3 ones have been recorded, the ten has not been carried over to the tens column. In the same way, in the tens column, $8 + 7 = 15$ (and the carried ten should make 16) but the 10 tens have not been carried to the hundreds column.

Students should now work in pairs to complete the activities on page 32 in the Student Book. Refer students to the speech bubble, which reminds them to estimate and use known facts to help them to spot errors. It is important for students to check and correct the errors in the questions on page 32. This will help them to spot their own errors in the future.

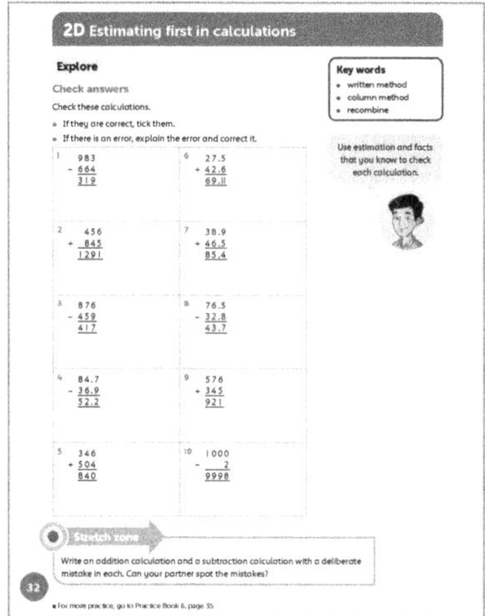

## Differentiation

**Supporting:** Help students to focus on spotting errors in the calculations that add two 3-digit numbers by modelling how to estimate.

**Consolidating:** Encourage students to estimate to spot errors and then to explain what the errors are.

**Extending:** Encourage students to create their own sets of calculations containing errors for partners to work on using estimation to check whether the answers are approximately correct.

**Stretch zone:** *Write an addition calculation and a subtraction calculation with a deliberate mistake in each. Can your partner spot the mistakes?*

Students should estimate the answers to their partner's calculations first so they know what the answer is likely to be close to. This will make it easier to spot the mistake.

You can ask students to create 'incorrect' questions for their parents to 'correct' at home.

 ## Reflection time

There are five errors in the calculations on page 32. In each case, ask a different student to come to the front and explain the error. Then ask them to calculate the answer correctly and to model the method to the rest of the class. *How did estimating help you to identify which calculations contained errors?*

**Practice Book:** Students complete Practice Book page 35. They can do this directly after the Main activity, as homework, or as the focus of a separate mathematics session to help students consolidate their learning and build fluency.

Students use their estimating skills to write calculations, using a given set of digits, to get an answer as close as possible to a target number.

| Differentiated outcomes | |
|---|---|
| **All students** | should use estimation, and other methods, to spot errors with support. |
| **Most students** | will use estimation, and other methods, to spot errors. |
| **Some students** | may use estimation, and other methods, to spot errors and create their own deliberate mistakes in calculations. |



**Student Book page 32**

1   correct

2   1301 – the carried '1' from the ones into the tens was not added.

3   correct

4   47.8 – the ones and tens have been recorded as the difference between the digits but the wrong way round, for example in the ones, 7 – 9 is not 2.

5   850 – the carried '1' from the ones into tens was not added.

6   70.1 – tenths total has been entered as two digits in one column and not carried into the ones.

7   correct

8   correct

9   correct

10  998 – the pattern has been extended one column too many into the thousands.

**Practice Book page 35**

Students choose their own numbers from digit cards to make additions and subtractions with answers that fit the challenges. Check that students' answers to their calculations are correct.

Stretch zone: For example, $1 + 3 + 5 + 7 + 9 = 25$, or $35 – 9 – 1 = 25$.

## 2E Which operation?

**Discover**   Student Book page 33 • Practice Book page 36

### Specific learning focus

- Solve addition and subtraction problems using numbers with the same and different numbers of decimal places, including amounts of money, but choosing the correct operation.

### Global skills

- **Creative skills:** problem solving

### Key vocabulary

- total, difference, how much longer/bigger/smaller?, how much is left?

### Resources

- mini whiteboards and markers
- metre rules or measuring tapes



Make links between money, other real-life measures contexts and the vocabulary of decimals. Use 'tenths' and 'hundredths'.

 ## Introductory activity

Ask pairs to write down as many examples as they can of the use of decimals in real life (for example money, length, volume, mass). Ask a student to measure your height. Ask another student to measure the height of the door. Ask pairs of students to calculate the difference between your height and the height of the door. Then ask pairs to measure objects in the classroom. *Make a list of objects with a total length as near to 2 metres as possible.*

 ## Main activity

Look together at page 33 of the Student Book. Display on the IWB, if possible. Refer students to the Think back question. Encourage students to look for key words in problems to help them decide which operation to use.

Write this problem on the board:

*Two sides of a garden need new fence. One side is 28.45 m long and the other side is 14.95 m long. What is the total length of fencing that needs to be bought?*

Ask students to identify the key word that tells them which operation to choose. They should be able to say that 'total' tells them to add. Encourage students to estimate the total length first: $30 + 15 = 45$ m. Now ask them, using their whiteboards, to use a written method to calculate the total. Ask one student to share their method and answer and see whether the rest of the class agree.

Now ask students to solve this problem: 'Find the difference in the lengths of the two sides of the garden.' Ask students to identify the key word that tells them which operation to choose. They should be able to say that 'difference' tells them to subtract. They should estimate the answer first before calculating. Ask a different student to share their method and answer, then ask whether anyone had a more efficient method.

Encourage students to complete the word problems on page 33 of the Student Book in pairs. Select two pairs to share their solutions with the class in Reflection time.

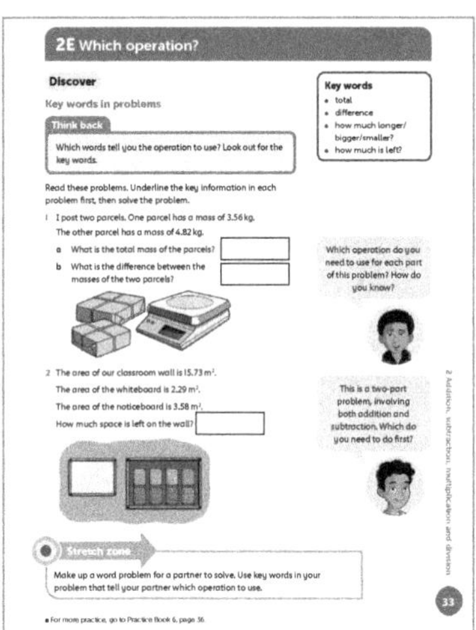

## Differentiation

**Supporting:** Help students to understand the word problems and form the correct calculations.

**Consolidating:** Encourage students to create word problems of their own that use adding or subtracting measures.

**Extending:** Encourage students to create two-step word problems using measures.

**Stretch zone:** *Make up a word problem for a partner to solve. Use key words in your problem that tell your partner which operation to use.*

As an extra activity, students can use real-life data from magazines or newspapers to ask problems of one another and their parents or other adults at home. Remind them to include the key words of addition and subtraction in their problems.



## Reflection time

Ask one pair to share their strategies for the two problems you selected earlier for the rest of the class. Ask another pair to come to the front of the class to solve the problems on the board. Encourage the rest of the class to support them. Discuss the strategies used and ask the class whether they can solve the problems using efficient mental methods. Look back to the problems in the Student Book and ask students to say which words in the problems indicated which operations to choose to solve them.

**Practice Book:** Students complete Practice Book page 36. They can do this directly after the Main activity, as homework, or as the focus of a separate mathematics session to help students consolidate their learning and build fluency.

Students need to select the correct information from word problems in order to solve the problems. The problems require them to use addition, subtraction or both.

| Differentiated outcomes | |
| --- | --- |
| **All students** | should solve the word problems by choosing the correct operations with support. |
| **Most students** | will solve the word problems by choosing the correct operation and set simple word problems. |
| **Some students** | may explain their methods clearly including choosing the correct operation and set two-step problems of their own. |

## Answers

### Student Book page 33

**1 a** 8.38 kg

**b** 1.26 kg

**2** 9.86 m$^2$

### Practice Book page 36

**1** $48.34

**2** No, the game costs $1.45 more than you have.

**3** $28.81

Stretch zone: Check that students' word problems are suitable and have been calculated correctly. Check that the addition or subtraction 'clue' words are included in the word problems.

# 2E Which operation?

## Explore   Student Book page 34 • Practice Book page 37

### Specific learning focus

- Solve addition and subtraction problems using numbers with the same and different numbers of decimal places, including amounts of money, but choosing the correct operation.

### Global skills

- **Creative skills:** problem solving
- **Real-world skills:** interpreting information

### Key vocabulary

- total, difference, how much longer/bigger/smaller? how much is left?

### Resources

- mini whiteboards and markers

### Language support

Help students with interpreting the word problems by highlighting the key words and numbers they need for the calculations. Support them by encouraging them to look for question words that can help, for example, '**How much**?', 'What is the total?' and so on.

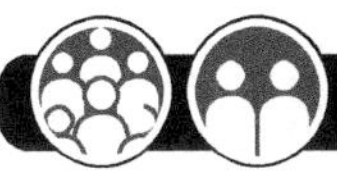 **Introductory activity**

Write this problem on the board:

*Rachael wants to buy a boat. She has saved $1347. If the price of the boat is $2580, how much money does she still have to save?*

Ask students to discuss in pairs and identify the question words for the calculation. (How much?) Ask whether the calculation is an addition or a subtraction. *How did you decide? What words in the question tell you whether it is an addition or subtraction?*

Ask pairs to solve the problem on their whiteboards using a mental or written method of their choice, and to check their answers using an inverse operation.

 **Main activity**

Discuss with students the things they need to remember when solving word problems.

- Read the question carefully.
- Identify the numbers in the question that they will use for the calculation.
- Identify the key words in the question that tell them what sort of thing they are calculating.

- Identify the question words that suggest whether it will be an addition or subtraction.
- Choose the correct strategy for solving the problem, then complete the calculation using a suitable mental or written method, then check using an inverse operation.
- Check that the answer makes sense with the question.

Now ask students to complete the word problems on page 34 of the Student Book. They should work individually but then compare answers with a partner.

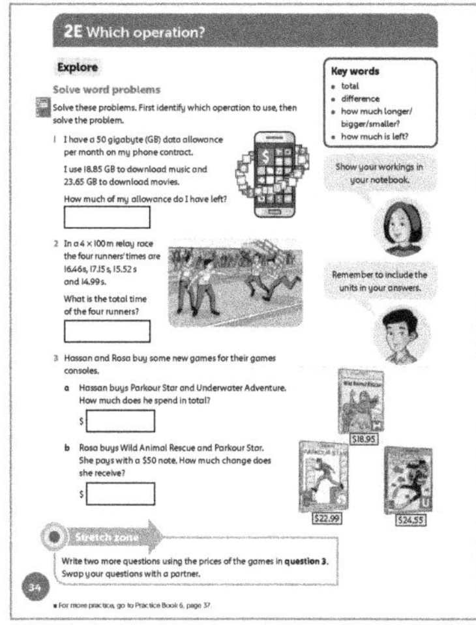

## Differentiation

**Supporting:** Help students to understand the word problems and form the correct calculations.

**Consolidating:** Encourage students to create word problems of their own that use addition or subtraction.

**Extending:** Encourage students to create two-step word problems with addition and subtraction.

**Stretch zone:** *Write two more questions using the prices of the games in question 3. Swap your questions with a partner.*

Ensure that students have included the key words in their problems that tell their partner whether they need to add or subtract (or both!).

 **Reflection time**

Ask some students who have completed the Stretch zone to share problems they have created for the rest of the class to work out. Make sure they have worded the problems in a clear way and considered the operation that needs to be used to solve the word problem. Other students can work out the problems and describe their strategies.

**Practice Book:** Students complete Practice Book page 37. They can do this directly after the Main activity, as homework, or as the focus of a separate mathematics session to help students consolidate their learning and build fluency.

Students need to select the correct information from the word problems in order to solve them. Students need to use addition, subtraction or both to solve these problems.

| Differentiated outcomes | |
| --- | --- |
| **All students** | should solve the word problems by choosing the correct operations with support. |
| **Most students** | will choose the correct operations to solve the word problems and set simple word problems. |
| **Some students** | may explain their methods clearly including choosing the correct operations and set two-step problems of their own. |



### Student Book page 34

**1** 7.5 GB

**2** 64.12 seconds (or 1 minute and 4.12 seconds)

**3 a** $47.54     **b** $8.06

### Practice Book page 37

**1** $4.34

**2** $20.55

**3** $11.26

Stretch zone: $43 + $40+ $56 = $139, so the headteacher needs to pay $11, which is less.

# 2F Multiplying 3- and 4-digit numbers

**Discover**   Student Book pages 35–36 • Practice Book page 38

## Specific learning focus

- Multiply 2- or 3-digit numbers by single- or 2-digit numbers.
- Multiply pairs of multiples of 10.

## Global skills

- **Creative skills:** problem solving

## Key vocabulary

- partitioning, grid method, column method

## Resources

- mini whiteboards and markers
- digit cards 0–9

## Language support

Model and use the language of place value – hundreds, tens, ones, partition, exchange, recombine – while explaining the calculation strategies. Correct any mistakes with language that students make.

### Introductory activity

Ask three students to come to the front of the class. Choose three digit cards to make a 3-digit number. Write the number on the board. Ask the students at the front of the class to **partition** the number on their whiteboards. They then stand in order at the front of

the class to show the result on their whiteboards. For example: 216 is shown as 200 + 10 + 6. Repeat with a 3-digit number with two decimal places. For example: $2.65 is shown as 2 + 0.6 + 0.05

### Main activity

Look together at page 35 of the Student Book. Display on the IWB, if possible. Refer students to the Think back text to remind them of the different strategies they can use to multiply larger numbers, depending on the numbers involved, or the strategy they feel most confident with. Model the **grid method** as follows. Remind the class that they have seen this method in the previous year. *We are going to calculate 216 × 52. So, we draw the following grid.*

| × | 50 | 2 | |
| --- | --- | --- | --- |
| 200 | | | |
| 10 | | | |
| 6 | | | |

Work through the multiplication step-by-step, reminding students that the grid represents the partitioning of the numbers until the grid is complete. So, the grid represents:

$(200 \times 50) + (200 \times 2) + (10 \times 50) + (10 \times 2) + (6 \times 50) + (6 \times 2)$

First, estimate the answer: 50 × 200 is larger than 10 000. Next, carry out the multiplications to fill each cell. Then total all the cells to find the answer by adding across the cells, and finally by adding the final column.

Ask students to work on the activities on pages 35–36 of the Student Book in pairs. Tell them to check each other's answers for questions 2 to 5. You may choose to group the less-confident students together so that you can model the grid method for them.

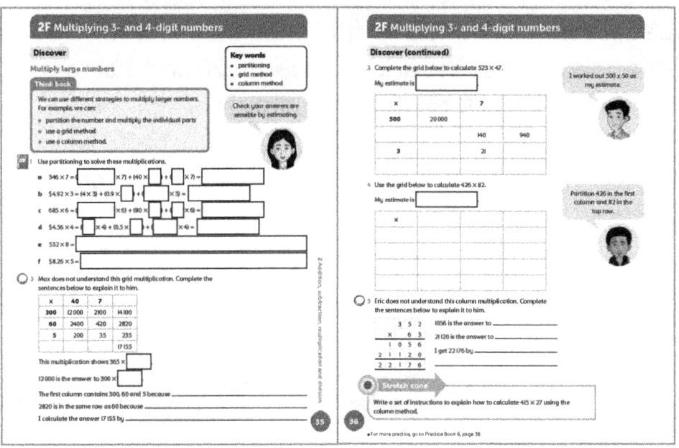

## Differentiation

**Supporting:** Work with less-confident students to support them in using the grid method.

**Consolidating:** Ask students how the **column method** relates to the grid method.

**Extending:** Ask students to explain the column method step by step.

**Stretch zone:** *Write a set of instructions to explain how to calculate 415 × 27 using the column method.*

You could give students a template such as:

First, I …
Then, I …
Finally, I …

For example: First, I multiply 415 × 7 and the answer is 2905.
Then, I multiply 415 × 20 and the answer is 8300.
Finally, I add these two products together and I get 11 205.

As a follow-up activity, students work in pairs to choose five digit cards to create another similar calculation. One student explains the method. The other student follows their instructions and completes the calculation on their whiteboard.

 **Reflection time**

Write 415 × 27 on the board. Ask confident students to explain how to carry out the calculation using both the grid method and a column method. Follow the explanation on the board. Make sure that you follow the instructions correctly.

**Practice Book:** Students complete Practice Book page 38. They can do this directly after the Main activity, as homework, or as the focus of a separate mathematics session to help students consolidate their learning and build fluency.

Students use partitioning to solve 3- and 4-digit × single-digit calculations and then use the grid method to solve 3- and 4-digit × 2-digit calculations. Help students to partition the numbers for the grid method, then they solve the calculations independently.

## Differentiated outcomes

| | |
|---|---|
| **All students** | should use the grid method with support. |
| **Most students** | will use the grid method to carry out calculations. |
| **Some students** | may model the column method for other students. |

## Answers

### Student Book pages 35–36

**1**  **a**  $346 \times 7 = (300 \times 7) + (40 \times 7) + (6 \times 7)$
$= 2100 + 280 + 42 = 2422$

   **b**  $\$4.92 \times 3 = (4 \times 3) + (0.9 \times 3) + (0.02 \times 3)$
$= 12 + 2.7 + 0.06 = \$14.76$

   **c**  $685 \times 6 = (600 \times 6) + (80 \times 6) + (5 \times 6)$
$= 3600 + 480 + 30 = 4110$

   **d**  $\$4.56 \times 4 = (4 \times 4) + (0.5 \times 4) + (0.06 \times 4)$
$= 16 + 2 + 0.24 = \$18.24$

   **e**  $532 \times 8 = (500 \times 8) + (30 \times 8) + (2 \times 8)$
$= 4000 + 240 + 16 = 4256$

   **f**  $\$8.26 \times 5 = (8 \times 5) + (0.2 \times 5) + (0.06 \times 5)$
$= 40 + 1 + 0.3 = \$41.30$

**2**  $365 \times 47$

   $300 \times 40$

   … $300 + 60 + 5 = 365$

   … it is the total of $60 \times 40$ and $60 \times 7$. (Accept other correct answers.)

   … adding all the individual totals together. (Accept other correct answers.)

**3**  Estimate, for example, $500 \times 50 = 25\,000$

| × | 40 | 7 | |
|---|---|---|---|
| **500** | 20 000 | 3500 | 23 500 |
| **20** | 800 | 140 | 940 |
| **3** | 120 | 21 | 141 |
| | | | 24 581 |

**4**

| × | 80 | 2 | |
|---|---|---|---|
| **400** | 32 000 | 800 | 32 800 |
| **20** | 1600 | 40 | 1640 |
| **6** | 480 | 12 | 492 |
| | | | 34 932 |

**5**  $1056 = 352 \times 3$

   $21\,120 = 352 \times 60$

   I get 22 176 by adding the totals together.

| | | | |
|---|---|---|---|
| **1** | 3192 | **5** | 60 613 |
| **2** | 2445 | **6** | 224 928 |
| **3** | 18 144 | **7** | 13 886 |
| **4** | 32 286 | **8** | 34 536 |

**Stretch zone:** Write 400 then 70 then 2 down the right-hand side of the grid, then write 20 and 3 across the top of the grid. You then multiply for each cell: $400 \times 20 = 8000$, $400 \times 3 = 1200$, $70 \times 20 = 1400$, $70 \times 3 = 210$, $2 \times 20 = 40$, $2 \times 3 = 6$. Then you add all the totals together: $8000 + 1200 + 1400 + 210 + 40 + 6 = 10 856$.

---

# 2F Multiplying 3- and 4-digit numbers

## Explore   Student Book page 37 • Practice Book page 39

### Specific learning focus

- Multiply 3- or 4-digit numbers by 2- and 3- digit numbers.

### Global skills

- **Creative skills:** investigating
- **Interpersonal skills:** teamwork

### Key vocabulary

- product, column method

### Resources

- mini whiteboards and markers

### Language support

As students work on the activity, ask them to estimate the product before they calculate, to reinforce the vocabulary of multiplication.

Ask, for example:

- *What do you think the product will be?*
- *Will this product be larger or smaller than that product?*

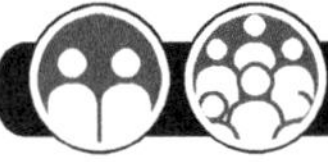 **Introductory activity**

Arrange students in pairs so that a more-confident student partners a less-confident student. Write these calculations on the board:

$45 \times 768 =$        $54 \times 768 =$

Ask pairs to discuss and decide which multiplication has the larger product. Take feedback and ask students to explain their answer.

Then write: $64 \times 587 =$     $58 \times 764 =$

*Which multiplication has the larger product?* This time, ask students to estimate a product to each calculation first. Ask students to explain their answer.

 **Main activity**

Put students into groups of four by doubling up pairs. Ask groups to work on the activity on page 37 of the Student Book. They can try different calculations on their whiteboards. Emphasise that they do not need to find all the possible products. Tell them to use their knowledge of place value to try to decide which calculations are most likely to give the largest products.

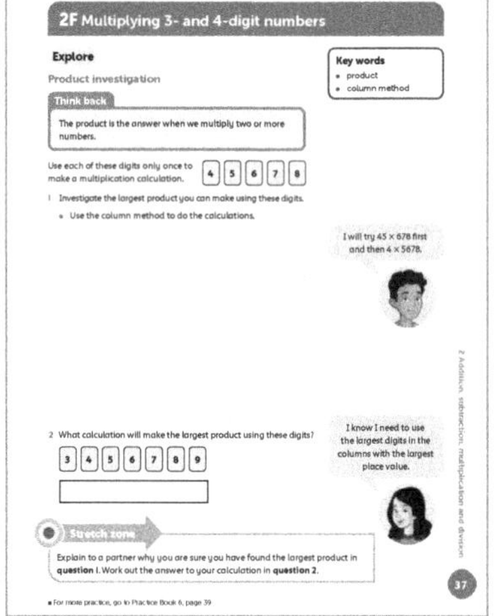

### Differentiation

All students should engage in the investigation together.

**Supporting:** Help students to record their calculations so that they can avoid repeating ones they have already done.

**Consolidating:** Ask students to explain why they think a pair of numbers will give the largest product.

**Extending:** Challenge students to find the smallest possible answer using the same digits and describe what they notice about the calculations.

**Stretch zone:** *Explain to a partner why you are sure you have found the largest product in question 1. Work out the answer to your calculation in question 2.*

When students are confident that they have found the largest product using the digits in question 2, ask them to find the next largest product. *Can you now find the difference between the two answers and explain the reason for this?*

 **Reflection time**

Ask groups to present their result to the rest of the class. Ask other students to comment on any differences in their solutions. *Did everyone get the same largest product? Can you explain how you know you have the largest product?*

**Practice Book:** Students complete Practice Book page 39. They can do this directly after the Main activity, as homework, or as the focus of a separate mathematics session to help students consolidate their learning and build fluency.

Students practise the column method, multiplying 3- and 4-digit numbers by 2-digit numbers. Encourage students to estimate their answers first before using a compact column method to solve each calculation.

| Differentiated outcomes | |
|---|---|
| **All students** | should find the largest products using given digits with support. |
| **Most students** | will find the largest products using given digits. |
| **Some students** | may find the largest products with given digits and explain their strategy. |

## Student Book page 37

**1** $864 \times 75 = 64\,800$ is the largest product.

**2** $9753 \times 864 = 8\,426\,592$

## Practice Book page 39

**1** 13 708

**2** 8505

**3** 61 854

**4** 67 494

Stretch zone: $2 \times 3214 = 6428$, so $20 \times 3214 = 64\,280$, $64\,280 + 3214 = 67\,494$

---

# 2G Long division

**Discover**  Student Book page 38 • Practice Book page 40

### Specific learning focus

- Divide 3-digit and 4-digit numbers by 2-digit numbers (no remainders) using the written method for long division.

### Global skills

- **Creative skills:** exploring

### Key vocabulary

- divisor, dividend, quotient

### Resources

- mini whiteboards and markers

### Language support

Students may need support in their explanations for this activity. Encourage them by asking questions, for example:

- *Why is that the first step?*
- *What is your estimate for the answer?*
- *What number facts do you know about that number?*
- *You know 10 times a number. How can you work out 15 times that number?*

 **Introductory activity**

Group students in pairs and try to group a more-confident student with a less-confident student. Write on the board: $912 \div 16$. Ask pairs to estimate the answer and write their estimates on their whiteboards. Give them two minutes to do this. Pick two or three different pairs of students to share their estimates. *How did you decide your estimate?* Compare the estimates and strategies and discuss these as a class.

Refer students to the Think back explanation and remind students that each number in the division calculation has a name. The number being divided is the **dividend**, the number you are dividing by is the **divisor** and the answer is the **quotient**. Check that they can pronounce quotient ('kwo-shunt').

Ask students to try to estimate the quotient to $672 \div 32$ and share the answers and strategies again.

 **Main activity**

Look together at page 38 of the Student Book. Display on the IWB, if possible. Remind students that division is a process of sharing. A division with no remainder means that the amount can be shared equally with nothing 'left over'. Model the method for doing the calculation by long division and say the phrases at each stage, using the example calculation from the Think back as follows.

$845 \div 13$

$$13\overline{)845}$$ How many thirteens in 8 (hundreds)?
We cannot do this so we move on.

$$\begin{array}{r} 6 \\ 13\overline{)845} \end{array}$$ Look at the first two digits (84 tens)
How many thirteens in 84 (tens)?
$6 \times 13 = 78$ so $60 \times 13 = 780$

$$\begin{array}{r} 6 \\ 13\overline{)845} \\ -780 \\ \hline 65 \end{array}$$ Subtract 780 from the dividend:
$845 - 780 = 65$

How many thirteens in 65?
$5 \times 13 = 65$

$$\begin{array}{r} 65 \\ 13\overline{)845} \\ 780 \\ \hline 65 \end{array}$$ $845 \div 13 = 65$

Ask students to work in pairs for the activity on page 38 the Student Book. Encourage them to explain the calculations to one another as they work through them. This gives them support with the calculations and the language. As they work, move around the room and ask, for example, *Why did you try that? How many of those go into that?*

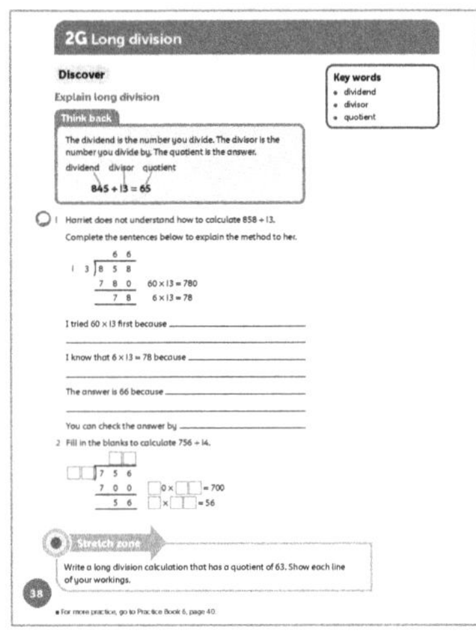

## Differentiation

**Supporting:** Work with less-confident students to support them in using the long written method.

**Consolidating:** Ask students to explain the long written method to you as they use it.

**Extending:** Ask students to explain the long written method step by step.

**Stretch zone:** *Write a long division calculation that has a quotient of 63. Show each line of your workings.*

Students could explain their method to their parents or other adults at home. They could also set their parents or other adults some practice calculations.

 ## Reflection time

Write $901 \div 17$ on the board. Ask one, more-confident, student to explain how to carry out the calculation, using long division. Write down each step as they describe it. Ask questions to make sure that students are clear about each step, for example:

- *Why do you write that digit in that column?*
- *What is the value of that digit?*

**Practice Book:** Students complete Practice Book page 40. They can do this directly after the Main activity, as homework, or as the focus of a separate mathematics session to help students consolidate their learning and build fluency.

Students solve long division calculations. There is a step-by-step reminder for students to work through first then to use as a template for their calculations. Remind students to estimate their answers first.

| Differentiated outcomes | |
|---|---|
| **All students** | should use the long division written method with support. |
| **Most students** | will use the long division written method to carry out calculations. |
| **Some students** | may explain the steps as they model the long division written method. |

## Answers

### Student Book page 38

**1** I tried $60 \times 13$ first because I know that $60 \times 12$ is 720.

I know $6 \times 13 = 78$ because $6 \times 12$ is 72.

The answer is 66 because $60 + 6$ is 66.

You can check the answer by multiplying 66 and 13.

**2**
$$\begin{array}{r} 54 \\ 14\overline{)756} \\ 700 \\ \hline 56 \end{array}$$
$50 \times 14 = 700$
$4 \times 14 = 56$

### Practice Book page 40

**1** 345

**2** 186

**3** 412

**4** 291

Stretch zone: Students choose their own numbers for the calculation. Check that they have used the long division method correctly.

# 2G Long division

## Explore  Student Book page 39 • Practice Book page 41

### Specific learning focus

- Divide 3-digit and 4-digit numbers by 2-digit numbers (no remainders) using the written method for long division.

### Global skills

- **Creative skills:** problem solving
- **Self-development skills:** reflecting on learning

### Key vocabulary

- divisor, dividend, quotient

### Resources

- mini whiteboards and markers

### Language support

A good way to organise the group work to support language development is to ask one of the group to describe the process and the rest to write down the solution. Tell the students who are writing not to write down anything they don't understand until the explanation is clarified. This ensures that the explanation is clear.

### Introductory activity

Write on the board the number fact $28 \times 16 = 448$. Ask students whether they can tell you the answers to these division calculations:

$$448 \div 28 = \qquad 448 \div 16 =$$

Discuss how the two division facts are related to the multiplication fact. Recall for students the idea of fact families.

Ask students to write on their whiteboards the fact families for $19 \times 17 = 323$ and $42 \times 18 = 756$.

### Main activity

Tell students you are thinking of a division calculation that has a quotient of 26. You don't know what the dividend and divisor are, but the dividend has three digits and the divisor has two digits. Ask students to use fact families to think of possible calculations.

You could give one example, such as $364 \div 14 = 26$, and ask how you knew that the dividend in this calculation was 364. (because you multiplied 26 by a 2-digit number)

Put together mixed-attainment groups of four students to work on the activity on page 39 of the Student Book. They will need to try different calculations. Emphasise that they do not need to find all the possible calculations, just those that meet the requirements of the question. Tell them to use their knowledge of place value to try to decide which calculations are most likely to give the correct answer and then to carry out the calculation using the long division method. They can use fact families to help them identify possible calculations.

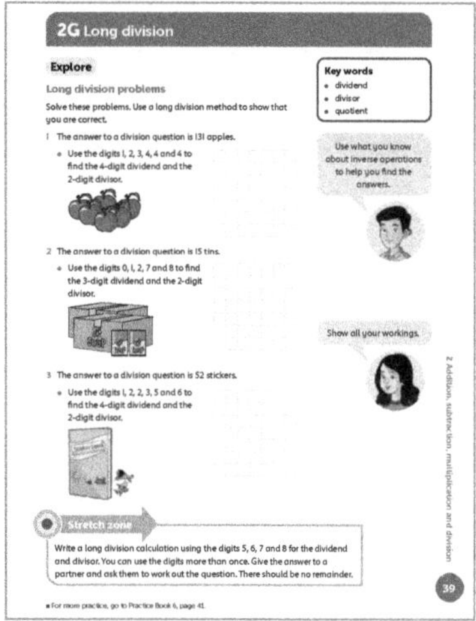

### Differentiation

All students should engage in the investigation together.

**Supporting:** Work with less-confident students to support them in using the long division written method to check the calculations.

**Consolidating:** Ask students to explain how they found the dividends for their calculations.

**Extending:** Challenge students to explain the connection between fact families and the calculations they chose.

**Stretch zone:** *Write a long division calculation using the digits 5, 6, 7 and 8 for the dividend and divisor. You can use the digits more than once. Give the answer to a partner and ask them to work out the question. There should be no remainder.*

Students can take their problems home to share with their parents and other adults at home.

### Reflection time

Ask groups to present their calculations to the rest of the class and explain how they were derived. Ask other students to comment on any differences in their solutions. At each step, ask students to try to tell you their thinking out loud.

**Practice Book:** Students complete Practice Book page 41. They can do this directly after the Main activity, as homework, or as the focus of a separate mathematics session to help students consolidate their learning and build fluency.

Students estimate the answers and then solve long division calculations using the written method.

| Differentiated outcomes | |
| --- | --- |
| **All students** | should derive a long division calculation for a given quotient with support. |
| **Most students** | will derive a long division calculation for a given quotient. |
| **Some students** | may explain how they derived long division calculations for a given quotient. |

## Answers

### Student Book page 39

Students make up their own calculations with the required answers. Check that their calculations are correct and that their methods are appropriate.

Possible answers:

**1** $3144 \div 24$

**2** $180 \div 12$

**3** $1352 \div 26$

### Practice Book page 41

Check that students' estimates are sensible.

**1** 32

**2** 63

**3** 580

**4** 382

Stretch zone: Students derive their own calculations that fit the question requirements. Check that they have made suitable calculations.

## 2H Short division

**Discover** Student Book page 40 • Practice Book page 42

### Specific learning focus

- Divide 3- and 4-digit numbers by 2-digit numbers using the written method for short division.

### Global skills

- **Creative skills:** exploring

### Key vocabulary

- written method, divisor, dividend, quotient

### Resources

- mini whiteboards and markers

### Language support

Students may need support in their explanations for this activity. Encourage them by asking questions, for example:

- *Why is that the first step?*
- *What is your estimate for the answer?*
- *What do you need to exchange?*
- *Where do you write the leftover tens?*

### Introductory activity

Ask students to say how many times 12 goes into 60.

Now ask them to use this fact to say how many times 12 goes into each of these numbers:

| 600 | 300 | 6000 | 3000 | 1500 | 150 |

For each number, students can work in pairs to discuss how many times 12 goes in then write their answers on their whiteboards. Then ask students to explain how they used one answer to help them find others. For example, ask, *If you know how many 12s go into 600, how can this help you find how many 12s go into 300?*

### Main activity

For this activity, keep the class together as you work through the first question on page 40 of the Student Book. Work through the problem on the board, talking through each step carefully, asking students to work out the missing numbers in each line and explain how they know. This gives them support with the calculations and the language.

Students should then use the short division written method to complete question 2 in the same way. As they work, move around the class and ask them what they are doing at each step and why.

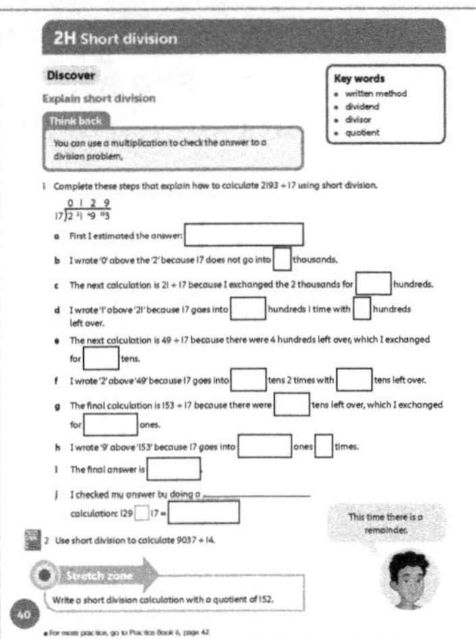

Students solve short division calculations. There is a step-by-step reminder for students to work through first then to use as a template for their calculations. Remind students to estimate their answers first.

| Differentiated outcomes | |
|---|---|
| **All students** | should use the short division written method with support. |
| **Most students** | will use the short division written method to carry out calculations. |
| **Some students** | may model the short division written method and explain the steps. |

## Differentiation

**Supporting:** Work with less-confident students to support them in using the short division written method.

**Consolidating:** Ask students to explain the short division written method to you as they use it.

**Extending:** Challenge students to explain the short division written method step by step.

**Stretch zone:** *Write a short division calculation with a quotient of 152.*

Students can use the fact families idea from 2G Explore to help them.

## Reflection time

Write on the board: $9037 \div 14$. Ask different students to explain how to carry out each step of the calculation. Write down each step as they describe it. Ask questions to make sure that students are clear about each step, for example:

- *Why do you write that digit in that column?*
- *What is the value of that digit?*

**Practice Book:** Students complete Practice Book page 42. They can do this directly after the Main activity, as homework, or as the focus of a separate mathematics session to help students consolidate their learning and build fluency.

## Answers

### Student Book page 40

**1 a** $2200 \div 20 = 110$     **f** 49, 15

  **b** 2     **g** 15, 150

  **c** 20     **h** 153, 9

  **d** 21, 4     **i** 129

  **e** 40     **j** multiplication calculation, $129 \times 17 = 2193$

**2** $9037 \div 14 = 645 \text{ r}7$

### Practice Book page 42

Check that students' estimates are sensible.

**1** 514     **3** 309

**2** 176     **4** 149

Stretch zone: $1807 \div 11 = 164 \text{ r}3$

# 2H Short division



### Specific learning focus

- Divide 3-digit and 4-digit numbers by 2-digit numbers (no remainders) using the written method for short division.

### Global skills

- **Creative skills:** problem solving
- **Self-development skills:** reflecting on learning

### Key vocabulary

- divisor, dividend, quotient

### Resources

- mini whiteboards and markers

### Language support

A good way to organise the group work to support language development is to ask one of the group to describe the process and the rest to write down the solution. The group can have a different student acting as 'scribe' for each problem. Make sure that the whole group understands each step before it is written down.

## Introductory activity

Ask students to write on their whiteboards the fact families for these calculations:

$27 \times 13 = 351$      $41 \times 28 = 1148$      $62 \times 14 = 868$

Students should then check the division statements in each family using short division.

For example, the fact family for $27 \times 13 = 351$ is:

$27 \times 13 = 351$   $13 \times 27 = 351$   $351 \div 13 = 27$   $351 \div 27 = 13$

Check:

$$13 \overline{)3\ {}^3 5\ {}^9 1} = 27$$

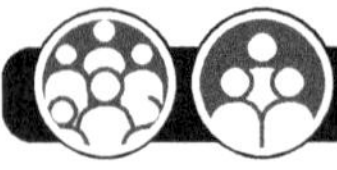

## Main activity

Tell students that you are thinking of a division calculation that has a quotient of 31. You don't know what the dividend and divisor are, but the dividend has three digits and the divisor has two digits. Ask students to use fact families to think of possible calculations.

You could give one example, such as $465 \div 15 = 31$, and ask how you knew that the dividend in this calculation was 465. (because you multiplied 31 by a 2-digit number)

Students should work in groups of four on the activities on page 41 of the Student Book. They will need to try different calculations. Emphasise that they do not need to find all the possible calculations, just those that meet the requirements of the question. Tell them to use their knowledge of place value to try to decide which calculations are most likely to give the correct answer and then to carry out the calculation using the short division written method. They can use fact families to help them identify possible calculations.

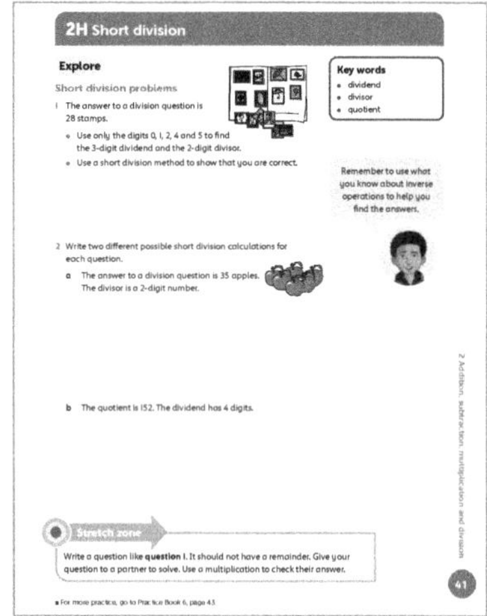

## Differentiation

All students should engage in the investigation together.

**Supporting:** Work with less-confident students to support them in using the short division written method to check the calculations.

**Consolidating:** Ask students to explain how they found the dividends for their calculations.

**Extending:** Challenge students to explain the connection between fact families and the calculations they chose.

**Stretch zone:** *Write a question like question 1. It should not have a remainder. Give your question to a partner to solve. Use a multiplication to check their answer.*

Students can take their problems home to share with their parents and other adults at home.

 Reflection time

Ask groups to present their calculations to the rest of the class and explain how they were derived. Ask other students to comment on any differences in their solutions. At each step, ask students to try to explain their thinking out loud.

**Practice Book:** Students complete Practice Book page 43. They can do this directly after the Main activity, as homework, or as the focus of a separate mathematics session to help students consolidate their learning and build fluency.

Students complete missing digit calculations to solve short division calculations. They should use estimation and knowledge of fact families to help them.

| Differentiated outcomes | |
| --- | --- |
| **All students** | should derive a short division calculation for a given quotient with support. |
| **Most students** | will derive a short division calculation for a given quotient. |
| **Some students** | may explain how they derived short division calculations for a given quotient. |



**Student Book page 41**

**1**  $420 \div 15 = 28$

**2**  Students make up their own calculations with the required answers. Check that students' calculations are correct and that their methods are appropriate. For example:

   **a**  $420 \div 12, 560 \div 16$    **b**  $2128 \div 14, 3192 \div 21$

**Practice Book page 43**

**1**  $819 \div 21 = 39$       **3**  $945 \div 15 = 63$

**2**  $884 \div 17 = 52$       **4**  $923 \div 13 = 71$

Stretch zone: Students find their own numbers, using their knowledge of divisibility rules for 9. Check that they have chosen suitable numbers and that they have used the short division written method to check.

# 21 Division with remainders

**Discover**  Student Book pages 42–43 • Practice Book page 44

## Specific learning focus

- Give an answer to division as a mixed number and as a decimal.

## Global skills

- **Creative skills:** investigating
- **Self-development skills:** reflecting on learning

## Key vocabulary

- divisor, dividend, remainder, quotient

## Resources

- mini whiteboards and markers
- digit cards 0–9

## Language support

Model the correct use of the language of division as discussed in the Introductory activity. For example, say:

- *I am dividing by …*
- *The remainder is …*
- *This is the same as …*

Choose a digit card. Write the number on the board. *Write a division calculation with an answer that has this number as a **remainder**.* For example, if you choose a 2, students write a division calculation with a remainder of 2. Give students, in pairs, two minutes to write their answers on their whiteboards. Ask pairs for their answers. *How did you work it out?*

Choose another digit card and repeat, each time asking students to explain how they knew that their answer would have the given remainder.



Write on the board the following calculations and model how to leave the remainder as a fraction. For example:

$17 \div 5 = 3 \text{ r } 2$   Since the divisor is 5, the remainder 2 can be written as $\frac{2}{5}$.

$34 \div 5 = 6 \text{ r } 4$   Since the divisor is 5, the remainder 4 can be written as $\frac{4}{5}$.

Repeat, this time modelling an answer as a decimal, for example:

17 divided by 5 is 3 remainder 2. $\frac{2}{5}$ is equivalent to $\frac{4}{10}$, or 0.4, so the answer is 3.4

34 divided by 5 is 6 remainder 4. $\frac{4}{5}$ is equivalent to $\frac{8}{10}$, or 0.8, so the answer is 6.8

Ask students to work in pairs on the activities on pages 42–43 of the Student Book. They can support one another and check answers. They write three calculations each for questions 1 to 3, then check answers. Check that students remember that 0.25 is equivalent to $\frac{1}{4}$.

You may choose to pair the least-confident students together so that you can model the idea of remainders using practical materials such as base-10 equipment or place-value counters. For example, when dividing 17 by 5, as in the example above, proceed as follows:

1 ten divided by 5 = 2.

7 ones divided by 5 = 1 remainder 2.

The 2 represents 2 of another group of 5, so $\frac{2}{5}$, which is equivalent to $\frac{4}{10}$, hence 0.4, making the total answer 3.4

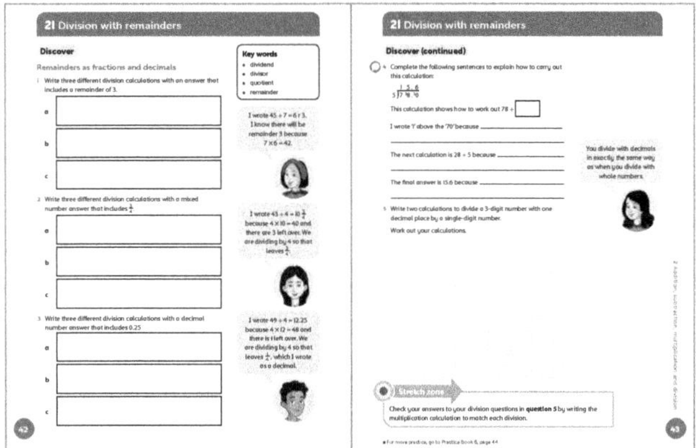

## Differentiation

**Supporting:** Model division with remainders using base-10 equipment or place-value counters.

**Consolidating:** Ask students to explain how they found the remainders.

**Extending:** Challenge students to write the same remainder in different forms.

**Stretch zone:** *Check your answers to your division questions in question 5 by writing the multiplication calculation to match each division.*

Ensure that students use their knowledge of fact families to write the correct corresponding multiplication calculations to check their answers.

 **Reflection time**

Ask one pair to give you their multiplication calculations from question 5. Ask the other students to write down the answer to the multiplication and the division that matches it. They can check that the calculation is correct by working out the long division.

**Practice Book:** Students complete Practice Book page 44. They can do this directly after the Main activity, as homework, or as the focus of a separate mathematics session to help students consolidate their learning and build fluency.

Students complete division calculations using short division. They write the remainder as a fraction and as a decimal. There is a reminder, on Practice Book page 44, of some of the most common fraction and decimal equivalents.

| Differentiated outcomes | |
| --- | --- |
| **All students** | should carry out calculations with support to make the given remainders. |
| **Most students** | will create calculations leaving remainders, using their knowledge of factors. |
| **Some students** | may explain how to write remainders as both fractions and decimals. |

## Answers

### Student Book pages 42–43

Answers will vary because students write their own calculations with the required remainders and instructions. Check that their calculations and instructions are correct.

Examples include:

**1** $94 \div 13 = 7 \text{ r } 3$

**2** $92 \div 16 = 5\frac{3}{4}$

**3** $78 \div 24 = 3.25$

**4** $78 \div 5$. I wrote 1 above the 70 because 5 goes into 70 ten times. The next calculation is $28 \div 5$ because 2 tens were left over from the first division. The final answer is 15.6 because 5 went into 28 five times with remainder 3, which became 30 tenths and 30 tenths divided by 5 is 6 tenths.

**5** Answers will vary because students choose their own numbers. Check that they have calculated correctly.

### Practice Book page 44

**1** $19 \text{ r } 3 = 19\frac{3}{4} = 19.75$    **6** $10 \text{ r } 8 = 10\frac{2}{3} = 10.66$

**2** $26 \text{ r } 2 = 26\frac{1}{3} = 26.33$    **7** $6 \text{ r } 2 = 6\frac{2}{9} = 6.22$

**3** $14 \text{ r } 1 = 14\frac{1}{3} = 14.33$    **8** $65 \text{ r } 2 = 65\frac{2}{5} = 65.4$

**4** $10 \text{ r } 3 = 10\frac{3}{8} = 10.375$    **9** $24 \text{ r } 7 = 24\frac{7}{8} = 24.875$

**5** $11 \text{ r } 3 = 11\frac{1}{2} = 11.5$    **10** $117 \text{ r } 6 = 117\frac{1}{2} = 117.5$

Stretch zone: For example: $1416 \div 12 = 118$

# 21 Division with remainders

## Explore
Student Book pages 44-45 • Practice Book page 45

### Specific learning focus
- Give an answer to a division problem as a rounded number including money contexts.

### Global skills
- **Creative skills:** problem solving
- **Real-world skills:** financial literacy
- **Self-development skills:** reflecting on learning

### Key vocabulary
- divisor, dividend, remainder, factor, quotient

### Resources
- real-life supermarket products (for example breakfast cereal) in different-sized packets, with their prices

### Language support
Use the language of problem solving throughout the activity. Ask, for example:
- *How do you know which calculation to use?*
- *How did you work that out?*
- *Is that answer reasonable?*
- *Do you need to round up or down?*

### Introductory activity
Divide the class into mixed-attainment groups of four to six students. Find an example of cereal that is sold in a local shop in different-sized packets. *Which packet of cereal is the best value?* Groups use division to work out how much each serving costs by dividing the cost of the packet by the number of servings it contains.

### Main activity
Present students with this problem:

*124 people are going on a day trip. They are travelling by minibus. Each minibus can carry 18 people. How many minibuses are needed for the trip?*

Give students some time to think about the calculation they need to do, then ask then to write the calculation and work out the answer. Let students share their thinking with the class. Help them by asking key questions, such as:
- *What division calculation did you choose to do?*
- *Was there an exact answer?*

- *Was there a remainder? What did you do with the remainder?*
- *How does the remainder affect the answer to the problem?*

Work through with the class the calculation $124 \div 18$ to get the answer 6 remainder 16. Ask students whether they can explain what the remainder means for the people going on the trip. Explain that in problems like this, the answer needs to be rounded up, so the trip requires 7 minibuses.

Students now work on the activities on pages 44–45 of the Student Book. The activity on page 45 builds on the ideas in the Introductory activity. Make sure that the more-confident students share their thinking carefully with other students who are less confident.

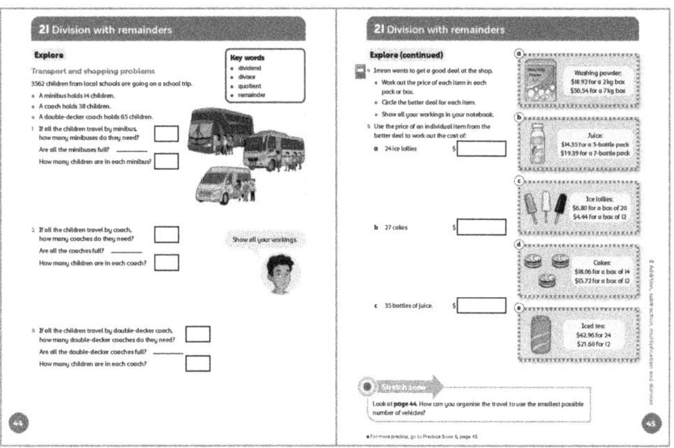

### Differentiation
**Supporting:** Help students to work out the calculation needed and check that they can answer the question correctly, depending on the context.

**Consolidating:** Ask students to interpret the answer in the context of the problem.

**Extending:** Challenge students to make up similar problems.

**Stretch zone:** *Look at page 44. How can you organise the travel to use the smallest possible number of vehicles?*

As a follow-up activity, you could ask students to visit local supermarkets to find other examples of 'shopping for best value' problems. They can bring their examples in for the class to solve in future lessons.

### Reflection time
Select a group to share their solution to the Introductory activity and then ask other groups to comment.

Ask students to explain to the rest of the class how they calculated the answers to questions 1 to 3. Can they explain why the answer needs to be rounded to the next highest whole number?

Discuss the answers to the questions on page 45. *Are bigger packets always better value for money? Why might someone choose to buy a smaller packet?* This could lead to a discussion about whether a smaller family might not be able to use a larger pack before the use-by date, or whether a family may not have the space to store larger packets.

**Practice Book:** Students complete Practice Book page 45. They can do this directly after the Main activity, as homework, or as the focus of a separate mathematics session to help students consolidate their learning and build fluency.

Students answer division problems involving money. They find the price per item, given the pack price and size. They round each answer to the nearest cent so that the answers are given in a realistic context.

| Differentiated outcomes | |
| --- | --- |
| **All students** | should solve the division word problems with support. |
| **Most students** | will solve the division problems and interpret the remainders. |
| **Some students** | may create their own similar problems that require the remainders to be rounded up. |



### Student Book pages 44–45

**1** 255 minibuses, not all full. 14 children in 254 minibuses and the remaining 6 in another minibus.

**2** 94 coaches, not all full. 38 children in 93 coaches and the remaining 10 children in another coach.

**3** 55 coaches, not all full. 65 children in 54 coaches and the remaining 52 children in another coach.

**4** **a** 7 kg box    **b** 7 bottles    **c** box of 20
   **d** box of 14   **e** 24 teas

**5** **a** $8.16    **b** $34.83    **c** $96.95

### Practice Book page 45

**1** $1.52

**2** $1.35

**3** 17 cents

**4** $1.40

**5** $1.60

Stretch zone: Each bottle costs $1.35, so you could sell them for $1.50 each.

---

# 2J Factors, multiples and primes

**Discover** Student Book pages 46–47 · Practice Book page 46

## Specific learning focus

- Recognise prime numbers up to 20 and find all prime numbers less than 100.
- Find factors of two-digit numbers.

## Global skills

- **Creative skills:** investigating
- **Self-development skills:** reflecting on learning

## Key vocabulary

- prime number, composite number

## Resources

- mini whiteboards and markers

## Language support

Ask students to read their answers saying:

- 'A prime number is …'
- 'A composite number is …'

Students could create a poster for display with the information completed.

 **Introductory activity**

Take the class outside if possible. Choose 24 students. Ask these students to stand in an array (a rectangular grid of rows and columns). They could form a 6 by 4 array, for example. Then ask them to stand in a new array. Repeat until they have found all the possibilities. These are $1 \times 24$, $2 \times 12$, $3 \times 8$, $4 \times 6$. Do the same activity with different numbers. Discuss which numbers have the most possibilities and which numbers have the fewest possibilities.

 **Main activity**

Look together at page 46 of the Student Book. Display on the IWB, if possible. Ask students to look at the Think back information. Remind them how arranging a number of items into an array helps you find the factors. Ask students to draw arrays to represent 20 on their whiteboards and to and find all the factors of 20.

Now ask them to find the factors of 17. *How many arrays can you make? How many factors does 17 have?* Agree that the only factors of 17 are 17 and 1, so the only array that can be made is a 17 × 1 (or a 1 × 17) array. Explain that a number with only two factors (itself and 1) is a **prime number**. *Can you find any more prime numbers smaller than 20?*

Then ask students, in pairs, to work through the activities on pages 46–47 of the Student Book. As students are working, ask them how they know they have found all the factors of each number and all the prime numbers to 100.

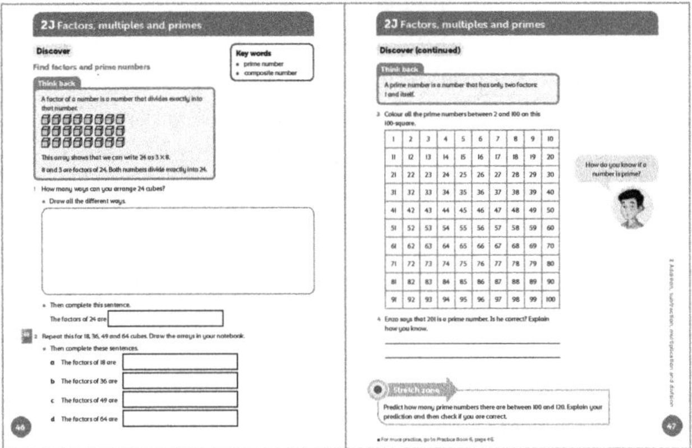

## Differentiation

**Supporting:** Support students in finding factors by modelling the process.

**Consolidating:** Ask students to explain what a factor is and what a prime number is.

**Extending:** Challenge students to find all the factors of all numbers to 30. Can they predict which will have the most factors before they find them all?

**Stretch zone:** *Predict how many prime numbers there are between 100 and 120. Explain your prediction and then check if you are correct.*

As an additional activity, you can ask students to research the history of prime numbers, including when they were first discovered.

## Reflection time

Ask students to work in pairs to complete the following sentences:

- A prime number is …
- A **composite number** is …
- A factor is …
- A multiple is …

Share three or four of these definitions. As a class, choose the class definitions. Write the class definitions and add them to the class students' glossary. You can also display the definitions in the classroom for students to refer to.

**Practice Book:** Students complete Practice Book page 46. They can do this directly after the Main activity, as homework, or as the focus of a separate mathematics session to help students consolidate their learning and build fluency.

Students find the lowest common multiple of two numbers, by first listing all the multiples of each number and then finding the multiples that they have in common.

<table>
<tr><td colspan="2">Differentiated outcomes</td></tr>
<tr><td>All students</td><td>should find prime numbers and understand what a factor and a multiple is, with support.</td></tr>
<tr><td>Most students</td><td>will find factors of numbers by drawing arrays.</td></tr>
<tr><td>Some students</td><td>may find factors of numbers using multiplication and division facts.</td></tr>
</table>

## Answers

### Student Book pages 46–47

**1** The factors of 24 are 1, 2, 3, 4, 6, 8, 12, 24.

**2 a** The factors of 18 are 1, 2, 3, 6, 9, 18.

   **b** The factors of 36 are 1, 2, 3, 4, 6, 9, 12, 18, 36.

   **c** The factors of 49 are 1, 7, 49.

   **d** The factors of 64 are 1, 2, 4, 8, 16, 32, 64.

**3** Check that students have followed the instructions for analysing the numbers in the 100-square.

Prime numbers: 2, 3, 5, 7, 11, 13, 17, 19, 23, 29, 31, 37, 41, 43, 47, 53, 59, 61, 67, 71, 73, 79, 83, 89, 97

**4** 201 is not a prime number – it is divisible by 3 because the digits add to a multiple of 3.

### Practice Book page 46

**1** 55

   multiples of 5: 5, 10, 15, 20, 25, 30, 35, 40, 45, 50, 55

   multiples of 11: 11, 22, 33, 44, 55

**2** 48

   multiples of 6: 6, 12, 18, 24, 30, 36, 42, 48

   multiples of 16: 16, 32, 48

**3** 91

   multiples of 7: 7, 14, 21, 28, 35, 42, 49, 56, 63, 70, 77, 84, 91

   multiples of 13: 13, 26, 39, 52, 65, 78, 91

**4** 114

   multiples of 6: 6, 12, 18, 24, 30, 36, 42, 48, 54, 60, 66, 72, 78, 84, 90, 96, 102, 108, 114

   multiples of 19: 19, 38, 57, 76, 95, 114

**5** 14

   multiples of 7: 7, 14

   multiples of 14: 14

**6** 24

   mulitples of 8: 8, 16, 24

   multiples of 12: 12, 24

Stretch zone: False. Multiplying two multiples gives a common multiple but not always the lowest. For example, for 4 and 8, the lowest common multiple is 8, but 4 × 8 = 32.

# 2J Factors, multiples and primes

## Explore
Student Book pages 48–49 • Practice Book page 47

### Specific learning focus
- Find prime factors of 2-digit numbers and common multiples.

### Global skills
- **Creative skills:** investigating
- **Self-development skills:** reflecting on learning

### Key vocabulary
- composite number, factor, multiple, prime factor, common multiple

### Resources
- mini whiteboards and markers
- 0–9 digit cards, class set

### Language support

Ask students to develop their own definitions of mathematical vocabulary to ensure that they develop their own mathematical understanding. Ask, for example:

- *What are the factors of 15?*
- *What are the common multiples of 4 and 10?*
- *Is 7 less than 15?*
- *Is 7 greater than 4?*
- *A factor is …*
- *A multiple is …*

Once students are confident with the term 'prime factor', they can add this to the class display they started in the previous lesson.

### Introductory activity

Remind students that multiples of numbers are 'numbers that appear in the times table of the number' – for example, multiples of 6 are 12, 18, 24 and so on. A **factor** of a number is a number that divides exactly into it. So the factors of 12 are 1, 2, 3, 4, 6, and 12 because you can divide 12 by each of these numbers without leaving any remainder.

Ask students to list all the factors of 3 and 7 below 30. Can they see any number that appears in both lists? 21 is in both lists of multiples, so 21 is the lowest **common multiple** of 3 and 7.

Repeat for factors of 6 and 18 below 40. Now there are two common multiples, 18 and 36, so 18 is the lowest common multiple of 6 and 18.

### Main activity

Look together at page 48 of the Student Book. Display on the IWB, if possible. Refer students to the Think back statement. Discuss how some numbers have just two factors (prime numbers) and others have more than two factors (composite numbers). Refer students to the definition of a '**prime factor**' below the Think back: 'A prime factor is a factor that is also a prime number.' Explain that every number can be broken down into its prime factors by a process of division, using a factor tree.

Look together at the worked example on page 48. Starting from 80, find two factors of 80 (2 and 40). Because 2 is prime, that branch of the tree stops. Now find two factors of 40 (4 and 10). Keeping finding two factors of each number until you reach prime factors. You then have all the prime factors of 80, so $80 = 2 \times 2 \times 2 \times 2 \times 5$.

Ask students to work in pairs, to complete the activities on pages 48–49 of the Student Book so that they can check one another's responses. Check that they understand the meaning of common multiples.

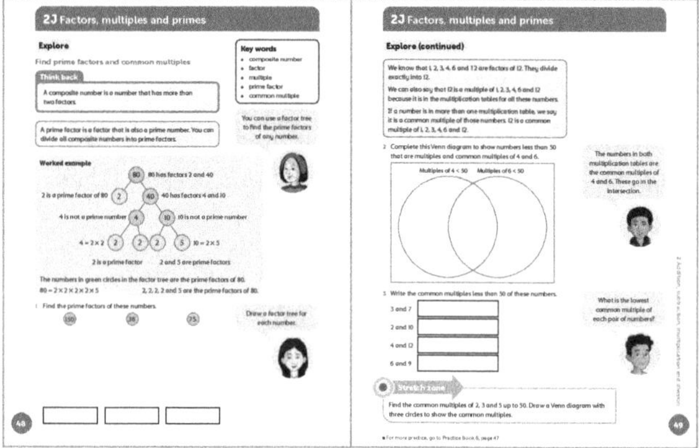

### Differentiation

**Supporting:** Work with students to complete the Venn diagram of common multiples. Use digit cards to support students so they can move them around on the Venn diagram.

**Consolidating:** Ask students to explain their reasoning when completing the Venn diagram.

**Extending:** Ask students to create new Venn diagrams to explore other common multiples.

**Stretch zone:** *Find the common multiples of 2, 3 and 5 up to 50. Draw a Venn diagram with three circles to show the common multiples.*

 **Reflection time**

Ask students to work in pairs to complete the following sentences:

- A factor is …
- A prime factor is…
- A lowest common multiple is…

Share three or four of these definitions. As a class, choose the best definitions. Write the class definitions on a poster to display and include examples.

**Practice Book:** Students complete Practice Book page 47. They can do this directly after the Main activity, as homework, or as the focus of a separate mathematics session to help students consolidate their learning and build fluency.

Students draw factor trees to find all the prime factors of given numbers. An example is given, which shows that there is more than one way to draw the factor tree for each number.

| Differentiated outcomes | |
| --- | --- |
| **All students** | should find prime factors and complete the Venn diagram with support. |
| **Most students** | will find prime factors and complete the Venn diagram. |
| **Some students** | may create their own Venn diagrams to explore other common multiples. |

## Answers

**Student Book pages 48–49**

**1** $150 = 2 \times 3 \times 5 \times 5$; $36 = 2 \times 2 \times 3 \times 3$; $75 = 3 \times 5 \times 5$

**2** Check that students have written numbers in the correct places on the Venn diagram.

Multiples of 4 are: 4, 8, 16, 20, 28, 32, 40, 44

Multiples of 6 are: 6, 18, 30, 42

Common multiples of 4 and 6 are: 12, 24, 36, 48

**3** Common multiples of 3 and 7 are: 21, 42

Common multiples of 2 and 10 are: 10, 20, 30, 40

Common multiples of 4 and 12 are: 12, 24, 36, 48

Common multiples of 6 and 9 are: 18, 36

### Practice Book page 47

While students are working, ask them to explain how they can use the factor trees to work out what the factors of the numbers are.

**1** $2 \times 2 \times 3 \times 5$      **6** $2 \times 3 \times 3 \times 5$

**2** $5 \times 17$      **7** $2 \times 2 \times 2 \times 2 \times 2 \times 2$

**3** $2 \times 2 \times 13$      **8** $3 \times 29$

**4** $2 \times 2 \times 3 \times 3$      **9** $2 \times 5 \times 5$

**5** $2 \times 2 \times 2 \times 3 \times 3$

Stretch zone: For example, 48 has a factor tree with 8 branches.

---

## 2K Order of operations

**Discover**   Student Book page 50 • Practice Book page 48

### Specific learning focus

- Use brackets in calculations to determine the order of operations.

### Global skills

- **Creative skills:** investigating, problem solving

### Key vocabulary

- brackets, order of operations

### Resources

- mini whiteboards and markers

### Language support

As they work through the calculations with brackets, encourage students to explain how they know which operations to carry out first. Model for them using examples by asking questions and making statements, such as:

- *In the calculation 5 + (3 × 2), which part must we work out first?*
- *Which calculation is done first in 24 – (5 × 4)?*

 **Introductory activity**

Write these calculations on the board:
$(6 + 3) \times 7$        $6 + (3 \times 7)$.

Ask students to calculate the answers. Students should have 63 and 27. Explain to the class that when there are **brackets** in a calculation, we work these out first, even though they may appear later in the calculation. If there are two sets of brackets, we can work them out in any order, but the brackets must be worked out before the rest of the calculation.

Write this example for students to try:
$(3 \times 4) + 15 - (2 \times 2)$.

Give them time to work this out and check with a partner before agreeing that the answer is 23.

## Main activity

Write these calculations on the board:

$(5 \times 2) + 3 \qquad 5 + (2 \times 3) \qquad 5 \times (2 + 3) \qquad (5 + 2) \times 3$

Ask students to work in pairs to work out which of these calculations has the answer 25. They should discuss each calculation and be able to explain why one makes 25 but the others do not. Share some of the explanations as a class.

Now ask students to write down the numbers 5, 4 and 6. They should combine these using different operations and brackets to make the following answers: 29, 14, 50, 6. For example, by calculating $(5 + 4) \times 6$, they could make 54. Listen to their discussions as they work and return to these during Reflection time.

Students then complete the activities on page 50 of the Student Book. As they work, listen to their discussions and ask questions about their answers, such as: *Why do you multiply those numbers first? Which part of the calculation will be done last?*

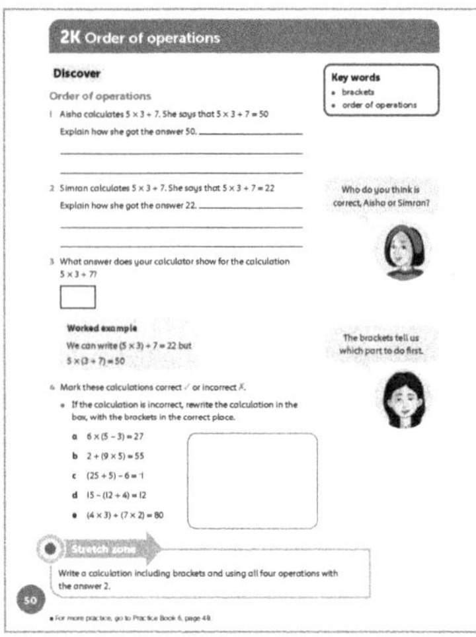

### Differentiation

**Supporting:** Help students by modelling the calculations, showing them which part to work out first.

**Consolidating:** Ask students to explain how they do each calculation.

**Extending:** Challenge students to find different possible answers by varying the position of the brackets in a calculation.

**Stretch zone:** *Write a calculation including brackets and using all four operations with the answer 2.*

Students can use whole numbers, negative numbers, decimals or fractions.

### Reflection time

Share as a class the calculations using 5, 4 and 6 from the Main activity. What different calculations did students make? Did they make each of the answers on the board? Can they explain their reasoning for how they placed the brackets?

Answers are: $29 = (6 \times 4) + 5$, $14 = (4 \times 5) - 6$, $50 = (6 + 4) \times 5$, $6 = (5 - 4) \times 6$

**Practice Book:** Students complete Practice Book page 48. They can do this directly after the Main activity, as homework, or as the focus of a separate mathematics session to help students consolidate their learning and build fluency.

Students use given digits, brackets and all four operations to write calculations. They then change the position of the brackets (but not the digits) to create a calculation with a different answer.

| Differentiated outcomes | |
| --- | --- |
| **All students** | should solve calculations including brackets with support. |
| **Most students** | will solve calculations including brackets. |
| **Some students** | may use brackets to make different calculations with the same numbers to get different answers. |

## Answers

### Student Book page 50

**1** $5 \times (3 + 7) = 50$

**2** $(5 \times 3) + 7 = 22$

**3** 22

**4 a** $(6 \times 5) - 3$    **b** $(2 + 9) \times 5 = 55$    **c** correct
   **d** correct    **e** $4 \times (3 + 7) \times 2 = 80$

### Practice Book page 48

Students choose their own numbers and operations. Check that they have calculated their answers correctly.

Stretch zone: Largest answer $= 52 \times 43 = 2236$

Smallest answer $= (2 - 3) - 54 = {}^{-}55$

# 2K Order of operations

## Explore    Student Book page 51 · Practice Book page 49

### Specific learning focus
- Use BIDMAS to determine the order to carry out operations in calculations.

### Global skills
- **Creative skills:** exploring

### Key vocabulary
- brackets, order of operations, BIDMAS, indices, index

### Resources
- mini whiteboards and markers

### Language support
Students should be supported with using the language associated with order of operations. Model for them the process of using BIDMAS to solve calculations. Ask questions such as:
- *Does a division come before a subtraction?*
- *Which comes first, an addition or a power?*

 **Introductory activity**

Write these expressions on the board:

$2^2$   $3^2$   $4^2$   $5^2$   $2^3$   $3^3$   $4^3$   $10^2$   $10^3$

Ask students to recall what these numbers represent. Students may remember that they are square and cube numbers. Ask students to work in pairs, using their whiteboards,  to work out the value of each expression. Choose students to say what each one is, and how they worked it out.

Explain that square and cube numbers are called powers of a number so, for example, $2^3$ is called '2 to the power 3'. The power numbers are also called **indices**, or a single one is called an **index**.

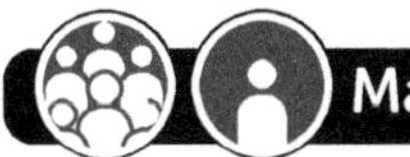 **Main activity**

Write $4^3 - 13 \times 4$ on the board.

Ask students to calculate the answer, then share their answers and methods with the class. Some students may have an answer of 204, others may have 12.

Look together at page 51 of the Student Book. Display on the IWB, if possible. Introduce students to **BIDMAS** and explain that it helps us remember the order we need to do the operations in a calculation (the **order of operations**). The following summarises what BIDMAS stands for.

*B is for brackets, so do any brackets first.*

*I is for indices, so work out the powers next.*

*D is for division, M is for multiplication – these operations can be done next, in any order.*

*A is for addition, S is for subtraction – these operations are done last, in any order.*

So, for the calculation $4^3 - 13 \times 4$, there is one power (index), a subtraction and a multiplication. Using the rules of BIDMAS, we calculate the index first, then the multiplication and finally the subtraction. $4^3 = 4 \times 4 \times 4 = 64$, and $13 \times 4 = 52$, giving us finally $64 - 52 = 12$.

Students should now complete the activity on page 51, trying to make calculations using 1, 2, 3 and 4 and any operations to make all the numbers from 1 to 20. Encourage them to use BIDMAS to make sure that they complete the calculations correctly.

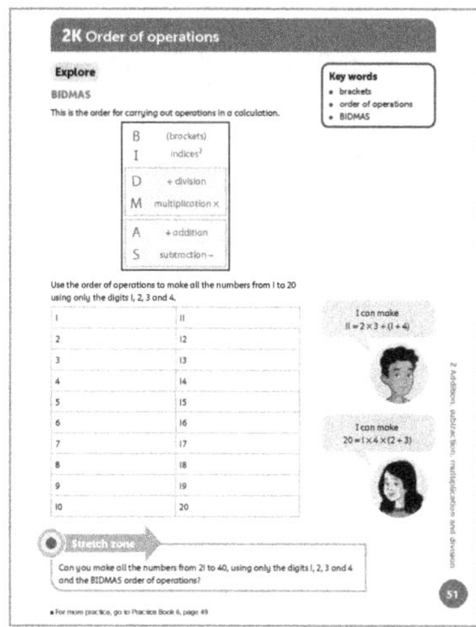

### Differentiation

**Supporting:** Help students apply BIDMAS by modelling it for them on calculations.

**Consolidating:** Ask students to explain how to solve a range of calculations using BIDMAS.

**Extending:** Challenge students to write some calculations of their own that require BIDMAS to solve and then swap them with a partner.

**Stretch zone:** *Can you make all the numbers from 21 to 40, using only the digits 1, 2, 3 and 4 and the BIDMAS order of operations?*

Students could use the calculations from numbers 1–20 to help them with the higher numbers.

 **Reflection time**

Enable students to share their solutions to finding calculations for the numbers 1–20 using 1, 2, 3 and 4. Where there is more than one solution offered for a number, compare them and ask the class to say which they think is easier to calculate.

**Practice Book:** Students complete Practice Book page 49. They can do this directly after the Main activity, as homework, or as the focus of a separate mathematics session to help students consolidate their learning and build fluency.

Students try to make answers from 1 to 20 using only the digits in the date 2021, and any of the operations, using the BIDMAS rules.

| Differentiated outcomes | |
|---|---|
| **All students** | should use BIDMAS to solve calculations with more than one operation with support. |
| **Most students** | will use BIDMAS to solve calculations with more than one operation. |
| **Some students** | may make calculations using more than one operation and solve them using BIDMAS. |

## Answers

### Student Book page 51

There are many possible answers to make the numbers from 1–20 using the digits 1, 2, 3, 4.

Examples are:

| | | | |
|---|---|---|---|
| **1** | $2 \times 3 - (1 + 4)$ | **11** | $2 \times 3 + 1 + 4$ |
| **2** | $4 - 3 + 2 - 1$ | **12** | $3 \times 4 \times (2 - 1)$ |
| **3** | $2 \times 3 - (4 - 1)$ | **13** | $3 \times 4 + 2 - 1$ |
| **4** | $2 \times 4 - (1 + 3)$ | **14** | $1 \times 4 \times 3 + 2$ |
| **5** | $(2 \times 4) - (1 \times 3)$ | **15** | $3 \times 4 + 2 + 1$ |
| **6** | $2 \times 4 - 3 + 1$ | **16** | $(1 + 3 + 4) \times 2$ |
| **7** | $3 \times (4 - 1) - 2$ | **17** | $(2 + 4) \times 3 - 1$ |
| **8** | $2 + 3 + 4 - 1$ | **18** | $(2 + 4) \times 3 \times 1$ |
| **9** | $2 \times 3 + (4 - 1)$ | **19** | $(2 + 4) \times 3 + 1$ |
| **10** | $1 + 2 + 3 + 4$ | **20** | $1 \times 4 \times (2 + 3)$ |

### Practice Book page 49

There are many possible answers to make the numbers from 1–20 using the digits 2, 0, 2, 1.

Examples are:

| | | | |
|---|---|---|---|
| **1** | $22 \times 0 + 1$ | **11** | $(20 \div 2) + 1$ |
| **2** | $(2 + 0) \div 2 + 1$ | **12** | $12 + (0 \div 2)$ |
| **3** | $2 \times 2 - 1 - 0$ | **13** | none |
| **4** | $2 \times 2 - (0 \times 1)$ | **14** | $12 + 2 + 0$ |
| **5** | $2 + 0 + 2 + 1$ | **15** | none |
| **6** | $(2 + 1) \times 2 + 0$ | **16** | none |
| **7** | $10 \div 2 + 2$ | **17** | $(20 - 2) - 1$ |
| **8** | none | **18** | $20 - (2 \div 1)$ |
| **9** | $(2 + 1)^2 + 0$ | **19** | none |
| **10** | $(20 \div 2) \times 1$ | **20** | $20 \div (2 - 1)$ |

Stretch zone: Answers will vary because students use their own birth date for the calculation.

# 2L Using the arithmetical laws



## Specific learning focus

- Know and apply the arithmetic laws as they apply to multiplication.

## Global skills

- **Creative skills:** investigating

## Key vocabulary

- associative law, commutative law, distributive law

## Resources

- digit cards 1–9
- calculators
- paper for notes
- mathematical dictionaries

## Language support

Writing their own definitions for the mathematical laws supports students in understanding the meanings of 'commutative', 'distributive' and 'associative'.

 **Introductory activity**

Write on the board: a(b + c). Choose three digit cards. Replace 'a' with the first digit you choose. Replace 'b' with the second digit you choose. Replace 'c' with the third digit you choose. For example: you choose 3, 5 and 2 so you write 3(5 + 2).

Tell the class that:

- this is the same as 3 × 7, which is 21
- you can also write this as (3 × 5) + (3 × 2) = 15 + 6 = 21.

Is this always true? Or is it only true sometimes? The class should work in mixed-attainment groups of four to six students to try different numbers in order to test the statement.

The activities in this unit look at similar questions. Remind students to try with fractions and negative numbers to test whether the statements are always true.

 **Main activity**

Ask students to work on the activities on page 52 of the Student Book in small groups. They can each test the statements with different numbers. Ask questions as they work. For example:

- *Is it true for negative numbers?*

- *Is it true for fractions?*
- *Is it true when there is a zero?*

Tell students to use mathematics dictionaries or the internet to support them in developing their definitions of the **associative law**, the **commutative law** and the **distributive law**. Remind them to use examples with different types of numbers to illustrate their definitions.

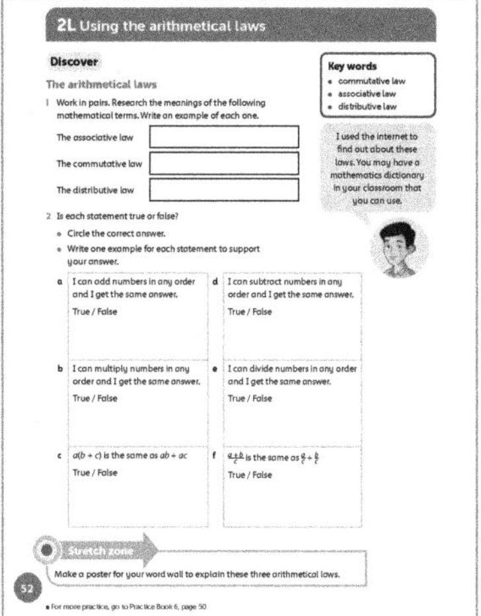

## Differentiation

All students should engage in the investigation together.

**Supporting:** Help students by modelling different forms of the same calculation.

**Consolidating:** Ask students to check their calculations using the rules of arithmetic.

**Extending:** Challenge students to make up calculations that are solved more easily by using one of the arithmetic laws.

**Stretch zone:** *Make a poster for your word wall to explain these three arithmetical laws.*

You could ask students to explore the rules of associativity, commutativity and distributivity.

 **Reflection time**

Ask the groups to share their definitions and give examples. As a class, agree on a single definition for each law. Create a class poster to display. You can refer to this poster in future lessons.

**Practice Book:** Students complete Practice Book page 50. They can do this directly after the Main activity, as homework, or as the focus of a separate mathematics session to help students consolidate their learning and build fluency.

Students solve multiplication calculations. They apply the commutative law to make the calculations easier to do, by multiplying in a different order.

| Differentiated outcomes | |
| --- | --- |
| **All students** | should test number statements about the laws of arithmetic with support. |
| **Most students** | will test number statements about the laws of arithmetic. |
| **Some students** | may use the laws of arithmetic to explain number statements. |

## Answers

### Student Book page 52

1 Check that students' definitions of the associative, commutative and distributive laws are correct and that their examples are appropriate, for example as follows.

The associative law: it doesn't matter how the numbers are grouped in an addition or multiplication calculation; the answer will always be the same, for example:

$(3 + 10) + 5 = 18$

$3 + (10 + 5) = 18$

The commutative law: it doesn't matter in which order the numbers are added or multiplied; the answer will always be the same, for example:

$4 \times 8 = 32$

$8 \times 4 = 32$

The distributive law: when multiplying a number, you can partition the numbers, multiply the separate parts and then add all the products together, for example:

$24 \times 7$

$= 20 \times 7 + 4 \times 7$

$= 140 + 28$

$= 168$

2 a True: $3 + 4 + 7 = 7 + 3 + 4$

  b True: $6 \times 2 \times 5 = 5 \times 6 \times 2$

  c True: $4 \times (2 + 3) = (4 \times 2) + (4 \times 3)$

  d False: $7 - 2$ is not the same as $2 - 7$

  e False: $10 \div 2$ is not the same as $2 \div 10$

  f True: $(8 + 2) \div 5 = (8 \div 5) + (2 \div 5)$

### Practice Book page 50

| | | | |
| --- | --- | --- | --- |
| 1 | 1700 | 7 | 8400 |
| 2 | 2600 | 8 | 1900 |
| 3 | 1260 | 9 | 240 |
| 4 | 900 | 10 | 2900 |
| 5 | 1600 | 11 | 3800 |
| 6 | 380 | | |

Stretch zone: For example, three single-digit numbers: $2 \times 3 \times 4$, $2 \times 4 \times 3$, $3 \times 2 \times 4$, $3 \times 4 \times 2$, $4 \times 2 \times 3$, $4 \times 3 \times 2$ (6 ways).

If splitting numbers into factors, there will be more ways, for example: $12 \times 24 \times 16 =$

  $3 \times 4 \times 6 \times 4 \times 4 \times 4$

  or $2 \times 6 \times 3 \times 8 \times 1 \times 16$

  or $1 \times 12 \times 2 \times 12 \times 4 \times 4$

For each of these examples the numbers can be written in any order, producing numerous ways to reorder the multiplication.

# 2L Using the arithmetical laws

 Student Book pages 53–54 • Practice Book page 51

## Specific learning focus

- Know and apply the arithmetic laws as they apply to multiplication.

## Global skills

- **Creative skills:** problem solving

## Key vocabulary

- associative law, commutative law, distributive law

## Resources

- none needed

## Language support

- Students should be encouraged to refer back to their definitions of commutative, associative and distributive from 2L Discover as they work through the multiplication calculations.

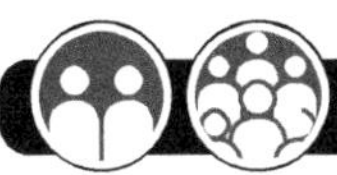 ## Introductory activity

Group students in pairs so that one of the pair is confident in using mental methods for multiplication. Look together at pages 53–54 of the Student Book. Display on the IWB, if possible. As a class, look at each of the worked examples. For each law, ask one student to explain what is happening in each worked example. *Do you think the calculation is easier to do now?*

Ask students to start the activities on pages 53–54 of the Student Book, completing questions 1a and 1b, 2a and 2b, 3a and 3b. They should be encouraged to share strategies with one another or model strategies for less-confident students. Ask students to stop and discuss their working after each pair of questions. For each law of arithmetic, ask, *How does applying the law help? Does everyone agree that it does help? Could you apply a different law?* After five minutes, bring the class together. Ask individual students to select a question that they found much easier to answer after applying the relevant arithmetic law. Do the rest of the class agree that this was a good example of how the arithmetic law made the question easier to answer?

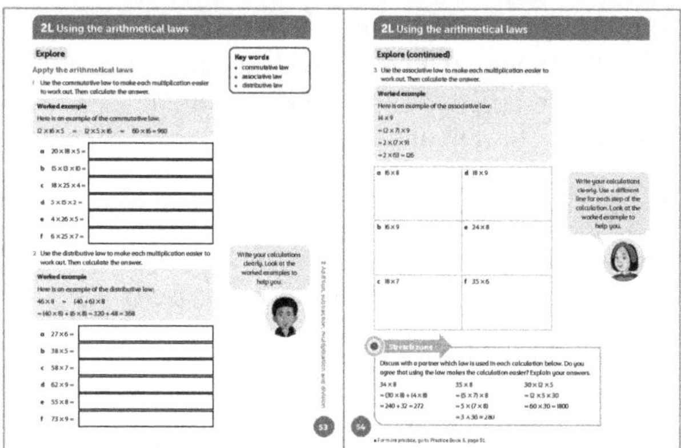

## Main activity

Ask students to complete the rest of the questions from the Student Book in pairs. When they have finished, ask them to make up calculations to test on a friend. The calculations should be similar to those completed, so they can be solved by using one of the arithmetic laws.

### Differentiation

**Supporting:** Help students by modelling the different laws of arithmetic.

**Consolidating:** Ask students to devise their own calculations that can be completed using one of the arithmetic laws.

**Extending:** Ask students to devise more complex calculations that can be solved more easily by using the arithmetic laws.

**Stretch zone:** *Discuss with a partner which law is used in each calculation below. Do you agree that using the law makes the calculation easier? Explain your answers.*

$34 \times 8 = (30 \times 8) + (4 \times 8) = 240 + 32 = 272$

$35 \times 8 = (5 \times 7) \times 8 = 5 \times (7 \times 8) = 5 \times 56 = 280$

$30 \times 12 \times 5 = 12 \times 5 \times 30 = 60 \times 30 = 1800$

The first calculation uses the distributive law, the second the associative law and the third the commutative law. Students should say whether or not they agree that these make the calculations easier, and explain their decisions.

 ## Reflection time

As a class, share three or four of the best calculations that students have written. Ask the student who devised each one to answer then explain: *Why does it work?*

**Practice Book:** Students complete Practice Book page 51. They can do this directly after the Main activity, as homework, or as the focus of a separate mathematics session to help students consolidate their learning and build fluency.

Students solve multiplication calculations. They apply the distributive law to make the calculations easier to do, by partitioning one of the numbers, multiplying the separate parts and then adding the products together.

| Differentiated outcomes | |
| --- | --- |
| **All students** | should carry out calculations, using the arithmetic laws, with support. |
| **Most students** | will carry out calculations, using the arithmetic laws, and write their own calculations for others to solve. |
| **Some students** | will carry out calculations, using the arithmetic laws, and write their own complex calculations that are more easily solved using one of the arithmetic laws. |

## Answers

### Student Book pages 53–54

**1 a** $20 \times 5 \times 18 = 100 \times 18 = 1800$

  **b** $15 \times 10 \times 13 = 150 \times 13 = 1950$

  **c** $4 \times 25 \times 18 = 100 \times 18 = 1800$

  **d** $5 \times 2 \times 15 = 10 \times 15 = 150$

  **e** $5 \times 4 \times 26 = 20 \times 26 = 520$

  **f** $6 \times 25 \times 7 = 150 \times 7 = 1050$

**2 a** $27 \times 6 = (20 \times 6) + (7 \times 6) = 120 + 42 = 162$

  **b** $38 \times 5 = (30 \times 5) + (8 \times 5) = 150 + 40 = 190$

  **c** $58 \times 7 = (50 \times 7) + (8 \times 7) = 350 + 56 = 406$

  **d** $62 \times 9 = (60 \times 9) + (2 \times 9) = 540 + 18 = 558$

  **e** $55 \times 8 = (50 \times 8) + (5 \times 8) = 400 + 40 = 440$

  **f** $73 \times 9 = (70 \times 9) + (3 \times 9) = 630 + 27 = 657$

**3 a** $16 \times 8 = (2 \times 8) \times 8 = 2 \times (8 \times 8) = 2 \times 64 = 128$

  **b** $16 \times 9 = (2 \times 8) \times 9 = 2 \times (8 \times 9) = 2 \times 72 = 144$

  **c** $18 \times 7 = (2 \times 9) \times 7 = 2 \times (9 \times 7) = 2 \times 63 = 126$

  **d** $18 \times 9 = (2 \times 9) \times 9 = 2 \times (9 \times 9) = 2 \times 81 = 162$

  **e** $24 \times 8 = (2 \times 12) \times 8 = 2 \times (12 \times 8) = 2 \times 96 = 192$

  **f** $35 \times 6 = 35 \times (3 \times 2) = (35 \times 3) \times 2 = 105 \times 2 = 210$

### Practice Book page 51

**1** $74 \times 3 = (70 \times 3) + (4 \times 3) = 210 + 12 = 222$

**2** $86 \times 5 = (80 \times 5) + (6 \times 5) = 400 + 30 = 430$

**3** $29 \times 6 = (20 \times 6) + (9 \times 6) = 120 + 54 = 174$

**4** $36 \times 7 = (30 \times 7) + (6 \times 7) = 210 + 42 = 252$

**5** $82 \times 4 = (80 \times 4) + (2 \times 4) = 320 + 8 = 328$

**6** $99 \times 7 = (90 \times 7) + (9 \times 7) = 630 + 63 = 693$

**7** $26 \times 8 = (20 \times 8) + (6 \times 8) = 160 + 48 = 208$

**8** $37 \times 8 = (30 \times 8) + (7 \times 8) = 240 + 56 = 296$

**9** $46 \times 7 = (40 \times 7) + (6 \times 7) = 280 + 42 = 322$

**10** $37 \times 8 = (30 \times 8) + (7 \times 8) = 240 + 56 = 296$

**11** $82 \times 6 = (80 \times 6) + (2 \times 6) = 480 + 12 = 492$

Stretch zone: $29 \times 6$ is easier done as $(30 \times 6) - 6$

# 2 Addition, subtraction, multiplication and division

## Connect  Student Book page 55

### Big idea

We can use mental or written strategies to solve calculations. The best strategy to use depends on the numbers. I have learned to find the most efficient strategy.

### Global skills

- **Creative skills:** problem solving
- **Interpersonal skills:** teamwork
- **Self-development skills:** reflecting on learning

### Key vocabulary

- addition, subtraction, multiplication, division, tenths, closest to

### Resources

- mini whiteboards and markers
- digit cards 1–9

### Language support

Use the vocabulary of place value, for example:

- *What is the value of that digit?*
- *What does that digit represent?*
- *Tens, hundreds, thousands, tenths? Why did you place that digit there?*
- *What is the number bond to 10 for that digit?*

## Introductory activity

Copy the grid for question 1 on page 55 of the Student Book onto the board. Divide the class into mixed-attainment groups of four or five students. Ask groups to play this game against one another. Explain the object of the game, which is to use digits 1–9 to create three numbers, each with one decimal place, then use these numbers in an addition calculation to get an answer as close to 100 as possible. Give groups two minutes to discuss their strategy before you choose the first digit card.

Choose a digit card. Ask groups to decide where to place the digit. They must place each digit after the card is chosen. After a digit is placed, they cannot move it. After nine choices of digit cards, groups add up their totals. The winner is the group with the answer nearest to 100. Ask, *What strategy did you use?* Play this game and the other games on page 55 of the Student Book as many times as you think useful.

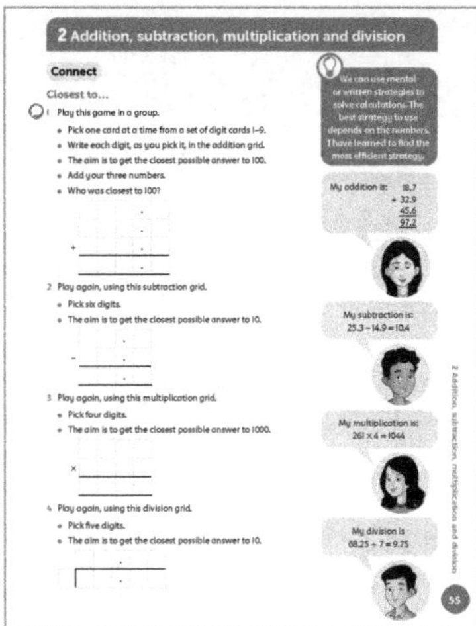


## Main activity

Ask the class to play against one another in groups of four or five students, taking it in turns to choose a digit card. Play one of the games with each group. Note the strategies that students use both to place the digits and to carry out the calculations.

Copy the grid for question 2 on page 55 of the Student Book onto the board. Choose six digit cards. Ask the groups to work out where to place the digits to give (a) the best possible result, (b) the worst possible result.

### Differentiation

Students should work in mixed-attainment groups to support one another in playing the game.

## Reflection time

After students have played the games, discuss the strategies they used when placing their digits. How did they try to ensure their digit was placed to give them the best possible outcome? How did it change when using different operations?

| Differentiated outcomes | |
|---|---|
| **All students** | should participate in the game. |
| **Most students** | will help to devise the best strategy to win each game. |
| **Some students** | may support other students in their group and provide alternative strategies. |

## Answers

### Student Book page 55

**1** For example: 15.9 + 36.8 + 47.2 = 99.9

**2–4** Answers will vary because students use selected digit cards to make the numbers for the calculations. Check that students' answers are correct.

# 2 Addition, subtraction, multiplication and division



## Global skills

- **Interpersonal skills:** communication
- **Self-development skills:** reflecting on learning

## Student Book

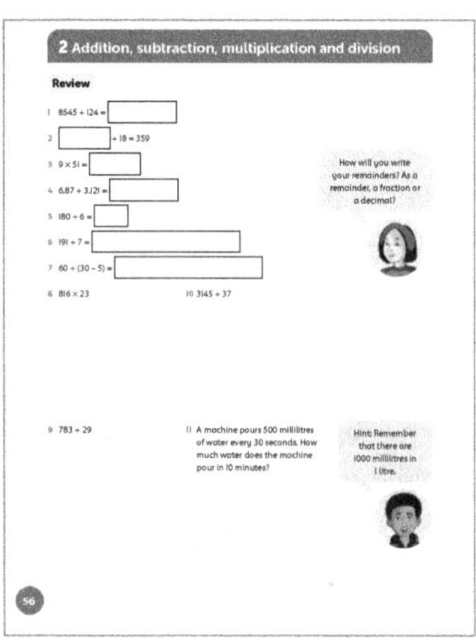

With young students, assessment activities are most effective when carried out as an everyday classroom activity. Watch as students calculate with large and small numbers and use both mental and written methods and the laws of arithmetic. Listen to check that they understand how to estimate and how to check answers using inverse operations.

Write 'Calculation strategies' in the middle of the board. Draw several arrows from this. Ask students to tell you the calculation strategies from this unit, for example: near multiples, commutative law, associative law, distributive law. Put these at the end of the arrows. Ask students to give examples of calculations that they can use each strategy to answer.

Ask students questions to assess their understanding, for example:

- *Which laws of arithmetic can you tell me about?*
- *Which is the best strategy for multiplying 2 × 43 × 5?*
- *When do you use rounding in solving problems?*

Ask students to work individually to complete the questions on page 56 of the Student Book to show what they now know about calculation strategies. Encourage them to show their working for each.

## Answers

### Student Book page 56

| | |
|---|---|
| **1** 8669 | **7** 2.4 or $2\frac{2}{5}$ or 2 r10 |
| **2** 341 | **8** 18 768 |
| **3** 459 | **9** 27 |
| **4** 9.991 | **10** 85 |
| **5** 30 | **11** 10 000 millilitres or 10 litres |
| **6** $27\frac{2}{7}$ or 27 r2 | |

## Practice Book

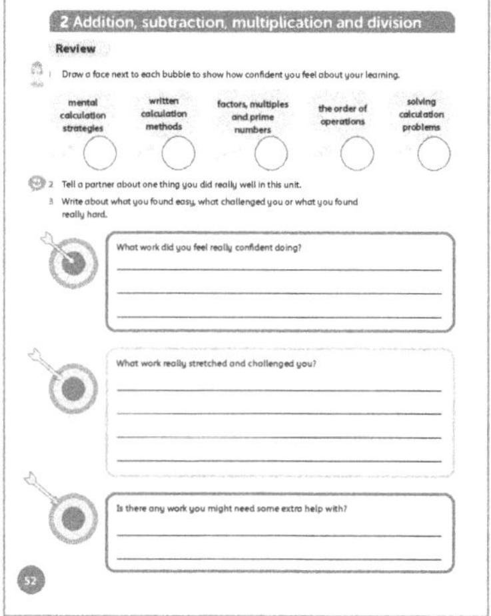

It is appropriate to complete this Practice Book Review as a whole-class discussion. You may choose to keep a record of the class discussion or a copy of the Review page for your own records. The Review provides an opportunity for students to reflect on their learning from the unit, to discuss any areas of mathematics that they feel went particularly well, and any areas that they feel less confident about. Ensure that all students have a copy of the Student Book as a reminder of the areas of mathematics that they have worked on in this unit.

Allow students plenty of time for discussion before asking them to complete the Practice Book page individually, and then, if appropriate, to share their responses with the rest of the class. If students complete this self-assessment at home, encourage them to discuss this with adults.

Make a note of areas that students still feel unsure about. As the four rules of arithmetic are essential in many of the other units, you can revisit this regularly. You can also build addition, subtraction, multiplication and division into everyday practice.

## Additional material

There are additional end-of-unit assessments available on the *Oxford Owl for School* website.

# 3 Fractions, decimals and percentages

## Overview

### Big idea

The most important idea for students to understand in this unit is how to carry out arithmetic using fractions, decimals and percentages.

With fractions, it is important that students have a clear understanding of fraction notation and how to manipulate fractions when adding, subtracting, multiplying or dividing. Students have already encountered equivalent fractions, and their enhanced knowledge of common multiples will enable them to convert easily between equivalent fractions, as well as modify them for easier use in calculations.

This unit further develops students' understanding of the equivalence between fractions, decimals and percentages. For example, $\frac{3}{4}$ is equivalent to 0.75 and to 75%. Students can use this knowledge of equivalence, when solving problems, to decide which fractional form is the best to use in a particular context.

Students will gain a deeper understanding, in this unit, of fractions greater than 1. These can be written as either improper fractions or mixed numbers. For example, $\frac{7}{4}$ is an improper fraction. The equivalent mixed number is $1\frac{3}{4}$. Students will use this understanding to solve problems involving mixed numbers.

### Look out for

- **Students who confuse the operations when calculating with fractions.** Use a range of visual representations and concrete material to help students relate the operation to different contexts. For example, use shape diagrams divided into equal parts with some parts shaded for each calculation, and use practical materials such as cubes in one, two or three different colours.

- **Students who struggle to know where to place the decimal point when adding, subtracting, multiplying or dividing decimals.** Reinforce the place value of each digit being multiplied and emphasise the language of tenths, hundredths and so on. Encourage students to estimate the answer first, by rounding each number to the nearest whole. They can then check their answer against their estimate to make sure that the decimal point is in the correct place.

### Possible misconceptions

- **Students think that all numbers that include fractions must be less than 1.** Use real contexts to help students to understand the idea of mixed numbers, for example $2\frac{1}{2}$ pizzas.

- **Students do not understand how dividing by a fraction can make a number larger.** Use the language of division to help students understand. For example, when dividing by $\frac{1}{2}$ ask 'How many halves are there in…?'

### Key vocabulary

- fraction, numerator, denominator, equivalent, equivalent fraction, simplest form
- improper fraction, mixed number
- common multiple, common denominator, bar model
- half, third, quarter, sixth, eighth, three-quarters, five-eighths, tenths, hundredths, thousandths
- one decimal place, two decimal places, round, to the nearest tenth, to the nearest whole, ascending/descending order, out of 100
- greater than, >, less than, <, equivalent to,
- decimal, decimal fraction, decimal equivalent, percentage, per cent, %
- price, reduction, proportion

## Coverage in lessons

| Learning objective | E | 3A | 3B | 3C | 3D | 3E | 3F | 3G | 3H | 3I | 3J | 3K | 3L | C | R |
|---|---|---|---|---|---|---|---|---|---|---|---|---|---|---|---|
| Use common factors to simplify fractions; use common multiples to express fractions in the same denomination. | | ✓ | ✓ | ✓ | | | | | | | | | | ✓ | ✓ |
| Compare and order fractions, including fractions > 1. | | ✓ | ✓ | ✓ | | | | | | | | | | | ✓ |
| Add and subtract fractions with different denominators and mixed numbers, using the concept of equivalent fractions. | | | ✓ | | ✓ | | | | | | | | | ✓ | ✓ |
| Multiply simple pairs of fractions, writing the answer in its simplest form (for example $\frac{1}{4} \times \frac{1}{2} = \frac{1}{8}$). | | | | | | ✓ | | | | | | | | ✓ | ✓ |
| Divide proper fractions by whole numbers (for example $\frac{1}{3} \div 2 = \frac{1}{6}$). | | | | | | | ✓ | | | | | | | | ✓ |
| Associate a fraction with division and calculate decimal fraction equivalents (for example 0.375) for a simple fraction (for example $\frac{3}{8}$). | | | | | | | | ✓ | | | | | ✓ | | ✓ |
| Identify the value of each digit in numbers given to three decimal places and multiply and divide numbers by 10, 100 and 1000 giving answers up to three decimal places. | | | | | | | | ✓ | ✓ | | | | | | ✓ |
| Multiply 1-digit numbers with up to two decimal places by whole numbers. | | | | | | | | | ✓ | ✓ | | | | | ✓ |
| Use written division methods in cases where the answer has up to two decimal places. | | | | | | | | | ✓ | | ✓ | ✓ | | ✓ | ✓ |
| Solve problems which require answers to be rounded to specified degrees of accuracy. | ✓ | | ✓ | | | | | ✓ | ✓ | ✓ | ✓ | ✓ | | ✓ | ✓ |
| Recall and use equivalences between simple fractions, decimals and percentages, including in different contexts. | ✓ | ✓ | ✓ | | | | | ✓ | | | | | ✓ | ✓ | ✓ |

 **Unit 3 Fractions, decimals and percentages**

# 3 Fractions, decimals and percentages

## Engage — Student Book page 57

### Big question

- How do I use fractions, decimals and percentages to solve mathematical problems in everyday contexts?

### Global skills

- **Creative skills:** problem solving
- **Interpersonal skills:** communication
- **Self-development skills:** reflecting on learning

### Key vocabulary

- percentage, %, out of 100, fraction

### Resources

- class set of local newspapers or magazines that students read at home

### Language support

Make a display using the percentages that appear in newspapers and magazines. Include all key vocabulary for reference during the unit.

##  Introductory activity

Look together at page 57 of the Student Book. Display on the IWB, if possible. Read the questions in the speech bubble together. Ask students what they know about percentages and where else they have seen them.

Ask students to work in pairs. Give each pair a newspaper or magazine. Ask them to find as many examples of the percentage sign (%) as they can in newspapers or magazines. Then ask each pair to share with the whole class one example from their publication. Try to share a wide range of contexts in which the percentage sign is used. You can normally find a % in headlines, adverts and news stories. Each time a student gives an example of a percentage, write it on the board so that at the end of the feedback you have a range of percentages.

Now ask students to talk to their partner about the percentages that they can see on the board. For example, they may know that '50% is equivalent to a half' and that 'you find 10% by dividing by 10'.

##  Main activity

Ask each pair to select three examples of percentages used in their newspaper, magazine or from the Student Book page. Then ask pairs to use their examples to write a word problem. They must be able to solve this problem themselves. Then ask students to work in groups of six to solve one another's problems. At the end of the session, ask each group to select their group's 'best' problem. Ask each group to justify their choice.

### Differentiation

**Supporting:** Support students in creating word problems with the percentages they know.

**Consolidating:** Encourage students to create word problems using a range of percentages.

**Extending:** Challenge students to create two-step word problems using a range of percentages.

###  Reflection time

Each group should explain their favourite problem and justify why they selected it. Ask one student from each group to model the solution at the front of the classroom for the whole class.

# 3A Equivalent fractions



### Specific learning focus

- Compare fractions with the same denominator and related denominators.

### Global skills

- **Creative skills:** problem solving

### Key vocabulary

- equivalent fraction, simplest form

### Resources

- mini whiteboards and markers

### Language support

As you work with students, ask them, for example:

- *What does the numerator/denominator show?*
- *Which fraction is equivalent?*
- *How can we simplify that fraction?*

It is important for students to say the fractions out loud, especially halves, quarters and eighths.

 **Introductory activity**

Ask pairs of students to draw three rectangles 20 cm long on their whiteboards. Ask them to divide the first into 5 equal parts, the second into 10 equal parts and the third into 20 equal parts. Ask pairs to discuss what the fractions should be called and how they should be written.

Now ask students to colour two parts on the first rectangle. *What fraction does this show?* Now can they colour an equivalent amount on the other two rectangles and write down what fraction each is showing?

*What fraction is unshaded on the first rectangle? What **equivalent fractions** are unshaded on the other two rectangles?*

 **Main activity**

Look together at page 58 of the Student Book. Display on the IWB, if possible. Ask students to look at the box at the top of the page. Ask them to discuss in pairs what equivalent fractions they can see.

Put students in mixed-attainment pairs so that the more-confident students can support the less-confident students. This also means that all students hear the correct vocabulary modelled and use appropriate vocabulary. Point out the activity on page 58 and ask, *How is the top row of the table filled in?* Ask individual students to share their reasoning with the rest of the class. Ask students to work in pairs to complete the table. They discuss the solutions together. To challenge pairs, offer them difficult examples to fill the last two rows in the table such as six twentieths or eight twenty-fourths. Check individual students' understanding of 'equivalence' and '**simplest form**' as they complete the table.

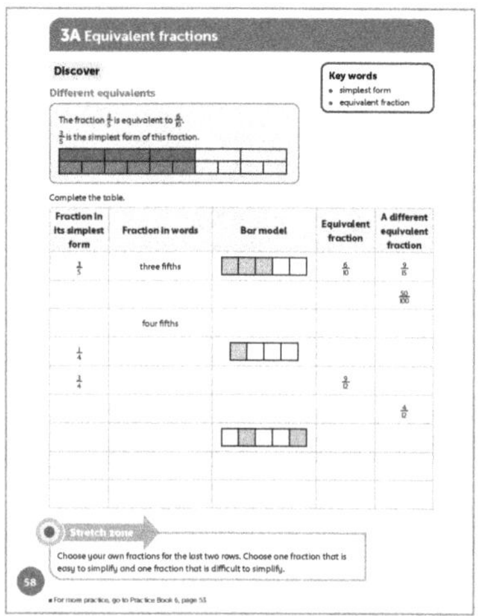

## Differentiation

**Supporting:** Support students by encouraging them to use drawings of equivalent fractions first.

**Consolidating:** Ask students to explain how they decide which representations are correct for each set of equivalent fractions.

**Extending:** Encourage students to find more equivalent fractions for each one in the table and explain how they found them.

**Stretch zone:** *Choose your own fractions for the last two rows. Choose one fraction that is easy to simplify and one fraction that is difficult to simplify.*

Can students select a difficult fraction and challenge their partner to complete the row of the table?

 **Reflection time**

Ask students to share their reasoning as they completed each row of the table. In each case, ask whether anyone had alternative equivalent fractions to those shared. Can students explain why they are the same?

Discuss the fractions used in the last two rows of the table and how the equivalent forms were made.

**Practice Book:** Students complete Practice Book page 53. They can do this directly after the Main activity, as homework, or as the focus of a separate mathematics session to help students consolidate their learning and build fluency.

Students complete bar models to check whether pairs of fractions are equivalent or not. *How do your bar models help you to check whether the two fractions are equivalent? Can you tell me another equivalent fraction for that pair? Can you tell me another?*

| Differentiated outcomes | |
| --- | --- |
| **All students** | should find equivalent fractions with support. |
| **Most students** | will calculate equivalent fractions by multiplying or dividing. |
| **Some students** | may use their knowledge of equivalence to find a greater number of equivalences to a given fraction. |

## Answers

### Student Book page 58

| Fraction in its simplest form | Fraction in words | Equivalent fraction | A different equivalent fraction |
| --- | --- | --- | --- |
| (Provided) $\frac{3}{5}$ | three fifths | $\frac{6}{10}$ | $\frac{9}{15}$ |
| $\frac{1}{2}$ | one half | $\frac{2}{4}$ | $\frac{50}{100}$ |
| $\frac{4}{5}$ | four fifths | $\frac{8}{10}$ | $\frac{16}{20}$ |
| $\frac{1}{4}$ | one quarter | $\frac{2}{8}$ | $\frac{5}{20}$ |
| $\frac{3}{4}$ | three quarters | $\frac{9}{12}$ | $\frac{15}{20}$ |
| $\frac{1}{3}$ | one third | $\frac{2}{6}$ | $\frac{4}{12}$ |
| $\frac{2}{5}$ | two fifths | $\frac{4}{10}$ | $\frac{6}{15}$ |

The last two rows contain students' own fractions. Check that each row is correct and that all the columns in one row contain an equivalent fraction.

### Practice Book page 53

**1** equivalent

**2** not

**3** equivalent

**4** not

**5** equivalent

Stretch zone: Students should refer to multiplying the numerator and denominator by the same number, or dividing them both by the same number.

# 3A Equivalent fractions



## Specific learning focus

- Order fractions by using equivalent fractions.

## Global skills

- **Creative skills:** problem solving

## Key vocabulary

- simplest form, equivalent fraction

## Resources

- mini whiteboards and markers
- six 3 cm-wide strips of A4 paper, cut across the width, for each student

## Language support

Model the key vocabulary carefully as students convert between equivalent fractions. Say, for example:

$\frac{8}{12}$ is equivalent to $\frac{16}{24}$ because 16 is double 8 and 24 is double 12.

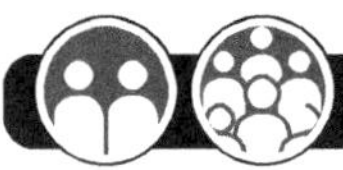 **Introductory activity**

Ask pairs of students to draw diagrams on their whiteboards to work out the answer to the question, *Which is larger: $\frac{3}{8}$ or $\frac{2}{5}$?* Students should be encouraged to use common multiples to make equivalent fractions so they can be compared by their common denominators. $\frac{3}{8} = \frac{15}{40}$ and $\frac{2}{5} = \frac{16}{40}$, so $\frac{2}{5}$ is larger. Share responses and diagrams when pairs have solved the problem. Repeat for $\frac{3}{5}$ and $\frac{2}{3}$.

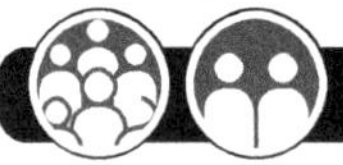 **Main activity**

Look together at page 59 of the Student Book. Display on the IWB, if possible. Point out the Think back information. You can suggest that, when sorting a set of fractions to be ordered, one way to start is to sort the fractions that are less than $\frac{1}{2}$ from those that are greater than $\frac{1}{2}$. For comparing fractions with simple multiples of denominators, use known equivalents such as the one in the example: $\frac{3}{4} = \frac{6}{8}$ so $\frac{3}{4} < \frac{7}{8}$.

Ask students to discuss in pairs why $\frac{7}{8} > \frac{5}{6}$. They can see it on the number line but ask them to justify it by using equivalent fractions or fractions with a common denominator. Share their responses and discuss which denominator is easier to use, as some students may have used 48 (8 × 6) but some may have recognised that

24 can also be used, being 3 × 8 and 4 × 6. Students then work in pairs to use equivalent fractions to place sets of fractions in order on a number line in the activities on page 59. As they work, listen to their discussions and prompt them with questions, for example, *How can you find a common denominator? Which equivalent fraction can help you here?*

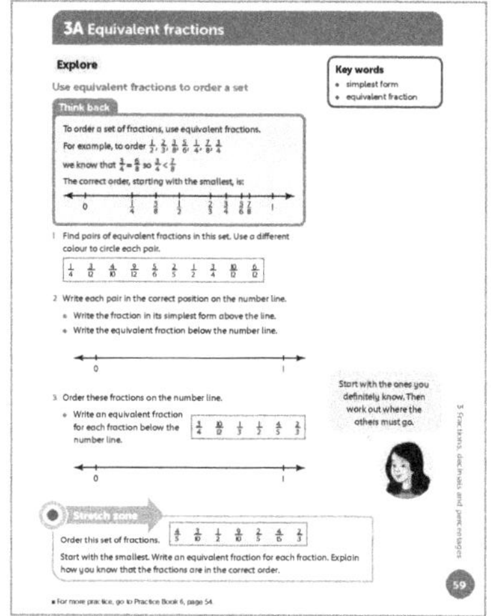

## Differentiation

**Supporting:** Help students by modelling the process for finding common denominators.

**Consolidating:** Ask students to explain their reasoning when comparing fractions.

**Extending:** Encourage students to sort fractions mentally first, using their knowledge of equivalence.

**Stretch zone:** *Order this set of fractions: $\frac{4}{5}, \frac{3}{10}, \frac{1}{2}, \frac{9}{10}, \frac{2}{5}, \frac{4}{15}, \frac{2}{3}$.*

*Start with the smallest. Write an equivalent fraction for each fraction. Explain how you know that the fractions are in the correct order.*

Students may write each fraction as an equivalent fraction with 30 as the denominator to help them. Alternatively, they may compare pairs of fractions to order the fractions one at a time. The correct order is: $\frac{4}{15}, \frac{3}{10}, \frac{2}{5}, \frac{1}{2}, \frac{2}{3}, \frac{4}{5}, \frac{9}{10}$.

 **Reflection time**

Look together at the set of fractions in the Stretch zone. Ask more-confident students to explain how they went about sorting and ordering the fractions. Ask them to describe how they compared pairs of fractions and which ones were easier to compare and why.

**Practice Book:** Students complete Practice Book page 54. They can do this directly after the Main activity, as homework, or as the focus of a separate mathematics session to help students consolidate their learning and build fluency.

Students write four different equivalent fractions for a given fraction. They need to divide the numerator and denominator of the starting fraction to find simpler fractions for each one.

| Differentiated outcomes | |
|---|---|
| **All students** | should order fractions on number lines using equivalent fractions with support. |
| **Most students** | will use equivalent fractions and common denominators to order fractions. |
| **Some students** | may order fractions mentally using their knowledge of equivalence. |

## Answers

### Student Book page 59

$\frac{1}{4}$ and $\frac{3}{12}$, $\frac{4}{10}$ and $\frac{2}{5}$, $\frac{9}{12}$ and $\frac{3}{4}$, $\frac{5}{6}$ and $\frac{10}{12}$, $\frac{1}{2}$ and $\frac{6}{12}$

Check that the number line has been marked accurately and that the fractions are in this order:

$\frac{1}{4}\left(\frac{3}{12}\right)$, $\frac{2}{5}\left(\frac{4}{10}\right)$, $\frac{1}{2}\left(\frac{6}{12}\right)$, $\frac{3}{4}\left(\frac{9}{12}\right)$, $\frac{5}{6}\left(\frac{10}{12}\right)$

$\frac{1}{3}, \frac{1}{2}, \frac{2}{3}, \frac{3}{4}, \frac{4}{5}, \frac{10}{12}$ – check that the number line is marked correctly and that students have written a correct equivalent fraction for each one, for example:

$\frac{2}{6}, \frac{2}{4}, \frac{4}{6}, \frac{6}{8}, \frac{8}{10}, \frac{5}{6}$

Various answers are possible. Here are examples for each:

| | Fraction | Equivalent fractions | | | |
|---|---|---|---|---|---|
| 1 | $\frac{100}{200}$ | $\frac{50}{100}$ | $\frac{25}{50}$ | $\frac{5}{10}$ | $\frac{1}{2}$ |
| 2 | $\frac{75}{100}$ | $\frac{3}{4}$ | $\frac{6}{8}$ | $\frac{30}{40}$ | $\frac{60}{80}$ |
| 3 | $\frac{40}{60}$ | $\frac{20}{30}$ | $\frac{10}{15}$ | $\frac{2}{3}$ | $\frac{6}{9}$ |
| 4 | $\frac{200}{500}$ | $\frac{100}{250}$ | $\frac{50}{125}$ | $\frac{2}{5}$ | $\frac{4}{10}$ |
| 5 | $\frac{160}{200}$ | $\frac{16}{20}$ | $\frac{8}{10}$ | $\frac{4}{5}$ | $\frac{40}{50}$ |
| 6 | $\frac{140}{200}$ | $\frac{14}{20}$ | $\frac{7}{10}$ | $\frac{70}{100}$ | $\frac{35}{50}$ |
| 7 | $\frac{90}{240}$ | $\frac{9}{24}$ | $\frac{3}{8}$ | $\frac{6}{16}$ | $\frac{30}{80}$ |
| 8 | $\frac{80}{280}$ | $\frac{8}{28}$ | $\frac{4}{14}$ | $\frac{2}{7}$ | $\frac{20}{70}$ |

Stretch zone: A fraction is in its simplest form when the numerator and denominator have no common factor except 1.

# 3B Mixed numbers and improper fractions

## Discover
Student Book page 60 · Practice Book page 55

### Specific learning focus

- Change an improper fraction to a mixed number.

### Global skills

- **Creative skills:** problem solving

### Key vocabulary

- improper fraction, simplest form, mixed number

### Resources

- mini whiteboards and markers
- cubes
- three chairs
- six bars of chocolate, each made up of six chunks

### Language support

It is important for students to share their solutions with other students and to explain their ideas. This allows them to use their mathematical language. Ask other groups to comment on any differences between the presented solution and their solution. Model key vocabulary in your response, for example:

- improper fraction, mixed number (*What sort of fraction is that?*)
- half, third, sixth
- decimal fraction (*What is the equivalent decimal fraction?*)

 **Introductory activity**

Write $\frac{11}{5}$ on the board. Ask students to work in pairs to draw an image on their whiteboards that shows $\frac{11}{5}$. Ask them to do this without naming the fraction – ask pairs to discuss how you might say this fraction. Tell students that it is called an **improper fraction**, because the numerator is larger than the denominator.

Share the images that students have drawn on their whiteboards. Ask students how they think they should say the fraction and agree on 'eleven fifths'. Ask students how many wholes can be made from eleven fifths. Agree that $\frac{11}{5} = 2$ wholes and one fifth, or $2\frac{1}{5}$. Repeat for other improper fractions with a variety of denominators such as $\frac{14}{3}$ and $\frac{17}{6}$.

Look together at page 60 of the Student Book. Display on the IWB, if possible. Refer students to the Think back information. Discuss the example given there and make sure that students understand the difference between an improper fraction and a **mixed number**.

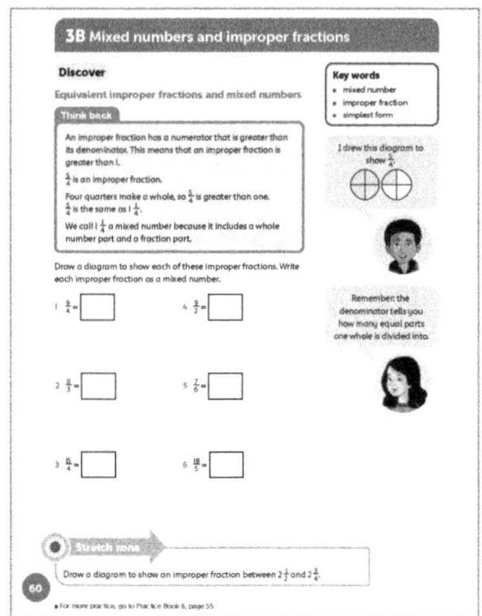

Main activity

Position three chairs at the front of the classroom. Place one chocolate bar on one chair, two bars on the next chair and three on the final chair. Ask ten students (you may decide to ask more) to come to the front of the class. They take it in turns to stand behind a chair. Tell them that they cannot change position after they have chosen a chair. When they are all behind a chair, they share the chocolate on that chair between them.

Write the fractions for each chair on the board. For example, there might be four students behind the first chair, so that is six chunks shared between four students, so they each get $\frac{6}{4}$ chunks. Once the fractions are written, discuss how to turn each fraction into an improper fraction, in this case $\frac{6}{4} = \frac{3}{2} = 1\frac{1}{2}$, so each student would get $1\frac{1}{2}$ chunks of chocolate. When you have completed the activity, students should work in groups to work out what the fairest solution to the problem is so that each person has a 'fair share'. Students need to offer a rationale for their solution as different groups will interpret 'fairness' differently.

Students then complete the activities on page 60 of the Student Book.

## Differentiation

**Supporting:** Work with students to model the use of cubes to turn improper fractions into mixed numbers.

**Consolidating:** Ask students to describe their strategies for changing improper fractions into mixed numbers and vice versa.

**Extending:** Challenge students to describe their mental strategies for changing improper fractions into mixed numbers and vice versa.

**Stretch zone:** *Draw a diagram to show an improper fraction between* $2\frac{1}{2}$ *and* $2\frac{3}{4}$.



Students should explain their choice of diagram and what it shows.

### Reflection time

Students should work together to agree on a definition for an improper fraction. Ask pairs to discuss how they can change an improper fraction into a mixed number and vice versa. Record the clearest explanation on a poster that you can display for the rest of the unit.

**Practice Book:** Students complete Practice Book page 55. They can do this directly after the Main activity, as homework, or as the focus of a separate mathematics session to help students consolidate their learning and build fluency.

Students practise converting improper fractions to mixed numbers and vice versa. Encourage them to use cubes or draw diagrams to help them.

| Differentiated outcomes | |
| --- | --- |
| **All students** | should convert improper fractions to mixed numbers with support. |
| **Most students** | will convert improper fractions to mixed numbers. |
| **Some students** | explain how they convert improper fractions to mixed numbers and vice versa. |

## Answers

### Student Book page 60

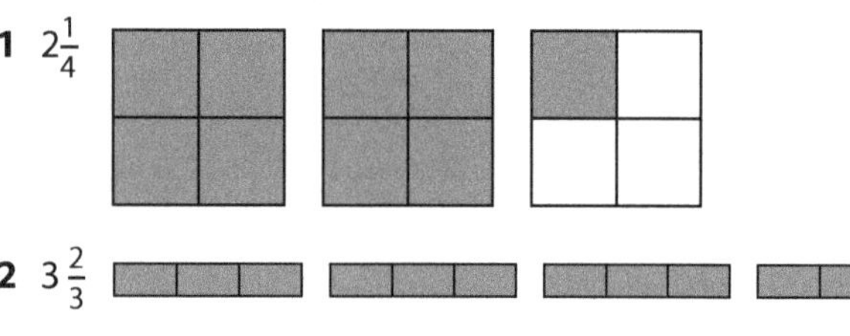

**1** $2\frac{1}{4}$

**2** $3\frac{2}{3}$

**3** $3\frac{3}{4}$

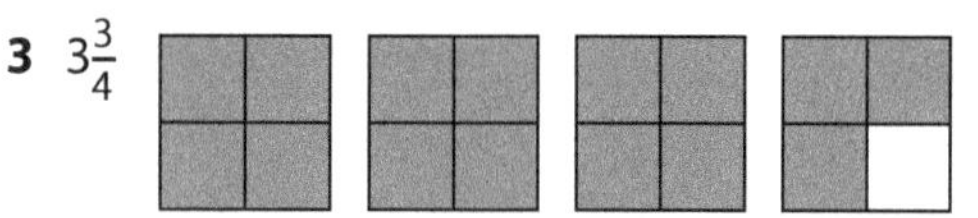

**4** $4\frac{1}{2}$

**5** $1\frac{1}{6}$

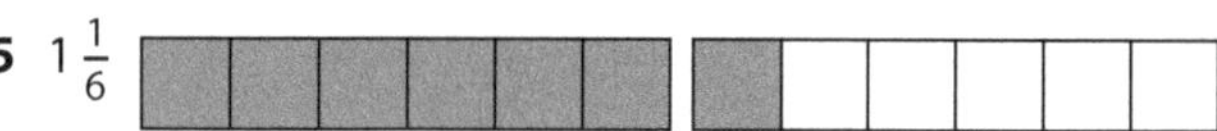

**6** $3\frac{3}{5}$

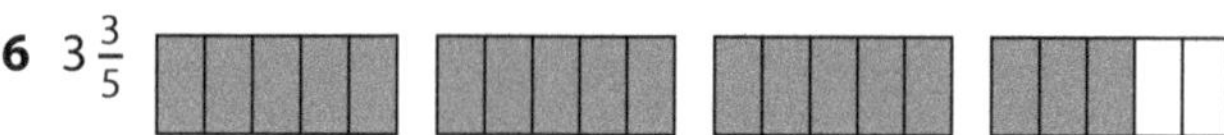

**1** $4\frac{1}{2}$  **5** $2\frac{1}{4}$  **9** $\frac{13}{4}$  **13** $\frac{19}{10}$

**2** $4\frac{1}{4}$  **6** $1\frac{5}{6}$  **10** $\frac{33}{8}$  **14** $\frac{81}{8}$

**3** $3\frac{1}{5}$  **7** $3\frac{1}{8}$  **11** $\frac{35}{8}$  **15** $\frac{34}{5}$

**4** $4\frac{3}{5}$  **8** $3\frac{1}{10}$  **12** $\frac{47}{9}$  **16** $\frac{42}{11}$

**Stretch zone:** Answers will vary because students choose their own numbers. Check that the improper fraction and mixed number are equivalent and that the diagram represents the number correctly.

# 3B Mixed numbers and improper fractions

## Explore
Student Book pages 61–62 • Practice Book page 56

### Specific learning focus

- Change improper fractions to mixed numbers to solve problems.

### Global skills

- **Creative skills:** problem solving

### Key vocabulary

- improper fraction, mixed number

### Resources

- large sheets of paper marked with large circles, scissors
- mini whiteboards and markers

### Language support

Asking students to explain their ideas and share their strategies allows them to practise and use mathematical language. Model key words and phrases in your discussions with students, for example:

- improper number, mixed number
- order in ascending size
- greater than, less than.

 **Introductory activity**

Cut out four large circles. Tell students to imagine these are pizzas. Cut each pizza into quarters. Sit students in a circle around the pizzas. Ask the following questions. After each question, share the pizzas out so that students have a visual image of the fractions. Then put the pizzas in the centre again.

*There is one person. How much or how many pizzas do they have? (4)*

Give all of the pizzas to one student to demonstrate this.

*There are two people. How much or how many pizzas do they have each? (2)*

Give two pizzas to two students to demonstrate this.

*There are four people. How much or how many pizzas do they have each? (1)*

Give one pizza to four different students to demonstrate this.

For five people, ask the class how to cut four pizzas. (Each each person will get $\frac{4}{5}$ of a pizza, so cut four pizzas into fifths and give four slices to each person.) Repeat with six people. (Each person will get $\frac{4}{6}$ of a pizza so cut four pizzas into sixths and give out four slices to each person.)

 **Main activity**

Draw four cakes on the board. *How much of a cake will each of five people get if they get equal shares?* Ask students to discuss this in pairs, and then to draw a diagram on their whiteboards. Agree that four cakes shared between five people means they each get $\frac{4}{5}$ of a cake.

*Now imagine that there are seven people to share the four cakes. How much will each person get?* $\left(\frac{4}{7}\text{ of a cake}\right)$ *What about eight people sharing five cakes?* (They each get $\frac{5}{8}$ of a cake.) *Can you describe the pattern here? Can you use it to say how much each person gets if 15 cakes are shared between 19 people?*

Now ask students to work to solve the problems on pages 61 and 62 of the Student Book. Students work in pairs to support one another and to discuss their solutions. Encourage them to draw diagrams on their whiteboards help them. Ask them to explain their drawings to you. It is important that they try to represent the problems visually.

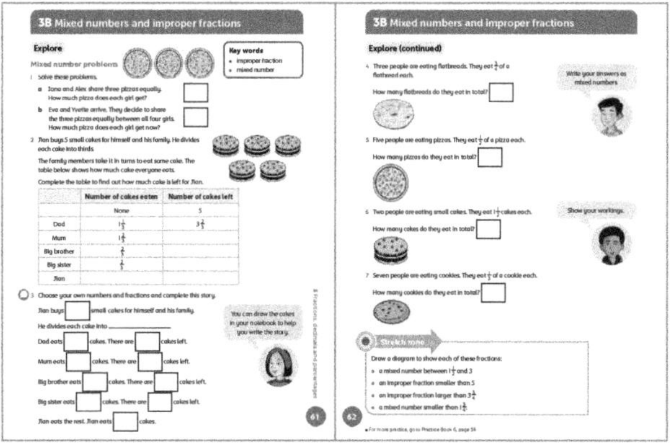

# Differentiation

**Supporting:** Help students to solve the problems using concrete resources.

**Consolidating:** Ask students to draw diagrams to help them solve the problems.

**Extending:** Challenge students to share their strategies for solving the problems.

**Stretch zone:** *Draw a diagram to show each of these fractions:*

- *a mixed number between $1\frac{1}{2}$ and 3*
- *an improper fraction smaller than 5*
- *an improper fraction larger than $3\frac{3}{4}$*
- *a mixed number smaller than $1\frac{3}{4}$.*

Students could see whether they can find more than one answer for each and represent them on diagrams.

## Reflection time

Ask three students to come to the front of the classroom and share their solutions to question 7 with the rest of the class. Ask them to explain their working very carefully. *Did you each use the same strategy? How did your strategies differ?* Ask the class, *Did anyone do the activity in a different way?*

**Practice Book:** Students complete Practice Book page 56. They can do this directly after the Main activity, as homework, or as the focus of a separate mathematics session to help students consolidate their learning and build fluency.

Students write mixed numbers and improper fractions to fit a given rule, such as 'a fraction between 8 and 9'.

| Differentiated outcomes | |
|---|---|
| **All students** | should solve the problems involving mixed numbers, using practical materials with support. |
| **Most students** | will solve the problems involving mixed numbers, using images and diagrams for support. |
| **Some students** | may solve problems involving mixed numbers without using images and diagrams. |



**Student Book pages 61–62**

**1 a** $\frac{3}{2}$ or $1\frac{1}{2}$ **b** $\frac{3}{4}$

**2**

| | Number of cakes eaten | Number of cakes left |
|---|---|---|
| | None | 5 |
| Dad: (Provided) | $1\frac{1}{3}$ | $3\frac{2}{3}$ |
| Mum | $1\frac{2}{3}$ | 2 |
| Big brother | $\frac{2}{3}$ | $1\frac{1}{3}$ |
| Big sister | $\frac{2}{3}$ | $\frac{2}{3}$ |
| Jian | $\frac{2}{3}$ | 0 |

**3** Check that students have completed the fraction story correctly.

**4** $\frac{9}{4}$ or $2\frac{1}{4}$

**5** $\frac{5}{2}$ or $2\frac{1}{2}$

**6** 3

**7** $\frac{7}{3}$ or $2\frac{1}{3}$

**Practice Book page 56**

Answers will vary because students choose their own mixed numbers and equivalent improper fractions to fit a given rule. Check that students' answers match the rules and show the equivalence of each mixed number and improper fraction.

Stretch zone: Check that students have ordered five of their fractions correctly.

# 3C Ordering fractions and mixed numbers

## Discover
**Student Book page 63 • Practice Book page 57**

### Specific learning focus

- Recognise and compare mixed numbers and improper fractions.

### Global skills

- **Creative skills:** investigating

### Key vocabulary

- mixed number, improper fraction

### Resources

- mini whiteboards and markers

### Language support

Model the key vocabulary carefully as students compare mixed numbers. Say, for example:

- $\frac{7}{2}$ is larger than $\frac{26}{8}$ because $\frac{7}{2}$ is $3\frac{1}{2}$ and $\frac{26}{8}$ is $3\frac{1}{4}$.

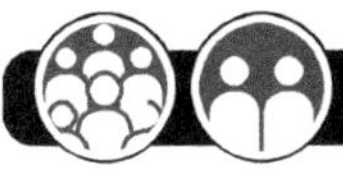 **Introductory activity**

Write on the board:

$$1\frac{3}{4} \qquad 3\frac{2}{5} \qquad 6\frac{3}{8}$$

Ask students to draw diagrams on their whiteboards to show what each number represents. They may suggest using a number line, full and part circles, or rectangles and part rectangles.

Record all three methods. Remind students that these numbers are called mixed numbers. A mixed number is made up of a whole number and a fraction.

Ask pairs to discuss when they use mixed numbers in everyday life. Share their responses.

 **Main activity**

Look together at page 63 of the Student Book. Display on the IWB, if possible. Ask students to look at the Think back statement. It reminds them to convert fractions to the same form when comparing them, either all as improper fractions or all as mixed numbers.

Show students these fractions:

$$\frac{9}{4}\,, \qquad 2\frac{2}{5} \qquad \frac{13}{6}$$

Ask students to work in pairs to decide which is the largest and which is the smallest of the three fractions. As they work, listen to their discussions about strategies.

After sufficient time, ask students to share their thinking. Ask them to give you all three fractions as mixed numbers first $\left(2\frac{1}{4}, 2\frac{2}{5}, 2\frac{1}{6}\right)$. *Which is the largest and which is the smallest? How do you know?* Students might say, for example, that $\frac{1}{4} > \frac{1}{6}$, but how do they compare the fraction with fifths to the others? They should recall the need to make common denominators. If they haven't already done this, give them some time to do so. Students work with two fractions at a time, so they may find that $\frac{9}{4} = \frac{45}{20}$ and $2\frac{2}{5} = \frac{48}{20}$, so this is the larger of these two. Repeat for the other pairings and finally use the results to put all three in the correct order.

Now ask students to complete the activities in the Student Book in pairs. As they work, ask them to read the mixed number to you. This allows you to check their pronunciation and understanding. Give students further practice by asking them to include additional numbers on the number line that have different denominators from those given. For example, for question 1 include $\frac{13}{20}$.

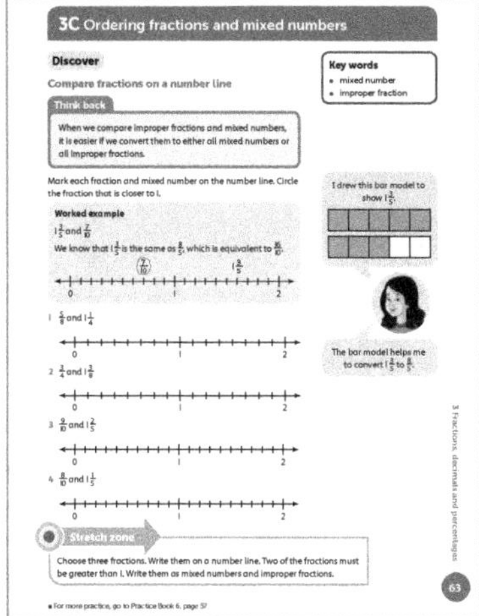

## Differentiation

**Supporting:** Work with students to model the use of the number line to compare fractions.

**Consolidating:** Ask students to support less-confident students in using the number line.

**Extending:** Ask students to use empty number lines to compare increasingly complex fractions.

**Stretch zone:** *Choose three fractions. Write them on a number line. Two of the fractions must be greater than 1. Write them as mixed numbers and improper fractions.*

Students could challenge each other to draw mixed fractions using diagrams and place them on number lines.

 **Reflection time**

Students share their results and explain how they found where to place each number on the line. Ask them to explain to the rest of the class their strategy for finding common denominators for the different questions.

**Practice Book:** Students complete Practice Book page 57. They can do this directly after the Main activity, as homework, or as the focus of a separate mathematics session to help students consolidate their learning and build fluency.

Students compare pairs of fractions saying which one is greater or less than the other. They can use a fraction wall to help them but will need to convert both fractions to the same denominator and show their working.

| Differentiated outcomes | |
|---|---|
| **All students** | should compare mixed numbers and improper fractions with support. |
| **Most students** | will compare mixed numbers and improper fractions. |
| **Some students** | may explain their strategy for comparing mixed numbers and improper fractions. |



**Student Book page 63**

**1** $1\frac{1}{4}$

**2** $\frac{3}{4}$

**3** $\frac{9}{10}$

**4** $\frac{8}{10}$ and $\frac{12}{10}$ are equally close to 1.

Check that the number lines are marked correctly.

**Practice Book page 57**

**1** $\frac{3}{4} > \frac{3}{8}$

**2** $\frac{2}{5} < \frac{5}{10}$,    $\frac{2}{5} < \frac{1}{2}$

**3** $\frac{8}{14} > \frac{3}{7}$,    $\frac{4}{7} > \frac{3}{7}$

**4** $\frac{9}{13} > \frac{4}{9}$

**5** $\frac{5}{15} < \frac{3}{6}$,    $\frac{1}{3} < \frac{1}{2}$

**6** $\frac{8}{16} > \frac{3}{8}$,    $\frac{1}{2} > \frac{3}{8}$

Stretch zone: Possible answers: $\frac{4}{7}, \frac{3}{5}, \frac{2}{3}$

---

## 3C Ordering fractions and mixed numbers

### Explore    Student Book page 64 • Practice Book page 58

#### Specific learning focus

- Order mixed numbers and improper fractions using a number line.

#### Global skills

- **Creative skills:** problem solving

#### Key vocabulary

- mixed number, improper fraction, numerator, denominator

#### Resources

- mini whiteboards and markers

#### Language support

Model the key vocabulary carefully as students order mixed numbers. Say, for example:

- $\frac{7}{2}$ is larger than $\frac{26}{8}$ because $\frac{7}{2}$ is $3\frac{1}{2}$ and $\frac{26}{8}$ is $3\frac{1}{4}$, but smaller than $\frac{29}{8}$, which is $3\frac{5}{8}$.



Write this sequence on the board for students to continue on their whiteboards:

$$\frac{3}{8} \quad \frac{7}{8} \quad \frac{11}{8} \quad \frac{15}{8} \quad \cdots \quad \cdots \quad \cdots \quad \cdots \quad \cdots$$

Ask them to complete the sequence and then turn the fractions greater than 1 into mixed numbers. *Can you describe the sequence? How much does it increase each time?*

Ask students to make up a sequence that starts with a fraction less than $\frac{1}{2}$ and increases by the same amount each time. Each student writes their sequence on their whiteboard, then swaps whiteboards with a partner and sees whether the partner can continue the sequence.



This lesson builds on 3C Discover, moving from comparing two mixed numbers or improper fractions to sorting and ordering three or more numbers and locating them on a number line. Write on the board:

$$\frac{4}{5} \qquad 1\frac{3}{10} \qquad 1\frac{1}{5} \qquad \frac{17}{10}$$

Draw a number line and mark the ends as 0 and 2, with 20 equal divisions between. Ask students to say what each division represents, counting in tenths from 0 to 2. Now ask them to decide where to place $\frac{4}{5}$ on the line. *What equivalent fraction can you make that will help you?* Agree that $\frac{4}{5} = \frac{8}{10}$, so it can be placed 8 divisions from 0.

Ask students to work in pairs to decide where to place the remaining fractions. Encourage them to turn mixed numbers into improper fractions and make the **denominators** into tenths to match the number line divisions. Look together at page 64 of the Student Book. Display on the IWB, if possible. Work through another example, using the worked example to check that each pair understands and completes the task correctly. Students then complete the tasks on page 64 of the Student Book.

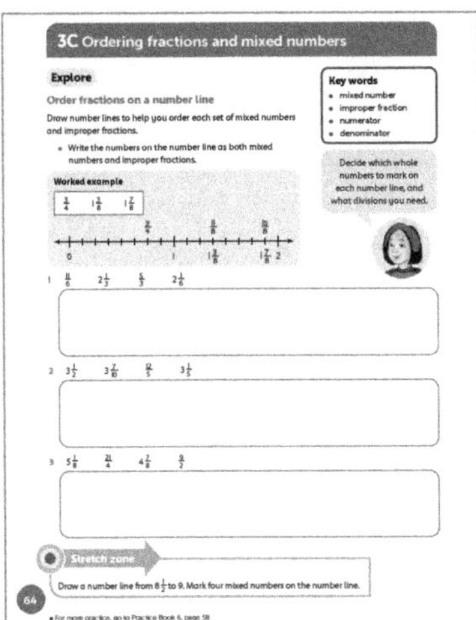

## Differentiation

**Supporting:** Work with students to model the use of the number line to order fractions.

**Consolidating:** Ask students to support less-confident partners in using the number line.

**Extending:** Ask students to use empty number lines to order increasingly complex fractions.

**Stretch zone:** *Draw a number line from $8\frac{1}{2}$ to 9. Mark four mixed numbers on the number line.*

Students should use their knowledge of equivalent fractions and mixed numbers to choose and place four numbers in this range.

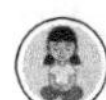 **Reflection time**

Students should share their results from the Student Book activity and explain how they decided where to place each number on the line. Ask them to explain their strategy for labelling the ends of the lines, then finding common denominators.

**Practice Book:** Students complete Practice Book page 58. They can do this directly after the Main activity, as homework, or as the focus of a separate mathematics session to help students consolidate their learning and build fluency.

Students choose four fractions between a given range, for example between 3 and 4. They label a number line and mark each of their fractions on the number line in the correct position.

| Differentiated outcomes | |
| --- | --- |
| **All students** | should order mixed numbers and improper fractions with support. |
| **Most students** | will order mixed numbers and improper fractions. |
| **Some students** | may explain their strategy for ordering mixed numbers and improper fractions. |

## Answers

### Student Book page 64

Check that the number lines are marked correctly with the numbers in the order:

**1** $\frac{5}{3}, \frac{11}{6}, 2\frac{1}{6}, 2\frac{1}{3}$

**2** $\frac{12}{5}, 3\frac{1}{5}, 3\frac{1}{2}, 3\frac{7}{10}$

**3** $\frac{9}{2}, 4\frac{7}{8}, 5\frac{1}{8}, 5\frac{1}{4}$

### Practice Book page 58

There are many possible answers. Make sure that students have chosen suitable fractions, labelled their number lines and marked appropriate fractions correctly.

Stretch zone: Students' choice of fractions will vary. Check that students have labelled their number lines and marked the appropriate fractions correctly.

# 3D Adding and subtracting fractions



### Specific learning focus

- Add and subtract fractions with different denominators.

### Global skills

- **Creative skills:** problem solving

### Key vocabulary

- numerator, denominator, common multiple, common denominator

### Resources

- Resource sheet 3.1: fraction wall
- mini whiteboards and markers

### Language support

Remind students that when adding, the denominators need to be the same and the answer has the same denominator. Reinforce this as they say their calculations, for example *two **sevenths** plus one **seventh** = three **sevenths***.

 **Introductory activity**

Write on the board: $\frac{7}{8} + \frac{3}{4}$. *What is the denominator of each fraction? Can you think of an equivalent fraction to $\frac{3}{4}$, using eighths?* $\left(\frac{3}{4} = \frac{6}{8}\right)$, so the calculation can be written as $\frac{7}{8} + \frac{6}{8}$

Draw two rectangles each divided into 8 equal parts. Ask a student to shade 7 of the parts in one colour, to represent $\frac{7}{8}$. Then ask another student to shade 6 more parts that represent $\frac{3}{4}$, or $\frac{6}{8}$. Finally, ask the class to say how many parts have been shaded altogether and what fractions of the two rectangles are now shaded. Emphasise the eighths in each fraction by saying 'seven *eighths* plus six *eighths* equals *thirteen eighths*'. Write this as $\frac{13}{8}$ and then ask a student to say what this is as a mixed number. $\left(1\frac{5}{8}\right)$

Repeat this for $\frac{9}{10} + \frac{3}{5}$.

 **Main activity**

Look together at page 65 of the Student Book. Display on the IWB, if possible. Point out the Think back worked example. Discuss how the fractions have been made into equivalent fractions having a **common denominator** so they can be added and subtracted. Use this as a further reminder that when adding or subtracting fractions,

students should make sure that the fractions have the same denominators, and they can use a fraction wall to find equivalent fractions so the denominators in the calculation are the same.

Now ask students to work on completing the activities on page 65 in pairs and listen to their discussions as they find equivalent fractions. Students should draw a bar model diagram for each calculation.

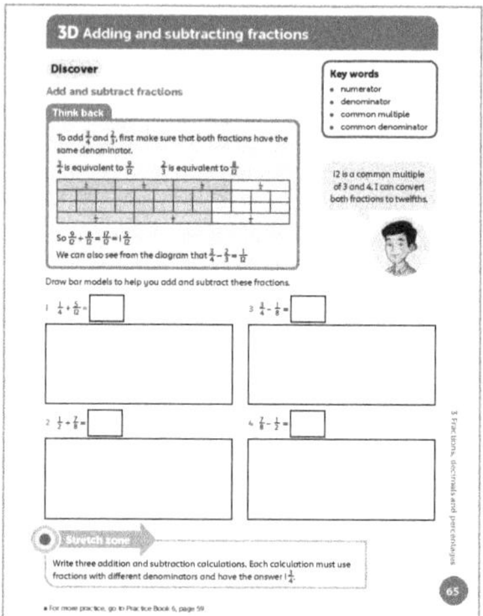

## Differentiation

**Supporting:** Help students to find equivalent fractions with common denominators for the calculations.

**Consolidating:** Ask students to explain how they found equivalent fractions with common denominators.

**Extending:** Challenge students to add or subtract more than two fractions with a variety of denominators.

**Stretch zone:** *Write three addition and subtraction calculations. Each calculation must use fractions with different denominators and have the answer $1\frac{3}{4}$.*

Encourage students to draw bar model diagrams to represent their calculations.

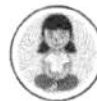 **Reflection time**

Ask several students to share their solutions to the Student Book activities. Ask students to work in pairs, using their whiteboards, to find further fraction additions and subtractions. They can use the fraction wall (Resource sheet 3.1) for support if necessary. Check their answers and understanding. *Do you think you could add any two fractions by finding equivalent fractions?*

**Practice Book:** Students complete Practice Book page 59. They can do this directly after the Main activity, as homework, or as the focus of a separate mathematics session to help students consolidate their learning and build fluency.

Students complete fraction additions to make totals as close to 1 as possible without going over. They complete the **numerator** and, in some cases, can choose their own numerator and denominator to complete the calculation.

## Answers

**Student Book page 65**

**1** $\frac{8}{12} = \frac{2}{3}$

**2** $\frac{11}{8}$

**3** $\frac{5}{8}$

**4** $\frac{3}{8}$

**Practice Book page 59**

**1** $\frac{1}{2} + \frac{3}{8} = \frac{7}{8}$

**2** $\frac{1}{2} + \frac{4}{10} = \frac{9}{10}$

**3** $\frac{1}{2} + \frac{5}{12} = \frac{11}{12}$

**4** $\frac{1}{2} + \frac{7}{16} = \frac{15}{16}$, for example.

**5** $\frac{1}{3} + \frac{3}{6} = \frac{5}{6}$

**6** $\frac{1}{3} + \frac{5}{9} = \frac{8}{9}$

**7** $\frac{1}{3} + \frac{7}{12} = \frac{11}{12}$

**8** $\frac{1}{3} + \frac{9}{15} = \frac{14}{15}$, for example.

**9** $\frac{1}{4} + \frac{5}{8} = \frac{7}{8}$

**10** $\frac{1}{4} + \frac{8}{12} = \frac{11}{12}$

**11** $\frac{1}{4} + \frac{11}{16} = \frac{15}{16}$

**12** $\frac{1}{4} + \frac{14}{20} = \frac{19}{20}$, for example.

**13** $\frac{1}{5} + \frac{7}{10} = \frac{9}{10}$

**14** $\frac{1}{5} + \frac{11}{15} = \frac{14}{15}$

**15** $\frac{1}{5} + \frac{15}{20} = \frac{19}{20}$

**16** $\frac{1}{5} + \frac{19}{25} = \frac{24}{25}$, for example.

Stretch zone: For example, $\frac{7}{20} + \frac{6}{10} = \frac{19}{20}$.

---

## 3D Adding and subtracting fractions

### Explore Student Book page 66 • Practice Book page 60

#### Specific learning focus

- Adding and subtracting fractions, including mixed numbers, with different denominators.

#### Global skills

- **Creative skills:** problem solving

#### Key vocabulary

- numerator, denominator, common denominator

#### Resources

- Resource sheet 3.1: fraction wall
- mini whiteboards and markers

#### Language support

Remind students that when subtracting, the denominators need to be the same and the answer has the same denominator. Reinforce this as they say their calculations, for example *seven **ninths** subtract two **ninths** = five ninths*.

 **Introductory activity**

Write on the board: $\frac{7}{8} - \frac{1}{4}$. *What is the denominator of each fraction? Can you think of an equivalent fraction to $\frac{1}{4}$ in eighths?* $\left(\frac{1}{4} = \frac{2}{8}\right)$, so the calculation can be written as $\frac{7}{8} - \frac{2}{8}\right)$

Draw a rectangle divided into 8 equal parts. Ask a student to shade 7 of the parts in one colour, to represent $\frac{7}{8}$. Then ask another student to cross out 2 parts that represent $\frac{1}{4}$, or $\frac{2}{8}$. Finally, ask the class to say how many parts are left from the original shading and what fraction of the rectangle is now shaded. Emphasise the eighths in each fraction by saying, *seven **eighths** subtract two **eighths** equals five **eighths***. Write this as $\frac{5}{8}$.

Repeat this for $\frac{9}{10} - \frac{1}{5}$.

 **Main activity**

Look together at page 66 of the Student Book. Display on the IWB, if possible. Point out the worked example. Discuss how the fractions have been made into equivalent fractions with a common denominator so they can be subtracted, and mixed numbers have been converted to improper fractions. Use this as a further reminder that when adding or subtracting fractions, these should have the same denominators, and that students can use a fraction wall to find equivalent fractions so the denominators in the calculation are the same.

Now ask students to work on completing the activities on page 65 in pairs and listen to their discussions as they find equivalent fractions and convert between mixed numbers and improper fractions.

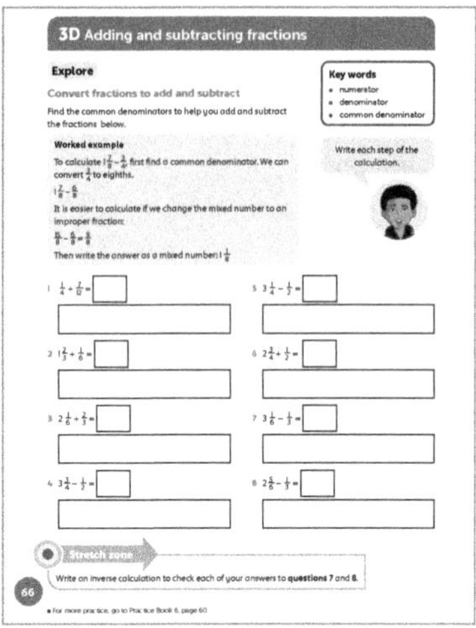

## Differentiation

**Supporting:** Help students to find equivalent fractions with common denominators for the calculations and convert mixed numbers to improper fractions when needed.

**Consolidating:** Ask students to explain how they found equivalent fractions with common denominators and converted mixed numbers to improper fractions.

**Extending:** Challenge students to add or subtract mixed numbers where both denominators need converting, for example $2\frac{1}{6} - \frac{5}{8}$.

**Stretch zone:** *Write an inverse calculation to check each of your answers to questions 7 and 8.*

Students should be confident to use inverse operations with fractions to check calculations.

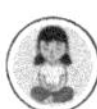 ## Reflection time

Ask several students to share their solutions to the Student Book activities. Ask students to work in pairs, using their whiteboards, to find further fraction and mixed number additions and subtractions. They can use the fraction wall (Resource sheet 3.1) for support if necessary. Check their answers and understanding. Ask students whether they could add or subtract any two mixed numbers by finding equivalent fractions.

**Practice Book:** Students complete Practice Book page 60. They can do this directly after the Main activity, as homework, or as the focus of a separate mathematics session to help students consolidate their learning and build fluency.

Students complete fraction subtraction with answers as close to 0 as possible without going below 0. They complete the numerator and, in some cases, can choose their own numerator and denominator to complete the calculation.

| Differentiated outcomes | |
| --- | --- |
| **All students** | should use equivalent fractions to add or subtract mixed numbers with support. |
| **Most students** | will use equivalent fractions to add or subtract mixed numbers. |
| **Some students** | may use equivalent fractions to add or subtract improper fractions or mixed numbers where both denominators need converting. |

## Answers

### Student Book page 66

**1** $\frac{10}{12} = \frac{5}{6}$

**2** $1\frac{5}{6}$

**3** $2\frac{5}{6}$

**4** $3\frac{1}{4}$

**5** $2\frac{3}{4}$

**6** $3\frac{1}{4}$

**7** $2\frac{5}{6}$

**8** $2\frac{1}{2}$

### Practice Book page 60

**1** $\frac{1}{2} - \frac{3}{8} = \frac{1}{8}$

**2** $\frac{1}{2} - \frac{4}{10} = \frac{4}{10}$

**3** $\frac{1}{2} - \frac{5}{12} = \frac{1}{12}$

**4** $\frac{1}{2} - \frac{7}{16} = \frac{1}{16}$, for example.

**5** $\frac{1}{3} - \frac{1}{6} = \frac{1}{6}$

**6** $\frac{1}{3} - \frac{2}{9} = \frac{1}{9}$

**7** $\frac{1}{3} - \frac{3}{12} = \frac{1}{12}$

**8** $\frac{1}{3} - \frac{4}{15} = \frac{1}{15}$, for example.

**9** $\frac{1}{4} - \frac{1}{8} = \frac{1}{8}$

**10** $\frac{1}{4} - \frac{2}{12} = \frac{1}{12}$

**11** $\frac{1}{4} - \frac{3}{16} = \frac{1}{16}$

**12** $\frac{1}{4} - \frac{4}{20} = \frac{1}{20}$, for example.

**13** $\frac{1}{5} - \frac{1}{10} = \frac{1}{10}$

**14** $\frac{1}{5} - \frac{2}{15} = \frac{1}{15}$

**15** $\frac{1}{5} - \frac{3}{20} = \frac{1}{20}$

**16** $\frac{1}{5} - \frac{4}{25} = \frac{1}{25}$, for example.

# 3E Multiplying fractions



## Specific learning focus

- Find a fraction of a fraction by multiplying.

## Global skills

- **Creative skills:** problem solving

## Key vocabulary

- numerator, denominator, simplest form

## Resources

- paper rectangles
- mini whiteboards and markers

## Language support

Remind students that when multiplying we sometimes refer to 'lots of', or just 'of'. Reinforce this as they say their calculations. For example, say, $\frac{1}{2} \times \frac{4}{5}$ is the same as finding $\frac{1}{2}$ of $\frac{4}{5}$, which is $\frac{2}{5}$.

## Introductory activity

Draw a series of shapes on the board – a circle, a square and a rectangle – and ask students to draw the same on their whiteboards. Now ask them to draw a straight line that cuts each shape in half and then shade one half of each shape. Now ask them to look at the half that has been shaded and draw a line that cuts that half into two equal parts. They should shade one of those parts in a different way. Ask, *How much of each shape has been shaded twice? How do you know? Is it the same fraction on each of your three shapes?* Agree that the doubly shaded part is $\frac{1}{4}$ of each shape in each case and express this as a sentence, *Half of a half equals a quarter.*

## Main activity

Draw a rectangle on the board and divide it into thirds vertically, like this:

Now draw a line that divides the rectangle into two halves horizontally, like this:

Ask students to notice what has happened to each third. They should see that each third has been divided into two halves, as well as the whole rectangle. Ask them to describe how much of the rectangle is taken up by one of the pieces. They may say that it is one of six equal pieces, so it is $\frac{1}{6}$ of the rectangle. They could also describe it as being 'half of a third'.

Write on the board: $\frac{1}{2}$ of $\frac{1}{3} = \frac{1}{6}$. Remind students that when we multiply, we sometimes say 'of' instead of 'times', so $\frac{1}{2}$ of $\frac{1}{3}$ is the same as $\frac{1}{2} \times \frac{1}{3}$, so $\frac{1}{2} \times \frac{1}{3} = \frac{1}{6}$.

Look together at page 67 of the Student Book. Display on the IWB, if possible. Ask students to look at the example at the top of the page. Here the rectangle is first divided into quarters vertically and three of the quarters $\left(\frac{3}{4}\right)$ are shaded in red. The rectangle is then divided into thirds horizontally, and one of the thirds is shaded blue. *How many pieces is the rectangle divided into?* Students should see that there are 12 equal pieces. *How many are shaded twice?* 3 of the 12 pieces are shaded twice (purple), so $\frac{1}{3}$ of $\frac{3}{4} = \frac{3}{12}$, or $\frac{1}{3} \times \frac{3}{4} = \frac{1}{4}$.

Use the comparison with finding a fraction of an amount. For example, if you had 3 cakes, then $\frac{1}{3}$ of that amount would be 1 cake. Here we had 3 quarters, and $\frac{1}{3}$ of that is 1 quarter.

Students should now complete the activities on page 67. For each question they should divide rectangles to show the fractions, either by folding paper or drawing rectangles.

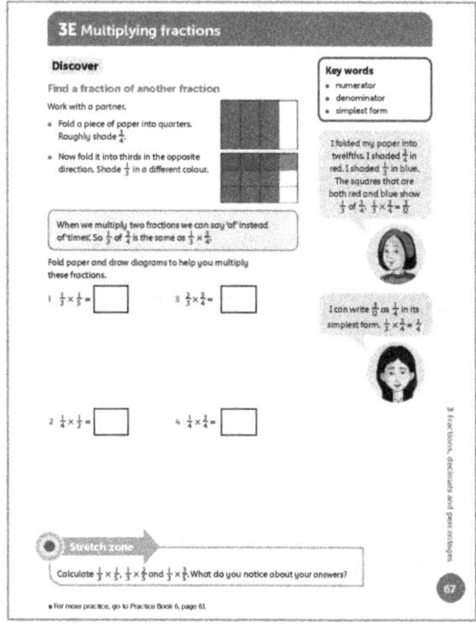

## Differentiation

**Supporting:** Help students by modelling the fractions on their rectangles or folded paper.

**Consolidating:** Ask students to explain how they find each fraction of a fraction.

**Extending:** Challenge students to think about finding the fractions of fractions mentally.

**Stretch zone:** *Calculate $\frac{1}{3} \times \frac{1}{5}$, $\frac{1}{3} \times \frac{2}{5}$ and $\frac{1}{3} \times \frac{3}{5}$. What do you notice about your answers?*

Students may notice that they develop a sequence of fifteenths when they find a third of a sequence of fifths. They might explore other similar series, for example a third of a series of quarters.



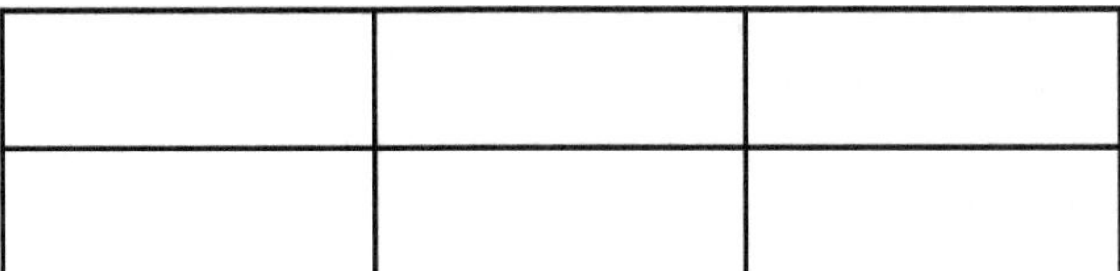

## Reflection time

Look together at the four questions on page 67 and ask different students to share their answers and explain their strategies. Ask how they decided to divide their rectangles (or fold their paper) and then to explain how they found the answer to each question from their rectangles. *Could you have found the answers just by looking at the numbers?*

**Practice Book:** Students complete Practice Book page 61. They can do this directly after the Main activity, as homework, or as the focus of a separate mathematics session to help students consolidate their learning and build fluency.

Students shade rectangles to show how they multiplied two fractions with different denominators.

| Differentiated outcomes | |
| --- | --- |
| **All students** | should multiply fractions by shading fractions of rectangles with support. |
| **Most students** | will multiply fractions by shading fractions of rectangles. |
| **Some students** | may begin to multiply fractions mentally by looking at the numbers in the fractions. |

# 3E Multiplying fractions

## Explore   Student Book page 68 · Practice Book page 62

### Specific learning focus

- Multiply two fractions using the numerators and denominators.

### Global skills

- **Creative skills:** problem solving

### Key vocabulary

- numerator, denominator, simplest form

### Resources

- mini whiteboards and markers

### Language support

Remind students that when multiplying fractions, the numerators can be multiplied and the denominators can be multiplied. It may be possible to write the answer in a simpler form. Reinforce this as they say their calculations.

## Answers

### Student Book page 67

**1** $\frac{1}{15}$   **3** $\frac{1}{2}$

**2** $\frac{1}{8}$   **4** $\frac{3}{16}$

### Practice Book page 61

**1** $\frac{1}{5}$   **4** $\frac{1}{3}$

**2** $\frac{2}{15}$   **5** $\frac{2}{5}$

**3** $\frac{4}{15}$   **6** $\frac{1}{12}$

Stretch zone: To multiply fractions, you can multiply the numerators together and multiply the denominators together, then write the fraction in its simplest form.

## Introductory activity

Write on the board:

$\frac{1}{5}$   $\frac{2}{5}$   $\frac{3}{5}$   $\frac{4}{5}$   $\frac{5}{5}$

Ask students to say which of these fractions they can halve easily. They may say the ones with even numerators, so $\frac{1}{2}$ of $\frac{2}{5}$ is $\frac{1}{5}$, $\frac{1}{2}$ of $\frac{4}{5} = \frac{2}{5}$ and so on.

Ask them to convert the fractions into equivalent fractions in tenths:

$\frac{2}{10}$   $\frac{4}{10}$   $\frac{6}{10}$   $\frac{8}{10}$   $\frac{10}{10}$

Now ask which ones can be halved easily. As all the numerators are now even, they can all be quickly halved.

Repeat with this sequence of sixths:

$\frac{1}{6}$   $\frac{2}{6}$   $\frac{3}{6}$   $\frac{4}{6}$   $\frac{5}{6}$   $\frac{6}{6}$

*How can you write them as equivalents so you can easily find one third of each fraction?* Some students may notice that multiplying the numerators and denominators by 3 will make equivalent fractions where one third of every numerator can then be found.

## Main activity

Look together at page 68 of the Student Book. Display on the IWB, if possible. Refer students to the worked example. Explain that the answer to $\frac{1}{3} \times \frac{3}{4}$ has been found by multiplying the two numerators together and then multiplying the two denominators together. The answer is then written in its **simplest form**.

Ask students to use the same method to find $\frac{1}{3} \times \frac{2}{3}$. They should agree this is $\frac{2}{9}$. Verify this by drawing a rectangle, dividing it into thirds vertically and shading two parts, then dividing the rectangle into thirds horizontally (as in 3E Discover) and shading one of the thirds. Finally, notice that 2 of the 9 parts have been double shaded, showing that $\frac{1}{3} \times \frac{2}{3} = \frac{2}{9}$.

Students now complete the activities on page 68 by using the method of multiplying numerators and then denominators. Remind them to write their final answer in its simplest form as an equivalent fraction.

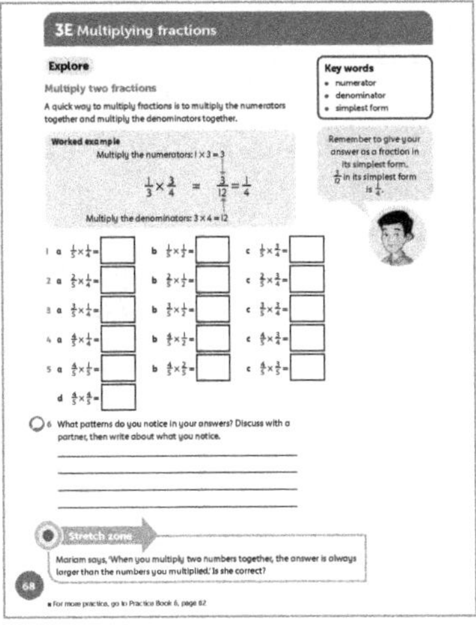

## Differentiation

**Supporting:** Model the method for students and use folded paper to verify the calculations.

**Consolidating:** Ask students to draw diagrams to check the calculation.

**Extending:** Challenge students to investigate multiplying three fractions and draw a diagram to check their answers.

**Stretch zone:** *Mariam says, 'When you multiply two numbers together, the answer is always larger than the numbers you multiplied.' Is she correct?*

Students should be able to use examples of finding a fraction of a fraction to show that this is not always true, for example $\frac{1}{2} \times \frac{1}{3} = \frac{1}{6}$.

 **Reflection time**

Look together at some of the questions on page 68 and ask different students to share their answers and explain their strategies. Ask how they were able to find the numerator and denominator of the answer and then how they were able to find the simplest form. Could they have found the answers just by looking at the numbers and finding the simplest form mentally?

**Practice Book:** Students complete Practice Book page 62. They can do this directly after the Main activity, as homework, or as the focus of a separate mathematics session to help students consolidate their learning and build fluency.

| Differentiated outcomes | |
|---|---|
| **All students** | should multiply fractions by multiplying the numerators and denominators with support. |
| **Most students** | will multiply fractions by multiplying the numerators and denominators. |
| **Some students** | may multiply three fractions and draw a diagram to check their answers. |

## Answers

### Student Book page 68

1 a $\frac{1}{20}$  b $\frac{1}{10}$  c $\frac{3}{20}$

2 a $\frac{1}{10}$  b $\frac{1}{5}$  c $\frac{3}{10}$

3 a $\frac{3}{20}$  b $\frac{3}{10}$  c $\frac{9}{20}$

4 a $\frac{1}{5}$  b $\frac{2}{5}$  c $\frac{3}{5}$

5 a $\frac{4}{25}$  b $\frac{8}{25}$  c $\frac{12}{25}$  d $\frac{16}{25}$

Students may notice that, before simplifying, answers in each set form a sequence.

### Practice Book page 62

1 a $\frac{1}{24}$  b $\frac{1}{12}$  c $\frac{1}{8}$

2 a $\frac{1}{6}$  b $\frac{1}{3}$  c $\frac{1}{2}$

3 a $\frac{3}{16}$  b $\frac{3}{8}$  c $\frac{9}{16}$

4 a $\frac{5}{24}$  b $\frac{5}{12}$  c $\frac{5}{8}$

5 a $\frac{1}{15}$  b $\frac{2}{15}$  c $\frac{1}{5}$

Stretch zone: Students may notice that the last answer in each set is the sum of the previous two answers, for example.

# 3F Dividing fractions

## Discover
**Student Book page 69 • Practice Book page 63**

### Specific learning focus

- Divide fractions by whole numbers, drawing a bar model.

### Global skills

- **Creative skills:** investigating

### Key vocabulary

- numerator, denominator, simplest form, bar model

### Resources

- mini whiteboards and markers

### Language support

- When dividing a fraction into equal parts, make sure that you emphasise the correct language. Use questions that prompt students to think about what fractions they are using and what is the whole in each case.

 **Introductory activity**

Draw half a circle on the board to represent half a cake. Explain that this much cake is left over and some friends are going to share it and finish it. *There are four friends who will share the half a cake, so what fraction of a whole cake will they each get?*

Allow students some time to discuss in pairs and then listen to their responses. On the drawing, divide half the cake into four equal parts. You can also draw in the missing half as well and show that there would have been four parts in that half too (perhaps use dotted lines for this half). Students can now see that the whole cake would have had eight equal pieces, so when they share the half-cake, each friend will be getting $\frac{1}{8}$ of the whole cake. Write this as $\frac{1}{2} \div 4 = \frac{1}{8}$.

*What fraction of a whole cake would each friend get if there are three friends sharing half a cake?* Ask them to draw the cake on their whiteboards and record the calculation. Agree that it is $\frac{1}{2} \div 3 = \frac{1}{6}$.

### Main activity

Write on the board: $\frac{1}{3} \div 2$

Draw a **bar model** to show a whole and three thirds like this:

| 1 whole | | |
|---|---|---|
| $\frac{1}{3}$ | $\frac{1}{3}$ | $\frac{1}{3}$ |

 **Unit 3** Fractions, decimals and percentages

Ask students to look at the calculation – it means to divide each third into two equal parts. Add another row to the model and show each third being equivalent to two equal parts:

| 1 whole | | |
|---|---|---|
| $\frac{1}{3}$ | $\frac{1}{3}$ | $\frac{1}{3}$ |
| | | |

*How many of these parts equal the whole? So, how much is each of those equal parts?* Agree that as there are 6 parts, then each part is $\frac{1}{6}$ of the whole. And each of the sixths is half of a third, so $\frac{1}{3} \div 2 = \frac{1}{6}$.

Students should now complete the tasks on page 69 of the Student Book. They should draw a bar model for each calculation to help them.

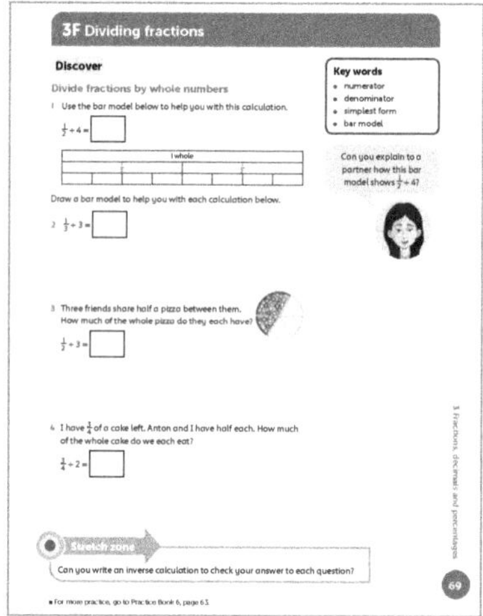

### Differentiation

**Supporting:** Help students construct the appropriate bar model for each question.

**Consolidating:** Ask students to explain how they created their bar model.

**Extending:** Challenge students to say how they could work out the calculations just by looking at the number, without using a diagram.

**Stretch zone:** *Can you write an inverse calculation to check your answer to each question?*

Students should be able to use a multiplication calculation to check their fraction divisions.

### Reflection time

Ask individual students to draw their diagrams on the board for the class to see. They should be encouraged to explain how they knew how many equal parts to show and how many to divide each part into. Ask other students to comment on each diagram and say what they understand by it.

**Practice Book:** Students complete Practice Book page 63. They can do this directly after the Main activity, as homework, or as the focus of a separate mathematics session to help students consolidate their learning and build fluency.

Students shade circle diagrams to show how to solve division of fractions by a whole number.

| Differentiated outcomes | |
| --- | --- |
| **All students** | should divide a fraction by a whole number with support. |
| **Most students** | will divide a fraction by a whole number using a bar model diagram. |
| **Some students** | may divide a fraction by a whole number with or without a diagram. |

## Answers

**Student Book page 69**

1  $\frac{1}{8}$

2  $\frac{1}{9}$

3  $\frac{1}{6}$

4  $\frac{3}{8}$

**Practice Book page 63**

1  $\frac{2}{9}$

2  $\frac{1}{9}$

3  $\frac{5}{18}$

4  $\frac{5}{12}$

5  $\frac{1}{4}$

6  $\frac{1}{6}$

Stretch zone: Students may notice that dividing a fraction by a whole number is the same as multiplying the denominator by that number and leaving the numerator as it is.

# 3F Dividing fractions

## Explore   Student Book page 70 · Practice Book page 64

### Specific learning focus

- Divide fractions by whole numbers using a bar model, and check with an inverse calculation.

### Global skills

- **Creative skills:** problem solving

### Key vocabulary

- numerator, denominator, simplest form

### Resources

- mini whiteboards and markers

### Language support

When dividing a fraction into equal parts, it is important to emphasise the correct language. Use questions that prompt students to think about what fractions they are using and what is the whole in each case. Encourage students to describe how they can use a multiplication to check their division calculation.

##  Introductory activity

Write this sequence on the board:

$\frac{1}{4} \div 2 \qquad \frac{1}{4} \div 3 \qquad \frac{1}{4} \div 4 \qquad \frac{1}{4} \div 5 \qquad \frac{1}{4} \div 6$

Ask students to draw on their whiteboards bar models to solve the first two calculations. The diagrams should show one whole, a row of quarters, then each quarter divided into two parts for the first calculation and into three parts for the second calculation, like this:

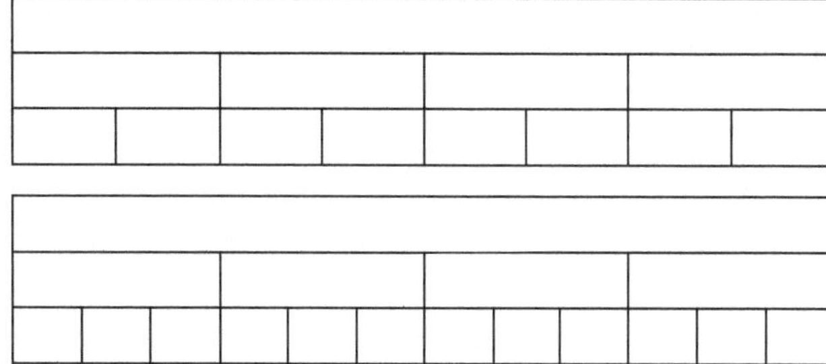

Once the answers to the first two have been established, $\left(\frac{1}{8} \text{ and } \frac{1}{12}\right)$, ask students whether they can predict the answers to the remaining calculations. *Can you spot a pattern? Is there a link between the denominator, the divisor and the answer?*

##  Main activity

As students to look again at the first calculation in the Introductory activity: $\frac{1}{4} \div 2 = \frac{1}{8}$. *What inverse operation could you use to check the answer?* They should recall that multiplication is the inverse of division and so the inverse calculation would be: $\frac{1}{8} \times 2 = \frac{1}{4}$. Ask a student to share their response with the class and check that everyone has the correct answer. Repeat with a different calculation from the Introductory activity, if necessary.

This lesson continues the work from 3F Discover, so students can now move on to the questions on page 70 of the Student Book. They should draw a diagram to help them if needed, and then write a multiplication to check their answer.

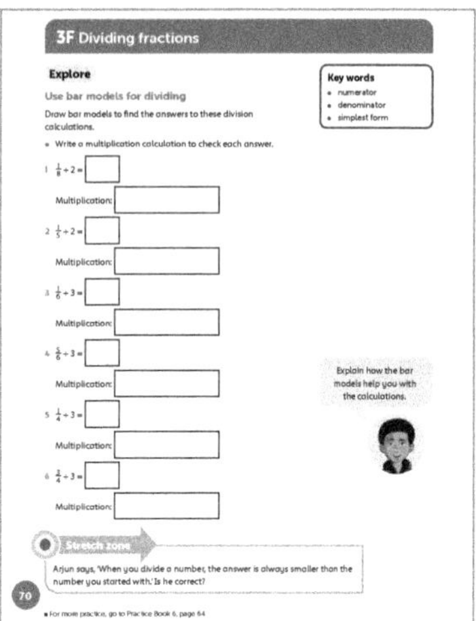

## Differentiation

**Supporting:** Help students to construct the appropriate bar model for each question.

**Consolidating:** Ask students to explain how they created their bar model diagrams.

**Extending:** Challenge students to explain how they knew which multiplication to use as a check.

**Stretch zone:** *Arjun says, 'When you divide a number, the answer is always smaller than the number you started with.' Is he correct?*

Students should see that if they divide a number by a whole number other than 1, then the answer is smaller. However, if they divide by a fraction between 0 and 1, the answer will be larger than the number they started with.

 **Reflection time**

Ask individual students to come to the board and draw their diagrams for the class to see. They should be encouraged to explain how they knew how many equal parts to show in each row of their diagram. Ask other students to comment on each diagram and say what they understand by it. Ask some students to explain how they chose which inverse calculation to write to check their answers.

**Practice Book:** Students complete Practice Book page 64. They can do this directly after the Main activity, as homework, or as the focus of a separate mathematics session to help students consolidate their learning and build fluency.

Students divide factions by whole numbers and draw a bar model to support their calculation.

| Differentiated outcomes | |
|---|---|
| **All students** | should divide a fraction by a whole number and check using an inverse calculation with support. |
| **Most students** | will divide a fraction by a whole number and check using an inverse calculation. |
| **Some students** | may divide a fraction by a whole number and explain why they use an inverse calculation as a check. |

## Answers

### Student Book page 70

**1** $\frac{1}{8} \div 2 = \frac{1}{16}$ $\qquad$ $\frac{1}{16} \times 2 = \frac{1}{8}$

**2** $\frac{1}{5} \div 2 = \frac{1}{10}$ $\qquad$ $\frac{1}{10} \times 2 = \frac{1}{5}$

**3** $\frac{1}{6} \div 3 = \frac{1}{18}$ $\qquad$ $\frac{1}{18} \times 3 = \frac{1}{6}$

**4** $\frac{5}{6} \div 3 = \frac{5}{18}$ $\qquad$ $\frac{5}{18} \times 3 = \frac{5}{6}$

**5** $\frac{1}{4} \div 3 = \frac{1}{12}$ $\qquad$ $\frac{1}{12} \times 3 = \frac{1}{4}$

**6** $\frac{3}{4} \div 3 = \frac{3}{12}$ $\qquad$ $\frac{3}{12} \times 3 = \frac{3}{4}$

### Practice Book page 64

**1** $\frac{1}{6}$

**2** $\frac{1}{9}$

**3** $\frac{1}{15}$

**4** $\frac{1}{3}$

**5** $\frac{2}{9}$

**6** $\frac{1}{6}$

### Stretch zone:

**1** There are 2 lots of $\frac{1}{6}$ in $\frac{1}{3}$

**2** There are 3 lots of $\frac{1}{9}$ in $\frac{1}{3}$

**3** There are 5 lots of $\frac{1}{15}$ in $\frac{1}{3}$

**4** There are 2 lots of $\frac{1}{3}$ in $\frac{2}{3}$

**5** There are 3 lots of $\frac{2}{9}$ in $\frac{2}{3}$

**6** There are 4 lots of $\frac{1}{6}$ in $\frac{2}{3}$

# 3G Fraction and decimal equivalents

## Discover
Student Book page 71 • Practice Book page 65

### Specific learning focus
- Recognise and use decimal and fraction equivalents.

### Global skills
- **Creative skills:** problem solving

### Key vocabulary
- fraction, decimal fraction, decimal equivalent

### Resources
- counting stick, sticky note
- mini whiteboards and markers
- calculators

### Language support

The key phrase in this lesson is 'What is the decimal equivalent?'. Students should be encouraged to complete sentences linking fractions and decimals, such as:

- The decimal equivalent of a half is zero point five.
- Three-quarters has a decimal equivalent of zero point seven five.

 Introductory activity

Hold up a counting stick and use your finger to make a line through the centre of the stick. Ask the class what each side of the stick is worth. $\left(\frac{1}{2}\right)$. Use a sticky note to label the centre of the counting stick $\frac{1}{2}$ and 0.5. Ask students to draw a number line on their whiteboards to represent the counting stick. *What will you label the ends of the number line?*

Now point to halfway between 0 and 0.5 on the counting stick. *How should we label this division?* Repeat with halfway between 0.5 and 1. Ask students to label their number lines on their whiteboards, using both the fraction and decimal form $\left(\frac{1}{4} \text{ and } 0.25 \text{ and } \frac{3}{4} \text{ and } 0.75\right)$.

Repeat, pointing to each of the ten marked divisions on the counting stick, asking students to draw a new number line on their whiteboards with ten divisions to represent the counting stick. *How much is each division worth?* (a tenth) Ask students to label each division. Check that students have correctly labelled the ends of the line 0 and 1, and correctly labelled the divisions $\frac{1}{10}$ and 0.1, $\frac{2}{10}$ and 0.2 and so on.

 Main activity

Ask each student to write on their whiteboard any fraction and its **decimal equivalent**. They may write simple fractions such as $\frac{1}{2} = 0.5$, $\frac{1}{4} = 0.25$ or $\frac{1}{10} = 0.1$. Write a selection on the board. Include some more complex fractions that any student has written, or add some, for example $\frac{11}{12}, \frac{11}{25}, \frac{64}{75}$.

Now give each student a calculator. Ask students to use it to divide 1 by 2. They will see the answer 0.5. Show the class that $\frac{1}{2}$ means 1 divided by 2. Show that dividing in this way gives the answer 0.5, and that you can use this to find the decimal equivalent of any fraction. Demonstrate this, using one of the more complex examples on your list.

Ask students, in pairs, to explore the other fractions listed on the board, noting any patterns they see in the fractions and their decimal equivalents.

Students now complete the activities on page 71 of the Student Book. Encourage them to find the number of squares, or part squares, to shade first. For example, they should be able to find $\frac{1}{5}$ of 10 squares easily, but for questions 5 and 6, they will need to divide 10 squares by 20 and by 25 to find out how many squares to shade (half a square for $\frac{1}{20}$ and just less than half a square for $\frac{1}{25}$).

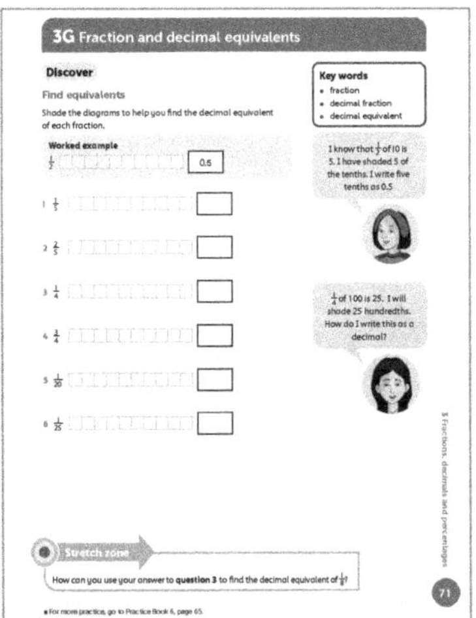

### Differentiation

**Supporting:** Help students to find decimal equivalents by modelling the division on the calculator.

**Consolidating:** Ask students to explain which decimal equivalents they can find without using a calculator, for example fifths, tenths, twentieths and those they know such as one quarter and three quarters.

**Extending:** Challenge students to use known equivalents to find new ones. For example, if $\frac{1}{20} = 0.05$, then what is $\frac{3}{20}$ as a decimal?

**Stretch zone:** *How can you use your answer to question 3 to find the decimal equivalent of $\frac{1}{8}$?*

Students should be able to use the decimal equivalent of $\frac{1}{4}$ and halve it to find the decimal equivalent for $\frac{1}{8}$.

Ask some students to share with the rest of the class something they learned this lesson. Check whether other students have anything to add.

You can ask students to look at the completed equivalents and use the **decimal fractions** to find new equivalent decimal fractions, for example:

$\frac{1}{5} = 0.2$, so $\frac{3}{5} = 0.6$, $\frac{4}{5} = 0.8$

$\frac{1}{20} = 0.05$, so $\frac{3}{20} = 0.15$, $\frac{7}{20} = 0.35$.

**Practice Book:** Students complete Practice Book page 65. They can do this directly after the Main activity, as homework, or as the focus of a separate mathematics session to help students consolidate their learning and build fluency.

Students match fractions to their decimal equivalents. They can use what they have learned in the Main activity to work out any decimal equivalents that are unfamiliar to them.

| Differentiated outcomes | |
| --- | --- |
| **All students** | should find the equivalent decimal fraction for a fraction using a calculator with support. |
| **Most students** | will find the equivalent decimal fraction for a fraction using a calculator. |
| **Some students** | may find equivalent decimal fractions using known equivalent decimals. |

## Answers

### Student Book page 71

**1** 0.2    2 squares shaded

**2** 0.4    4 squares shaded

**3** 0.25    $2\frac{1}{2}$ squares shaded

**4** 0.75    $7\frac{1}{2}$ squares shaded

**5** 0.05    $\frac{1}{2}$ square shaded

**6** 0.04    $\frac{2}{5}$ square shaded

### Practice Book page 65

**1** $\frac{3}{8} = 0.375$

**2** $\frac{1}{2} = 0.5$

**3** $\frac{5}{8} = 0.625$

**4** $\frac{3}{4} = 0.75$

**5** $\frac{1}{5} = 0.2$

**6** $\frac{3}{5} = 0.6$

**7** $\frac{9}{10} = 0.9$

**8** $\frac{1}{4} = 0.25$

**9** $\frac{1}{8} = 0.125$

Stretch zone: $\frac{3}{8} < \frac{1}{2}$ because $\frac{1}{2} = \frac{4}{8}$ and $\frac{3}{8} < \frac{4}{8}$, and $\frac{3}{8} = 0.375$, which is less than $\frac{1}{2}$, which is 0.5.

## 3G Fraction and decimal equivalents

### Explore    Student Book page 72 · Practice Book page 66

#### Specific learning focus
- Solve word problems using fractions and decimal equivalents.

#### Global skills
- **Creative skills:** problem solving

#### Key vocabulary
- fraction, decimal fraction, decimal equivalent

#### Resources
- large cards with digits 0 and 1 on them
- mini whiteboards and markers

### Language support

Writing problems and defining good questions helps to support language development.

Write a word bank on the board to help students to write questions including, for example:

- equivalent (What is the equivalent decimal?)
- greater than, less than, in between (Which fractions are between __ and __?)
- decimal, decimal fraction (Which of these decimal fractions is greater, 0.37 or 0.09?)

 **Introductory activity**

Stick a large card with 0 on it on the left-hand side of the classroom wall (as students look at it) and a large 1 on the right-hand side of the same wall. Ask a student to write any fraction or decimal between 0 and 1 on their whiteboard. Then ask that student to come to the front of the class and to stand in the correct place between the 0 and 1 cards. Ask another student to write down a different fraction between the first fraction and 1. Ask that student to come to the front of the class and to stand in the correct place (between the first student and the 1).

Repeat this until there are five students on the 'number line' at the front of the class. Repeat the activity. This time, ask for fractions between 0 and the previous student.

## Main activity

Write this problem on the board:

*In a class, there are 12 boys and 18 girls. What fraction of the class are girls?*

Ask students to tell you how many are in the class in total (30). *How many of the 30 are girls? What fraction is this?* Write down $\frac{18}{30}$. Ask students to tell you a fraction that is equivalent to $\frac{18}{30}$. Agree that $\frac{18}{30}$ in its simplest form is $\frac{3}{5}$.

Now ask what is an equivalent decimal to $\frac{3}{5}$? Students should recall that $\frac{3}{5} = 0.6$, or they could work this out from the fact that $\frac{3}{5} = \frac{6}{10}$.

*So, what fraction of the class are boys and what decimal is this?* Students should be able to say that $\frac{12}{30}$ are boys, which is equivalent to $\frac{2}{5}$ or 0.4.

Students then complete the problems on page 72 of the Student Book.

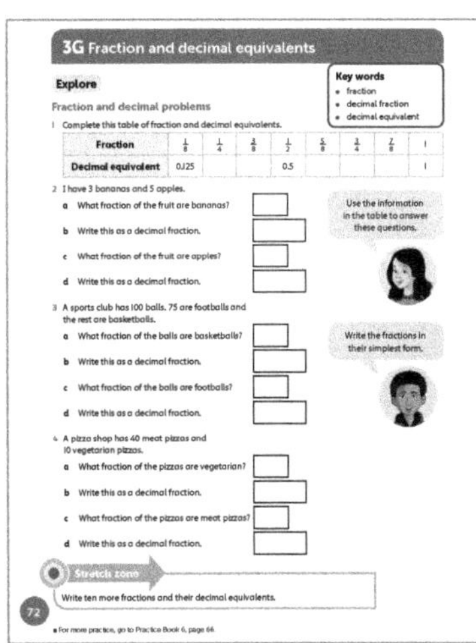

## Differentiation

**Supporting:** Help students to solve word problems by modelling the context in fractions and decimal equivalents.

**Consolidating:** Ask students to explain how they worked out which fraction was being represented in each question.

**Extending:** Challenge students to make up a word problem using fractions and decimal equivalents for a partner to solve.

**Stretch zone:** *Write ten more fractions and their decimal equivalents.*

Students could use fractions they know or work out new ones from known fractions.

## Reflection time

Ask different students to share their answers and explain their reasoning for the word problems they completed. Now ask pairs to write their own questions for the rest of the class. Select five of these questions for the class to answer.

**Practice Book:** Students complete Practice Book page 66. They can do this directly after the Main activity, as homework, or as the focus of a separate mathematics session to help students consolidate their learning and build fluency.

Students write decimal equivalents to some given fractions. They should show their working for each one, first simplifying the fraction to make it easier to convert into a decimal, and then writing the decimal.

| Differentiated outcomes | |
| --- | --- |
| **All students** | should solve fraction and decimal word problems with support. |
| **Most students** | will solve fraction and decimal word problems. |
| **Some students** | may write fraction and decimal word problems for others to solve. |

## Answers

### Student Book page 72

**1**

| Fraction | $\frac{1}{8}$ | $\frac{1}{4}$ | $\frac{3}{8}$ | $\frac{1}{2}$ | $\frac{5}{8}$ | $\frac{3}{4}$ | $\frac{7}{8}$ | 1 |
| --- | --- | --- | --- | --- | --- | --- | --- | --- |
| **Decimal equivalent** | 0.125 | 0.25 | 0.375 | 0.5 | 0.625 | 0.75 | 0.875 | 1 |

**2 a** $\frac{3}{8}$   **b** 0.375   **c** $\frac{5}{8}$   **d** 0.625

**3 a** $\frac{1}{4}$   **b** 0.25   **c** $\frac{3}{4}$   **d** 0.75

**4 a** $\frac{1}{5}$   **b** 0.2   **c** $\frac{4}{5}$   **d** 0.8

### Practice Book page 66

| | Fraction | Decimal |
| --- | --- | --- |
| | $\frac{1}{20}$ | 0.05 |
| 1 | $\frac{3}{20}$ | 0.15 |
| 2 | $\frac{6}{20}$ | 0.3 |
| 3 | $\frac{9}{20}$ | 0.45 |
| 4 | $\frac{10}{20}$ | 0.5 |
| 5 | $\frac{19}{20}$ | 0.95 |
| 6 | $\frac{3}{25}$ | 0.12 |
| 7 | $\frac{6}{25}$ | 0.24 |
| 8 | $\frac{10}{25}$ | 0.4 |
| 9 | $\frac{18}{25}$ | 0.72 |

Check that students' workings match the answers they have given.

Stretch zone: $0.3 < \frac{1}{2}$ because $\frac{1}{2} = 0.5$

# 3H Place value in decimals

## Discover
Student Book page 73 • Practice Book page 67

### Specific learning focus

- Read and represent 6-digit numbers with decimals to three decimal places.

### Global skills

- **Creative skills:** exploring

### Key vocabulary

- tenths, hundredths, thousandths, decimal places

### Resources

- mini whiteboards and markers
- digit cards 0–9

### Language support

Ask students questions such as:

- *What do the values of the digits in 471.632 represent? How do you know?*
- *Which is greater, 381.573 or 381.753? How do you know?*
- *When do you see numbers with three decimal places in real life?*

###  Introductory activity

Write 493.752 on the board. Ask pairs to discuss what they know about this number. Agree that it is a decimal number that contains whole numbers and decimal fractions. It has 4 hundreds, 9 tens, 3 ones, 7 tenths, 5 hundredths and 2 thousandths. Write the number in a place-value grid to clarify the value of each digit in the number. Explain that we could also write the number as a mixed number, using whole numbers and a fraction, that is: $493\frac{752}{1000}$.

| H | T | O | . | t | h | th |
|---|---|---|---|---|---|----|
| 4 | 9 | 3 | . | 7 | 5 | 2 |

Ask a student to come to the front of the class and select six digit cards. Students should use the digits chosen to make six different numbers that include three **decimal places**. For example, if they choose 1, 3, 4, 6, 7 and 9, some of the numbers they could make are:

479.163     691.437     364.917

Pairs should draw number lines from 0–1000 on their whiteboards and write these decimal numbers in the correct place on their number lines (or as accurately as they can). Ask a pair to describe how they decided where

to write each of the numbers. Discuss how accurately they can place the numbers using the decimal places on a number line from 0–1000. For example, could they show 371.469 and 371.496 on that number line? Why not? Then ask, *Can you draw a number line with different start and end values that would make placing these two numbers easier?* Demonstrate by drawing a number line from 371.4 to 371.5 and then writing these numbers on that line, as accurately as possible.

###  Main activity

Ask students to make a 6-digit number using six of the digit cards. They should write down the number on their whiteboards and then hold up their whiteboards so that everyone in the class can see each other's number.

Ask a series of questions for students to check against the number, for example:

- *Who has made the largest number?*
- *Who has made the smallest number?*
- *Whose number has most hundreds?*
- *Whose number has most tenths?*
- *Whose number is closest to 400?*

Students then complete the activity on page 73, using the digit cards to make a set of numbers and then answering questions about them.

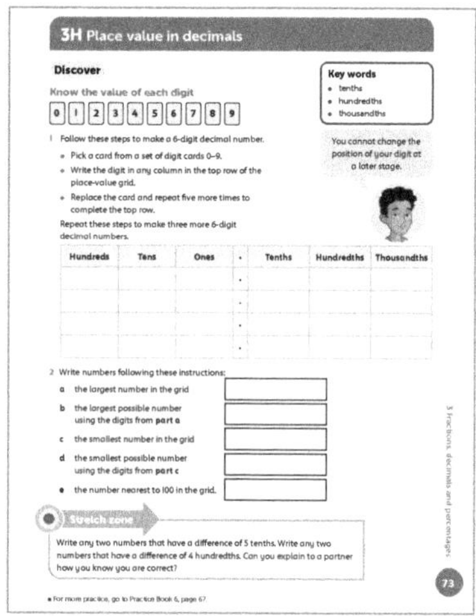

### Differentiation

**Supporting:** Help students to make 6-digit numbers with decimals to thousandths, and to write the numbers on a number line or in a place-value grid to aid understanding of the value of the numbers.

**Consolidating:** Ask students to explain their thinking as they make their 6-digit numbers and write them on a number line or in a place-value grid.

**Extending:** Challenge students to make 6-digit numbers to three decimal places that can all be placed on the same number line between consecutive whole numbers, for example between 284 and 285.

Students can use place-value grids to help with their explanations.

 **Reflection time**

Invite students to share the numbers they made with the class, explaining what the value of each digit is. See who in the class made the largest numbers and smallest numbers. Whose number was closest to 500? Whose was closest to 1000? (Continue with other examples.)

**Practice Book:** Students complete Practice Book page 67. They can do this directly after the Main activity, as homework, or as the focus of a separate mathematics session to help students consolidate their learning and build fluency.

Students annotate a Gattegno chart to show the value of digits in decimal numbers with up to three places of decimals.

| Differentiated outcomes | |
| --- | --- |
| **All students** | should make 6-digit numbers to three decimal places and explain the value of each digit with support. |
| **Most students** | will make 6-digit numbers to three decimal places and explain the value of each digit. |
| **Some students** | will make 6-digit numbers to three decimal places close to a given number. |

## Answers

### Student Book page 73

1 Students make their own numbers using the digit cards. Check that their numbers are recorded in the place-value grid correctly.

2 Check that students have identified the largest and smallest numbers from their set.

### Practice Book page 67

Check that students have correctly marked the numbers in coloured dots on their Gattegno charts.

Stretch zone: Students should describe the patterns they see on the chart. For example, they might say that in each column the number of zeros changes, or the same digit starts each number along the same row.

---

## 3H Place value in decimals

**Explore**   Student Book page 74 • Practice Book page 68

### Specific learning focus

- Divide numbers by 10, 100 or 1000.

### Global skills

- **Creative skills:** exploring

### Key vocabulary

- tenths, hundredths, thousandths

### Resources

- digit cards 0–9 per pair
- mini whiteboards and markers

### Language support

Encourage students to explain their strategies and use the vocabulary of place value: tenths, hundredths, thousandths. Ask, for example:

- *What happens to a number when you divide by 10?*
- *What happens to the ones digit when you divide by 100?*

 **Introductory activity**

Draw a place-value grid on the board and fill in the number 2375 as follows:

| Th | H | T | O | . | t | h | th |
| --- | --- | --- | --- | --- | --- | --- | --- |
| 2 | 3 | 7 | 5 | . | | | |

Tell students to copy this grid onto their whiteboards, leaving room to add extra rows. Ask students to tell you the value of each digit. Now ask them to recall what happens to the digits in a number when you divide it by 10. They should remember that the digits move one place to the right.

Ask students what happens to the 2 thousands, the 3 hundreds the 7 tens and the 5 ones when they are divided by 10. Ask them to show this in a row they add to the grid on their whiteboards (237.5).

Now ask students what happens to the digits in a number when you divide it by 100. They should remember that the digits move two places to the right.

Ask them what happens to the 2 thousands, the 3 hundreds the 7 tens and the 5 ones in the original number when they are divided by 1000. They write this in another row they add to the grid (23.75).

Finally, repeat for dividing 2375 by 1000 to finish with 2.375

**Unit 3 Fractions, decimals and percentages**   95

 **Main activity**

Ask students, in pairs, to complete the activities on page 74 of the Student Book. As they work, ask them to describe to each other the effect on the digits of dividing by 10, 100 or 1000 in each case. Ask them questions about digits in the numbers they have written, for example: *Your number has 4 hundreds. What happens to the value of this digit when you divide this by 1000?*

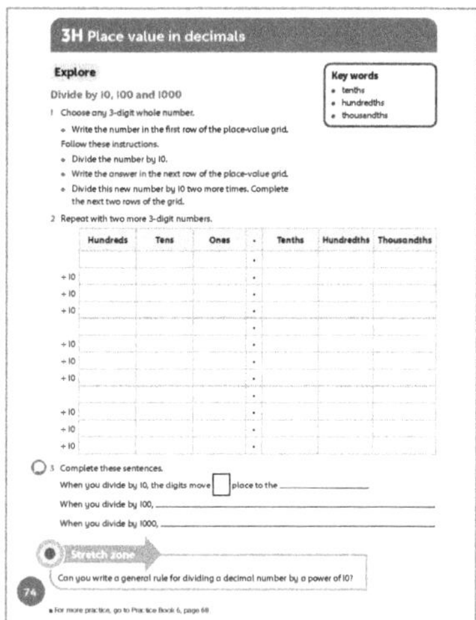

## Differentiation

**Supporting:** Help students to describe the effect of moving digits when dividing by 10, 100 or 1000.

**Consolidating:** Ask students to describe the patterns they notice as they divide.

**Extending:** Ask students for a general rule for dividing by 10, 100 and 1000.

**Stretch zone:** *Can you write a general rule for dividing a decimal number by a power of 10?*

Students may be able to say that the digits move to the right the same number of places as there are zeros in the number you are dividing by. For example, the digits move one place to the right if dividing by 10, two places for 100, three for 1000, four for 10 000 and so on.

 **Reflection time**

Select one or two students who have good understanding of dividing by powers of 10 and who explained their ideas well in their pairs. Ask them to share their explanation with the whole class.

Ask the class, *Can you give me an example that shows this statement is incorrect?*

'When you divide by 1000, you move digits three places to the left.'

**Practice Book:** Students complete Practice Book page 68. They can do this directly after the Main activity, as homework, or as the focus of a separate mathematics session to help students consolidate their learning and build fluency.

Students look at the effect that multiplying by a power of 10 has on a decimal number. They write the number in a place-value grid, then write the number again, moving the digits the correct number of places to the left, then complete the answer to the calculation.

| Differentiated outcomes | |
| --- | --- |
| **All students** | should divide by 10, 100 or 1000 with support. |
| **Most students** | will divide by 10, 100 or 1000. |
| **Some students** | may generalise the effect of dividing by powers of 10. |

## Answers

### Student Book page 74

**1–2** Answers will vary because students make their own numbers to divide. Check they have divided and written the results correctly in the grid.

**3** The digits move 1 place to the right.

The digits move 2 places to the right.

The digits move 3 places to the right.

### Practice Book page 68

**1** 532.7

**2** 857.21

**3** 957.21

**4** 12.3

**5** 123

Stretch zone: 0.1

# 31 Multiplying decimals

## Discover  Student Book page 75 • Practice Book page 69

### Specific learning focus

- Multiply decimals using place-value counters.

### Global skills

- **Creative skills:** exploring

### Key vocabulary

- tenths, hundredths, thousandths

### Resources

- place-value grid and counters
- mini whiteboards and markers

### Language support

Remind students of the language associated with place-value counters. Say, for example:

- *You can exchange 10 counters in the tenths for 1 counter in the ones.*
- *You can exchange 10 counters in the hundredths for 1 counter in the tenths.*
- *You can exchange 10 counters in the thousandths for 1 counter in the hundredths.*

 **Introductory activity**

Look together at page 75 of the Student Book. Display on the IWB, if possible. Draw students' attention to the worked example. Ask them to say how each digit in the number to be multiplied (2.313) has been represented in the first row of the place-value grid. For example, the 3 tenths have been represented by 3 counters in the tenths column.

Ask how the counters have been used to show that we are multiplying by 3. Students should see that there are three identical rows of counters. Then ask how the total number of counters show the final answer. For example, ask, *How may thousandths are there in total?* Students should be encouraged to say that '3 thousandths × 3 = 9 thousandths' and so on.

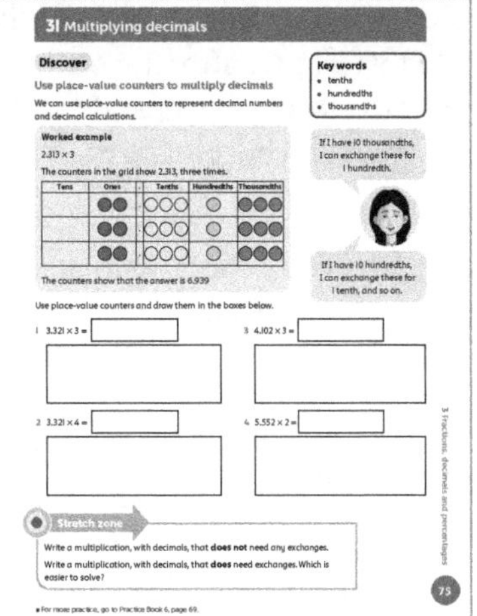

 **Main activity**

Extend the example used in the Introductory activity to consider 2.313 × 4. *How would the place-value grid be changed for multiplying by 4 instead of 3?* Students should see that a fourth identical row of counters would be needed.

*How many thousandths would there now be in total?* (12) *How can this be written down?* Students should be able to describe how to exchange 10 of the counters in the thousandths and place 1 counter in the hundredths. A similar exchange would be needed for the tenths. Ask students to complete the calculation for 2.313 × 4 on their whiteboards. Agree that the answer is 9.252

Students complete the activities on page 75 of the Student Book. They can use place-value grids and counters to help them and then draw these.

### Differentiation

**Supporting:** Model the exchange process for students when multiplying tenths, hundredths and thousandths.

**Consolidating:** Ask students to explain their method when multiplying tenths, hundredths and thousandths.

**Extending:** Challenge students to work mentally, multiplying tenths, hundredths and thousandths.

**Stretch zone:** *Write a multiplication, with decimals, that does not need any exchanges.*

*Write a multiplication, with decimals, that does need exchanges. Which is easier to solve?*

Students should provide appropriate examples. They are likely to say that calculations with no exchanges are easier.

 **Reflection time**

Discuss with students their solutions to the questions in the Student Book. Ask, for example, *How did you set out the place-value grid using the counters? What did you do first? What did you do next and why? How did you record the answer? Can you check your answer another way?*

**Practice Book:** Students complete Practice Book page 69. They can do this directly after the Main activity, as homework, or as the focus of a separate mathematics session to help students consolidate their learning and build fluency.

Students draw place-value counters in grids to show how to complete a multiplication of a decimal number. They add the value of all the counters in the grid. then write the answer to the calculation.

| Differentiated outcomes | |
|---|---|
| **All students** | should multiply decimals to thousandths by a single digit with support. |
| **Most students** | will multiply decimals to thousandths by a single digit using a place-value grid and counters. |
| **Some students** | may multiply decimals to thousandths mentally. |

## 31 Multiplying decimals

### Explore  Student Book page 76 • Practice Book page 70

#### Specific learning focus

- Multiply numbers with two decimal places using a written method.

#### Global skills

- **Creative skills:** problem solving

#### Key vocabulary

- tenths, hundredths, thousandths

#### Resources

- mini whiteboards and markers
- digit cards 1–9

#### Language support

Ask students questions to monitor their understanding of the methods, for example:

- *Can you explain to me the steps you took to multiply that decimal number by 3?*
- *Is there another strategy that you can use to find the answers to a multiplication calculation?*

## Answers

### Student Book page 75

**1** $3.321 \times 3 = 9.963$

**2** $3.321 \times 4 = 13.284$

**3** $4.102 \times 3 = 12.306$

**4** $5.552 \times 2 = 11.104$

### Practice Book page 69

**1** $1.242 \times 2 = 2.484$

**2** $13.422 \times 2 = 26.844$

**3** $13.113 \times 3 = 39.339$

Check that students have drawn the place-value counters in the grids correctly to represent the calculations.

Stretch zone: $7.35 \times 3 = 22.05$

 **Introductory activity**

Set this problem:

*Claire has five strips of material, each 3.43 metres in length. What length of material does she have altogether?*

Ask students to discuss with a partner how they could work out the answer. Take feedback. Invite students to show their methods on the board and let the class discuss how the calculation has been done. They may suggest using place-value counters as they have done previously, or they may suggest partitioning mentally, or perhaps a different method.

 **Main activity**

Look together at page 76 of the Student Book. Display on the IWB, if possible. Refer students to the worked example at the top of the page. Explain that this is the calculation for $2.72 \times 5$ and talk through with students how it has been completed. Each line of the working is the multiplication of 5 by each digit in turn, starting with 5 lots of the hundredths, then 5 lots of the tenths and so on. The final answer is found by adding these separate answers.

Write the calculation from the Introductory activity on the board: $3.43 \times 5$

Ask students to set out the calculation in the same way as the example, then work in pairs to complete the calculation.

```
      3 . 4 3
    ×   5
      0 . 1 5      (5 × 0.03)
      2 . 0 0      (5 × 0.4)
    1 5 . 0 0      (5 × 3)
    1 7 . 1 5
```

Take feedback and check that students have used the method correctly. Students now complete the calculations on page 76.

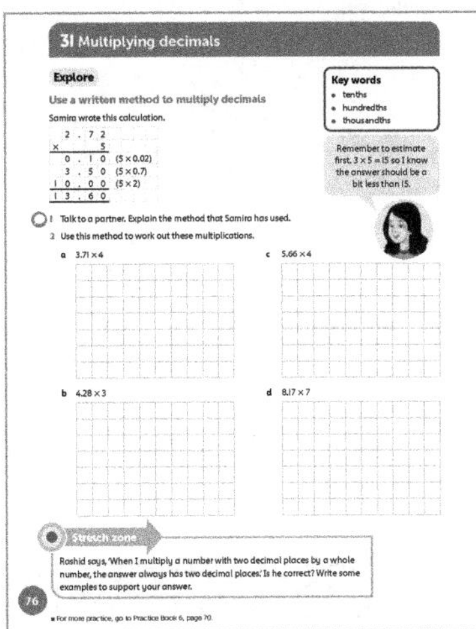

## Differentiation

**Supporting:** Help students to multiply numbers with two decimal places by single digits, using the expanded written method.

**Consolidating:** Ask students to explain how they multiply numbers with two decimal places by single digits using the expanded written method.

**Extending:** Challenge students to multiply numbers with three decimal places by single digits using the expanded written method.

**Stretch zone:** *Rashid says, 'When I multiply a number with two decimal places by a whole number, the answer always has two decimal places.' Is he correct? Write some examples to support your answer.*

Students should try to find an example where the number of hundredths in the answer is zero, for example $1.42 \times 5 = 7.10$

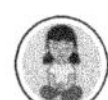 **Reflection time**

Take feedback from the activity, working through some of the examples as a class. Invite students to describe the method in their own ways. This will support other students. Sum up 'multiplying decimals' by saying: *You have learned to multiply decimals using place-value counters and using an expanded written method. You could also use a compact written method, for example:*

$$
\begin{array}{r}
3.43 \\
\times\ \ 5 \\
\hline
17.15 \\
2\ 1
\end{array}
$$

**Practice Book:** Students complete Practice Book page 70. They can do this directly after the Main activity, as homework, or as the focus of a separate mathematics session to help students consolidate their learning and build fluency.

Students solve multiplications of decimal numbers by single-digit whole numbers, using a written method. The example shown is the expanded written method but if students prefer to use the compact method, they can do so.

| Differentiated outcomes | |
|---|---|
| **All students** | should multiply a number with two decimal places by a single digit, using a written method, with support. |
| **Most students** | will multiply a number with two decimal places by a single digit, using a written method. |
| **Some students** | may multiply a number with three decimal places by a single digit, using a written method. |

## Answers

### Student Book page 76

**1** Students will discuss the method with partners.

**2 a** 14.84 **b** 12.84 **c** 22.64 **d** 57.19

### Practice Book page 70

**1** Estimate $2.5 \times 3 = 7.5$, answer 7.35

**2** Estimate $2 \times 5 = 10$, answer 8.75

**3** Estimate $6 \times 3 = 18$, answer 17.55

**4** Estimate $6 \times 6 = 36$, answer 35.1

**5** Estimate $3 \times 4 = 12$, answer 12.76

Stretch zone: 21.15 and 21.28

# 3J Dividing decimals

## Discover   Student Book page 77 • Practice Book page 71

### Specific learning focus

- Dividing decimals using place-value counters.

### Global skills

- **Creative skills:** exploring

### Key vocabulary

- tenths, hundredths, thousandths

### Resources

- place-value grid and counters
- mini whiteboards and markers

### Language support

Remind students of the language associated with place-value counters. Say, for example:

- *You can exchange 1 counter in the ones for 10 counters in the tenths.*
- *You can exchange 1 counter in the tenths for 10 counters in the hundredths.*
- *You can exchange 1 counter in the hundredths for 10 counters in the thousandths.*

### Introductory activity

Look together at page 77 of the Student Book. Display on the IWB, if possible. Point out the worked example. Ask students to explain how the dividend (3.369) and the divisor (3) have been represented, using the place-value grid and the counters.

Students should explain that the three rows in the grid represent the divisor (3) and that all the counters represent the dividend (3.369). You could count each set of counters to show that, for example, there are 3 ones in total, 3 tenths, 6 hundredths and 9 thousandths. The counters representing the dividend have been equally shared between the three rows. Then ask how the number of counters in one row shows the answer. For example, ask, *What is 9 thousandths divided by 3?* Students should be encouraged to say that '9 thousandths divided by 3 = 3 thousandths', which are shown in each row. Repeat for the hundredths, tenths and ones and agree that the answer is 1.123

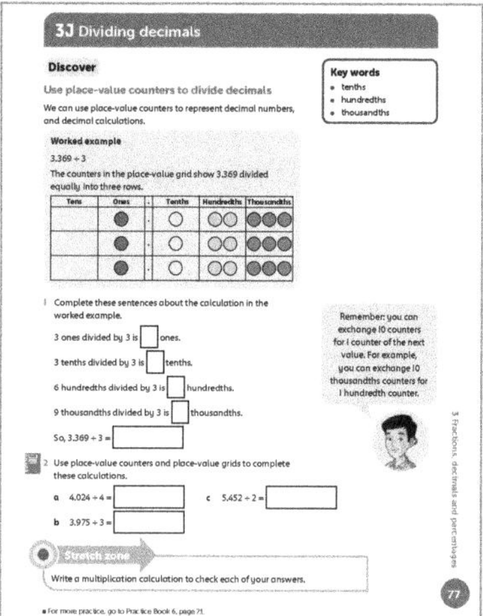

### Main activity

Extend the example used in the Introductory activity to consider $6.442 \div 2$. How would the place-value grid be changed for dividing by 2 instead of 3? Students should see that there should be two rows in the grid instead of 3 and that the counters should be shared equally between the two rows.

*How many thousandths do we start with?* (2) *So how many are there in one row?* (1) *How can this be written down?* Ask students to complete the calculation for $6.442 \div 2$ on their whiteboards. Agree that the answer is 3.221

Students complete the activities on page 77. They can use place-value grids and counters to help them and then draw these.

### Differentiation

**Supporting:** Model the exchange process for students when dividing tenths, hundredths and thousandths.

**Consolidating:** Ask students to explain their method when dividing tenths, hundredths and thousandths.

**Extending:** Challenge students to work mentally dividing tenths, hundredths and thousandths.

**Stretch zone:** *Write a multiplication calculation to check each of your answers.*

Students should be able to multiply their answers by single digits to get back to the dividend. For example, for question 2a they should write $4 \times 1.006$

### Reflection time

Discuss students' solutions to the questions in the Student Book. Ask, for example, *How did you set out the place-value grid using the counters? What did you do first? What did you do next and why? How did you record the answer? Can you check your answer another way?*

**Practice Book:** Students complete Practice Book page 71. They can do this directly after the Main activity, as homework, or as the focus of a separate mathematics session to help students consolidate their learning and build fluency.

Students draw place-value counters in grids to show how to complete a division of a decimal number by a single-digit number. They share the counters equally between the rows of the place-value grid and then add the value of the counters in one row of the grid to find the answer to the calculation.

| Differentiated outcomes | |
| --- | --- |
| **All students** | should divide decimals to thousandths by a single digit with support, using place-value counters. |
| **Most students** | will divide decimals to thousandths by a single digit using a place-value grid and counters. |
| **Some students** | may divide decimals to thousandths using place-value counters, or mentally. |

## 3J Dividing decimals

### Explore  Student Book page 78 · Practice Book page 72

#### Specific learning focus

- Divide numbers with two decimal places using a written method.

#### Global skills

- **Creative skills:** problem solving

#### Key vocabulary

- tenths, hundredths, thousandths

#### Resources

- mini whiteboards and markers
- digit cards 1–9

#### Language support

Ask students questions to monitor their understanding of the methods, for example:

- *Can you explain to me the steps you took to divide a number with two decimal places by 5?*
- *Is there another strategy that you can use to find the answers to a division calculation?*

## Answers

**Student Book page 77**

**1** 1 ones

   1 tenths

   2 hundredths

   3 thousandths

   1.123

**2 a** 1.006    **b** 1.325    **c** 2.726

**Practice Book page 71**

**1** 4.224

**2** 1.321

**3** 4.223

Stretch zone: Students should show the short division method for question 3.

 Introductory activity

Set this problem:

*Johnson has a wooden plank, 10.65 metres in length. If he cuts 5 shelves of equal length from the plank, how long will each shelf be?*

First ask students to estimate the answer. Explain that this will help them know where the decimal point goes in their answer. (10 ÷ 5 = 2) Ask students to discuss with a partner how they could work out the answer and make notes on their whiteboards. Take feedback. Invite students to show their methods on the board and let the class discuss how the calculation has been done. They may suggest using place-value counters as they have done previously, or they may suggest partitioning mentally, or perhaps a different method.

### Main activity

Look together at page 78 of the Student Book. Display on the IWB, if possible. Draw students' attention to the short division example at the top of the page. Explain that this is the calculation for 6.85 ÷ 5. Ask students to estimate the answer to this division (for example 5 ÷ 5 = 1, as given in the speech bubble). Next, talk through with students how the calculation has been completed. Each step of the working is the division by 5 of each digit in turn, starting with ones divided by 5, then tenths divided by 5 and so on. Each time, the remainder is carried forward to the next column and exchanged. Remind students to insert the decimal point in the answer.

Write on the board: 2.46 ÷ 6

Ask students to set out this calculation in the same way as the example, then work in pairs first to estimate and then to complete the calculation.

$$\begin{array}{r} 0 \, . \, 4 \, 1 \\ 6\,\overline{)\,2\, .\,{}^2\!4\, 6} \end{array}$$

Take feedback and check that students have used the method correctly. Students now complete the calculations on page 78.

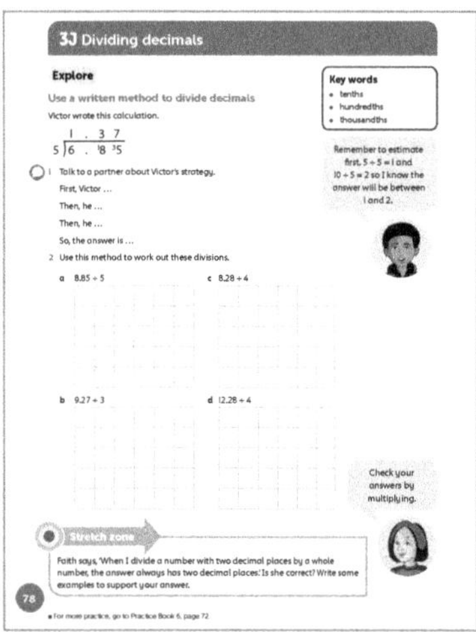

## Differentiation

**Supporting:** Help students to divide numbers with two decimal places by single digits, using a written method.

**Consolidating:** Ask students to explain how they divide numbers with two decimal places by single digits using a written method.

**Extending:** Challenge students to divide numbers with three decimal places by single digits using a written method.

**Stretch zone:** *Faith says, 'When I divide a number with two decimal places by a whole number, the answer always has two decimal places.' Is she correct? Write some examples to support your answer.*

Students should try to find examples where the number of thousandths in the answer is not zero, for example 6.45 ÷ 2 = 3.225

 ## Reflection time

Take feedback from the activity, working through some of the examples as a class. Invite students to describe the methods in their own ways. This will support other students.

**Practice Book:** Students complete Practice Book page 72. They can do this directly after the Main activity, as homework, or as the focus of a separate mathematics session to help students consolidate their learning and build fluency.

Students estimate and then complete division calculations involving decimal numbers with up to three decimal places. They use a short division written method each time.

| Differentiated outcomes | |
|---|---|
| **All students** | should divide a number with two decimal places by a single digit with support, using a written method. |
| **Most students** | will divide a number with two decimal places by a single digit using a written method. |
| **Some students** | may divide a number with three decimal places by a single digit using a written method. |

## Answers

### Student Book page 78

**1** First, Victor divided 6 ones by 5, which goes once, with 1 one left over. He exchanged this one for 10 tenths, which are written in the tenths column to make 18 tenths.

Then, he divided 18 tenths by 5 which goes 3 times with 3 tenths left over. He exchanged these 3 tenths for 30 hundredths, which are written in the hundredths column to make 35 hundredths.

Then, he divided 35 hundredths by 5 which goes 7 times.

So, the answer is 1.37

**2 a** 1.77  **b** 3.09  **c** 2.07  **d** 3.07

### Practice Book page 72

**1** Estimate 12 ÷ 3 = 4, answer 4.24

**2** Estimate 12 ÷ 2 = 6, answer 6.36

**3** Estimate 12 ÷ 4 = 3, answer 3.18

**4** Estimate 35 ÷ 5 = 7, answer 7.185

**5** Estimate 24 ÷ 6 = 4, answer 4.24

**6** Estimate 24 ÷ 3 = 8, answer 8.48

Stretch zone: The answer to question 6 is double the answer to question 5 because the divisor is halved.

# 3K Decimal problems

## Discover
Student Book page 79 • Practice Book page 73

### Specific learning focus
- Round decimals to the nearest tenth.

### Global skills
- **Creative skills:** problem solving

### Key vocabulary
- two decimal places, round, nearest tenth, one decimal place, nearest whole number

### Resources
- 0–9 digit cards, large set
- 0–9 digit cards, class set
- mini whiteboards and markers

### Language support
Model key vocabulary. Ask, for example:
- *What is that number approximately equal to?*
- *What is that number rounded up/down to the nearest whole?*
- *What is that number rounded up/down to the nearest tenth?*

 Introductory activity

Ask three students to come to the front of the class and each pick one of the large 0–9 digit cards. Ask them to stand in a line facing the rest of the class, holding out their cards so that everyone can see them. Ask the rest of the class to work in pairs to decide how many possible numbers can be made, using these three digits, that have two decimal places. For example, if the cards chosen were 4, 8 and 2 they could make: 4.82, 4.28, 2.84, 2.48, 8.24 and 8.42. Encourage students to work systematically to check that they have all possible arrangements of the numbers.

Explain that students are going to round all the numbers they made to the **nearest whole number** by looking at the number of tenths.

Working through the numbers one at a time, ask, for example, *What two whole numbers does 4.82 come between?*

Agree that 4.82 is between 4 and 5. Explain that, as we are rounding to the nearest whole number, we look at the digit one place to the right of this number, which is the tenths. *How many tenths are in 4.82?* Agree that it has 8 tenths. Ask students whether they can remember the rule for rounding, that is, '4 or less **round** down, 5 or more round up' *So will we round 4.82 down to 4 or up to 5?* Agree that, as 4.82 has 8 tenths, we need to round up. *So, 4.82 rounded to the nearest whole number is 5.*

 Main activity

Look together at page 79 of the Student Book. Display on the IWB, if possible. Refer students to the worked example. Explain that, this time, you are going to round numbers to the **nearest tenth**. Talk through the worked example, showing how to use a number line to see where the number lies and which tenth it is nearer to.

Choose three digits from a set of digit cards 0–9 and make a 3-digit number with two decimal places, for example 2.54. Ask students to round the number to the nearest tenth by looking to see which multiples of a tenth are either side of the number, in this case 2.50 and 2.60. As 54 tenths is nearer to 50 tenths than 60 tenths, the rounded number will be 2.50. Repeat this process several times with other digits.

Ask students to work on the activity on page 79 of the Student Book in pairs. Ask individual students, *What is the 'rule' for rounding to the nearest tenth?* Students may be able to explain that it is the same as when they were rounding to the nearest whole. That is, they look at the digit to the right of the tenths (the hundredths) and then apply the rule of 4 or less, round down; 5 or more, round up. Note which students have the clearest descriptions of their rules. These students can support you in Reflection time.

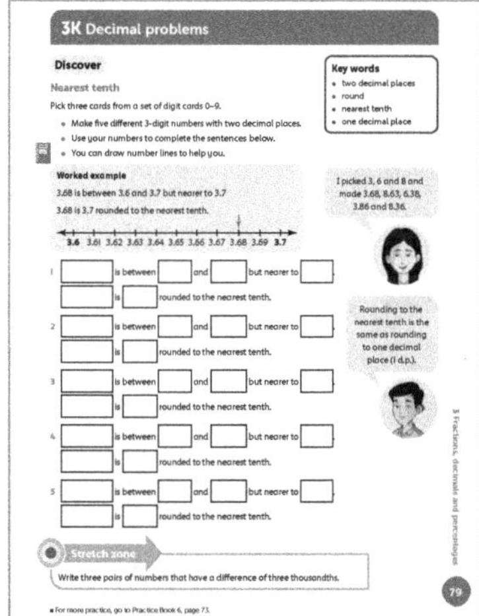

### Differentiation
**Supporting:** Model how to use a number line to support rounding to the nearest tenth.

**Consolidating:** Ask students to show you how they use a number line to support them with rounding numbers to the nearest tenth.

**Extending:** Ask students to explain the rule for rounding to the nearest tenth.

**Stretch zone:** *Write three pairs of numbers that have a difference of three thousandths.*

Students could write a number to three decimal places, then either add or subtract 3 thousandths, for example 3.245 and 3.248.

### Reflection time
Ask two students to work together to explain to the rest of the class what the 'rule' is for rounding numbers to the nearest tenth.

**Unit 3 Fractions, decimals and percentages** 103

Generate a 4-digit number with three decimal places using the large digit cards. Write this on the board. Ask students to round this to the nearest whole, nearest tenth, and then to the nearest hundredth. They should write their responses on their whiteboards. Wait, then ask all students to show you their whiteboards at the same time. Repeat with other numbers with three decimal places.

**Practice Book:** Students complete Practice Book page 73. They can do this directly after the Main activity, as homework, or as the focus of a separate mathematics session to help students consolidate their learning and build fluency.

Students look at reaction times of athletes, given to three decimal places. They write each of the times to the nearest second, the nearest tenth of a second and the nearest hundredth of a second. They then order the times, using the different rounded times, and compare the lists they produce. *Would you expect the times to be in the same order in each list?*

| Differentiated outcomes | |
| --- | --- |
| **All students** | should round numbers to the nearest whole and tenth with support. |
| **Most students** | will round numbers to the nearest whole and tenth using a number line. |
| **Some students** | may explain the rule for rounding decimals clearly. |

## Answers

### Student Book page 79

Students choose their own digit cards. Check that for each number they make, they have located it between the correct multiples of tenths to help them round it correctly to the nearest tenth.

### Practice Book page 73

| | Athlete | Reaction time (seconds) | Rounded to the nearest hundredth | Rounded to the nearest tenth | Rounded to the nearest second |
| --- | --- | --- | --- | --- | --- |
| **1** | Grace | 0.243 | 0.24 | 0.2 | 0 |
| **2** | Sobia | 0.476 | 0.48 | 0.5 | 0 |
| **3** | Zara | 0.673 | 0.67 | 0.7 | 1 |
| **4** | Rowan | 0.336 | 0.34 | 0.3 | 0 |
| **5** | Cam | 0.824 | 0.82 | 0.8 | 1 |
| **6** | Hana | 0.505 | 0.51 | 0.5 | 1 |

**7 a** G, R, S, H, Z, C     **b** G, R, S, H, Z, C     **c** G, R, S and H, Z, C     **d** G, R and S; H, Z and C

**8**  The order is the same for nearest hundredth but for tenths and seconds some athletes have the same time.

**9 a** 0.581     **b** 0.58

Stretch zone: Students may say that rounding makes the calculations easier, but it makes them less accurate.

## 3K Decimal problems

### Explore   Student Book page 80 • Practice Book page 74

#### Specific learning focus
- Doubling amounts of money to two decimal places.

#### Global skills
- **Creative skills:** problem solving
- **Real-world skills:** financial literacy

#### Key vocabulary
- tenths, hundredths, thousandths, two decimal places

#### Resources
- mini whiteboards and markers
- digit cards 0–9

### Language support

Explain to students the language of charity fundraising and donations. Ask students questions to support them with the doubling and rounding processes, for example:

- *How much is double that amount?*
- *Can you double the dollars and cents separately?*
- *How much is that rounded to the nearest ten dollars?*
- *How much is it rounded up to the next ten dollars?*

 **Introductory activity**

Ask a student to choose three digit cards and make an amount of money with two decimal places, for example $4.82. Ask the class to work in pairs to double this amount on their whiteboards. Agree that the new amount is $9.64.

Now students double that amount and then double twice more. Ask them to share the amounts they made, which should be:

$19.28     $38.56     $77.12

Now go back to the original amount and round it to the nearest 10 cents ($4.80). Double this amount four times and see how it compares with the first result.

The new amount after doubling four times is $76.80.

Repeat the activity with a new amount drawn from four digit cards. For example $17.67 doubled four times makes $282.72, but rounding to the nearest 10 cents and then doubling four times will give $283.20.

## Main activity

Look together at page 80 of the Student Book. Display on the IWB, if possible. Discuss with students the information in the top box, explaining how, each week, Business A will double the amount raised by a charity. Then look at the second box, which describes how, each week, Business B will always round up the amount raised to the next multiple of $10 and then double it.

Students should then work in pairs to calculate the donations each week and find the total amount of donations by both businesses over five weeks.

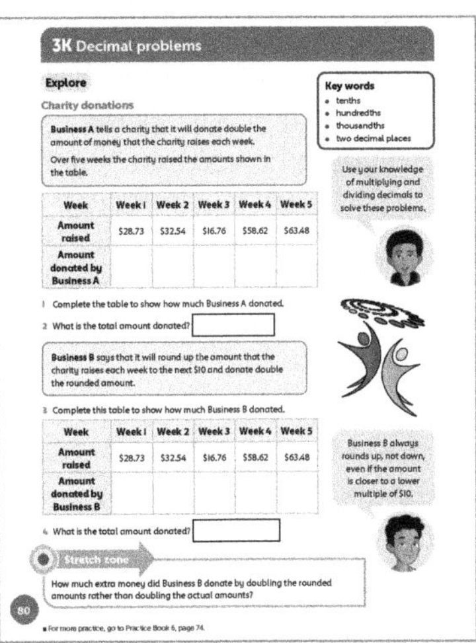

### Differentiation

**Supporting:** Help students with the doubling process if necessary and help them round up to the nearest $10 for questions 3 and 4.

**Consolidating:** Ask students to explain how they rounded the amounts before doubling.

**Extending:** Challenge students to calculate the amount raised by Business B if it rounded to the nearest dollar (rounding up or down) compared to always rounding up to the next dollar.

**Stretch zone:** *How much extra money did Business B donate by doubling the rounded amounts rather than doubling the actual amounts?*

Ask students to estimate first before calculating. Donating double the original amounts would be $400.26, but donating double the rounded amounts would be $440, so the difference is $39.74.

## Reflection time

Bring the class together and discuss the results. Ask pairs of students to share their answers and explain their method. Ask students to suppose that the two businesses waited until the end of the five weeks and then made their donations on the same basis as before. *How much would Business A donate?* Can students predict whether it would be more, less, or the same as donating weekly? *What about Business B?* (The amounts will be the same; it does not matter whether the money is donated week by week or totalled after five weeks.)

**Practice Book:** Students complete Practice Book page 74. They can do this directly after the Main activity, as homework, or as the focus of a separate mathematics session to help students consolidate their learning and build fluency.

Students round large amounts of money to the nearest hundredth of a million pounds. They then answer questions based on these rounded numbers.

| Differentiated outcomes | |
|---|---|
| **All students** | should round and double the amounts with support. |
| **Most students** | will round and double the amounts. |
| **Some students** | should round and double mentally and estimate the different totals by the different methods. |

## Answers

### Student Book page 80

**1**

| Week | Week 1 | Week 2 | Week 3 | Week 4 | Week 5 |
|---|---|---|---|---|---|
| **Amount raised** | $28.73 | $32.54 | $16.76 | $58.62 | $63.48 |
| **Amount donated by Business A** | $57.46 | $65.08 | $33.52 | $117.24 | $126.96 |

**2** $400.26

**3**

| Week | Week 1 | Week 2 | Week 3 | Week 4 | Week 5 |
|---|---|---|---|---|---|
| **Amount raised** | $28.73 | $32.54 | $16.76 | $58.62 | $63.48 |
| **Amount donated by Business B** | $60 | $80 | $40 | $120 | $140 |

**4** $440

**1**

| Company | A | B | C | D | E |
|---|---|---|---|---|---|
| Profit in £ | 2 361 525 | 3 652 590 | 2 587 325 | 4 143 565 | 1 999 250 |
| Profit written as '£x.xx million' | £2.36 million | £3.65 million | £2.59 million | £4.14 million | £2.00 million |

**2** E, A, C, B, D   **3** £2.14 million   **4** £14.74 million   **5** £7.77 million   **6** £0.828 million

Stretch zone:   £14 744 255

£14.74 million

Yes, it is the same as the answer for question 4.

---

## 3L Fractions, decimals and percentages

### Discover   Student Book page 81 · Practice Book page 75

#### Specific learning focus

- Find simple fractions and percentages of shapes and whole numbers.

#### Global skills

- **Creative skills:** problem solving
- **Real-world skills:** interpreting information
- **Self-development skills:** reflecting on learning

#### Key vocabulary

- fraction, decimal, percentage, equivalent

#### Resources

- mini whiteboards and markers
- counting stick marked into ten sections

#### Language support

Ask students to explain their ideas. Ask, for example:

- *How did you work out that a mark of 20 out of 80 is 25%?*

### Introductory activity

Hold up a counting stick marked into ten sections and say that the whole stick represents 100%. Count up in tens, calling out each section as a percentage: 10%, 20%, 30% and so on.

Now say that the stick represents 40 marks available on a test. Point to different positions on the stick and ask students to say what percentage it shows and what mark on the test. For example, ask, *If I got half the question correct* (point to halfway along the stick) *how many marks did I get and what percentage is that?* Agree that the halfway mark represents 20/40 (20 out of 40) or 50%. Repeat, pointing to the first division on the stick, which represents 4/40 marks, or 10% and so on.

### Main activity

Ask students to draw a rectangle and a circle on their whiteboards. Ask them to shade 50% of each shape then show you. They should have shaded half of each shape. *What fraction does this show?*

Repeat using different amounts. Sometimes ask students to shade a fraction and tell you the percentage, sometimes ask them to shade a percentage and tell you the equivalent fraction. Ask them to explain their strategy for finding the correct amount to shade. For example, for quarters they could have halved and halved again.

Ask students to complete the activities on page 81 of the Student Book, attempting questions individually then comparing answers with a partner. Ask some students to explain their ideas. For example, ask, *How did you work out that a mark of 20 out of 80 is 25%?* You can also ask them about marks that do not appear in the table. For example, ask, *My mark is 25. What percentage is that?*

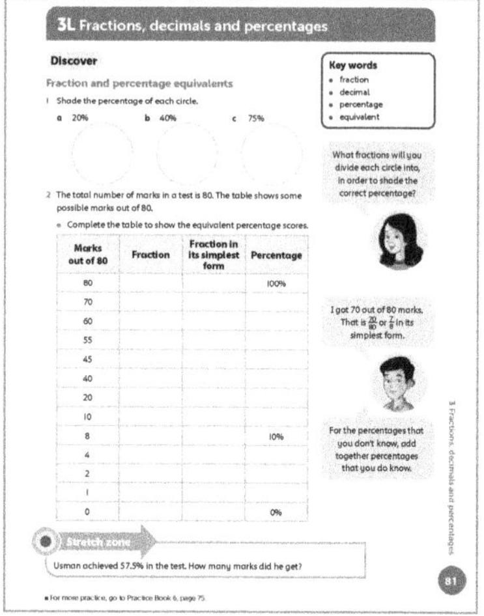

### Differentiation

**Supporting:** Ask students to write down all the equivalent fractions and percentages they know.

**Consolidating:** Ask students to explain their thinking as they complete the table.

**Extending:** Ask students to find as many different percentages as they can for a test marked out of 80.

**Stretch zone:** *Usman achieved 57.5% in the test. How many marks did he get?*

The answer is 46, which students can work out by adding the percentages for 40 marks (50%) 4 marks (5%) and 2 marks (2.5%).

You can set similar problems. For example, ask students to work out what 25% of the marks are or what the mark is after a 10% reduction.

##  Reflection time

Ask students to draw a square on their whiteboards. Then ask them to shade 25% of the square. Ask them to show you their answers at the same time. Repeat, asking for 33.3% shading, then 80% shading. Use any errors to clear up misconceptions about finding fraction and percentage equivalents.

**Practice Book:** Students complete Practice Book page 75. They can do this directly after the Main activity, as homework, or as the focus of a separate mathematics session to help students consolidate their learning and build fluency.

Students convert percentages to fractions and then find percentages for test scores, out of a possible score of 60. *If 6 out of 60 is 10%, what percentage is 3 out of 60? What about 9 out of 60?*

| Differentiated outcomes | |
| --- | --- |
| **All students** | should convert between fractions and percentages with support. |
| **Most students** | will convert between fractions and percentages. |
| **Some students** | may find percentages that do not appear in the table. |



## Student Book page 81

**1** Accept any accurate shading of $\frac{1}{5}$ of the first circle, $\frac{2}{5}$ of the second circle and $\frac{3}{4}$ of the third circle.

| Marks | % | Marks | % | Fraction | Simplest form | Fraction | Simplest form |
| --- | --- | --- | --- | --- | --- | --- | --- |
| 80 | 100% | 10 | 12.5% | $\frac{80}{80}$ | 1 | $\frac{10}{80}$ | $\frac{1}{8}$ |
| 70 | 87.5% | 8 | 10% | $\frac{70}{80}$ | $\frac{7}{8}$ | $\frac{8}{80}$ | $\frac{1}{10}$ |
| 60 | 75% | 4 | 5% | $\frac{60}{80}$ | $\frac{3}{4}$ | $\frac{4}{80}$ | $\frac{1}{20}$ |
| 55 | 68.75% | 2 | 2.5% | $\frac{55}{80}$ | $\frac{11}{16}$ | $\frac{2}{80}$ | $\frac{1}{40}$ |
| 45 | 56.25% | 1 | 1.25% | $\frac{45}{80}$ | $\frac{9}{16}$ | $\frac{1}{80}$ | $\frac{1}{80}$ |
| 40 | 50% | 0 | 0% | $\frac{40}{80}$ | $\frac{1}{2}$ | $\frac{0}{80}$ | $\frac{0}{80}$ |
| 20 | 25% | | | $\frac{20}{80}$ | $\frac{1}{4}$ | | |

## Practice Book page 75

**1** $\frac{1}{4}$    **5** $\frac{1}{5}$    **10** $\frac{7}{10}$    **15** 25%

**2** $\frac{3}{4}$    **6** $\frac{9}{10}$    **11** 90%    **16** 20%

**3** $\frac{1}{10}$    **7** $\frac{1}{2}$    **12** 75%    **17** 40%

**4** $\frac{2}{5}$    **8** $\frac{1}{3}$    **13** 70%    **18** 5%

   **9** $\frac{2}{3}$    **14** 50%

Stretch zone: 53

---

## 3L Fractions, decimals and percentages

### Explore
Student Book pages 82–83 • Practice Book page 76

### Specific learning focus

- Find simple percentages of shapes and whole numbers. Write fractions and decimal equivalents.

### Global skills

- **Creative skills:** problem solving

### Key vocabulary

- fraction, decimal, percentage, equivalent

### Resources

- Resource sheet 3.2: 100-square
- sheets of card, coloured pens, scissors

### Language support

Ask students to explain their reasoning. This allows them to practise language skills and show their understanding of key terms.

###  Introductory activity

Ask a student to choose any number from this list on the board: 24, 48, 60, 90, 128, 360, 1600, 3600

Ask students to write as many facts as they can about the number using percentages and equivalent fractions. For example, for 24: $\frac{3}{10}$ of 24 = 7.2, 15% of 24 = 3.6.

Repeat with other numbers.

Display this table of equivalences on the board:

| Fraction | $\frac{1}{2}$ | | $\frac{1}{5}$ | | | $\frac{9}{10}$ | | $\frac{1}{3}$ |
|---|---|---|---|---|---|---|---|---|
| Decimal | | | | 0.7 | | | 0.6 | |
| Percentage | | 75 | | | 25 | | | |

Ask students, in pairs, to complete the missing values, then to explain how they knew each one.

Look together at page 82 of the Student Book. Display on the IWB, if possible. Talk through the first example, noting how $\frac{1}{10}$ of the 100-square is shaded and the equivalent decimal and percentage of the shaded part are recorded. Ask students to explain why $\frac{1}{10}$ is the same as 0.1 and why this is the same as 10%, relating 'percentage' to 'per cent' or 'out of 100'.

Students then complete the examples on pages 82–83. For the game on page 83, they should copy the details in the table onto card then complete the missing cards before cutting them out.

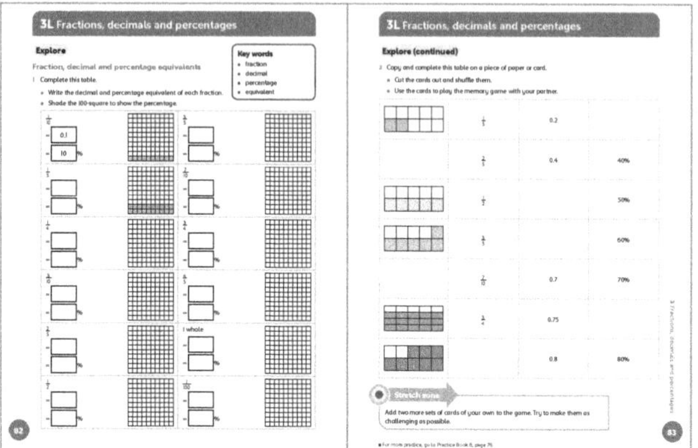

## Differentiation

**Supporting:** Give students 100-squares to shade in to support them in completing the table.

**Consolidating:** Encourage students to use their knowledge of equivalences to complete the table and to explain how they worked each one out.

**Extending:** Ask students to add extra rows to the matching cards game, to explore eighths as percentages.

**Stretch zone:** *Add two more sets of cards of your own to the game. Try to make them as challenging as possible.*

Students make the cards as difficult as they can, but must make sure that they have a correct set of shapes, fractions, decimals and percentages. For example, they could add a set of cards as follows: $\frac{3}{8} = 0.375$ or 37.5%, or $\frac{5}{8} = 0.625$ or 62.5%.

 **Reflection time**

Write the following fractions on separate pieces of paper:

$\frac{3}{4}$   $\frac{3}{8}$   $\frac{1}{4}$   $\frac{3}{5}$   $\frac{3}{10}$   $\frac{1}{10}$   $\frac{16}{20}$

Ask a student to come to the front of the class and choose a fraction. Ask them to write the equivalent percentage on the piece of paper. *How do you know this is correct?* Then ask the student to stand in the correct place on an imaginary number line. Repeat with six other students so that when seven students are at the front, the fractions and equivalent percentages are arranged in ascending order on the 'number line'.

**Practice Book:** Students complete Practice Book page 76. They can do this directly after the Main activity, as homework, or as the focus of a separate mathematics session to help students consolidate their learning and build fluency.

Students complete diagrams by choosing some different fractions and percentages of given numbers, and writing the answers, some of which involve decimals, for example: 10% of 24 is 2.4, $\frac{1}{3}$ of 24 is 8, $\frac{1}{20}$ of 24 is 0.12

| Differentiated outcomes | |
|---|---|
| **All students** | should convert fractions, decimals and percentages, using 100-squares for support. |
| **Most students** | will convert fractions, decimals and percentages, using facts that they do know to complete those they don't know. |
| **Some students** | may complete fractions, decimal and percentage conversions, using mental recall. |

## Answers

### Student Book pages 82–83

**1**

| (Provided) $\frac{1}{10} = 0.1 = 10\%$ = 10 squares shaded | $\frac{3}{5} = 0.6 = 60\%$ = 60 squares shaded |
|---|---|
| $\frac{1}{5} = 0.2 = 20\%$ = 20 squares shaded | $\frac{7}{10} = 0.7 = 70\%$ = 70 squares shaded |
| $\frac{1}{4} = 0.25 = 25\%$ = 25 squares shaded | $\frac{3}{4} = 0.75 = 75\%$ = 75 squares shaded |
| $\frac{3}{10} = 0.3 = 30\%$ = 30 squares shaded | $\frac{4}{5} = 0.8 = 80\%$ = 80 squares shaded |
| $\frac{2}{5} = 0.4 = 40\%$ = 40 squares shaded | 1 whole = 1 = 100% = 100 squares shaded |
| $\frac{1}{2} = 0.5 = 50\%$ = 50 squares shaded | $\frac{1}{100} = 0.01 = 1\%$ = 1 squares shaded |

**2** Students match the cards and create their own cards.

Answers will vary because students choose their own percentages, fractions and decimals of the numbers provided. Check that their calculations are correct.

Stretch zone: Check that students' amounts related to 0.75 are correct.

# 3 Fractions, decimals and percentages

## Connect  Student Book page 84

### Big idea

I can compare, order and find equivalent fractions, decimals and percentages to solve problems.

### Global skills

- **Creative skills:** problem solving
- **Real-world skills:** interpreting information
- **Interpersonal skills:** teamwork
- **Self-development skills:** reflecting on learning

### Key vocabulary

- percentage, price, reduction, proportion

### Resources

- large sheets of paper or access to presentation software to prepare a presentation

### Language support

You can use a word and phrase bank to support the presentations, for example:

- per cent, percentage, %
- 'The equivalent fraction is …'
- 'The decimal equivalent is …'

 **Introductory activity**

Write on the board:

$5            $16            $21

Ask students to find 10% and 5% of each of the three amounts. *How can you use these to help you find 35% of the amount?* Finally, ask students to reduce each price by 35%. *What is the new price? How did you work it out?*

 **Main activity**

Arrange the class into mixed-attainment groups of about four students for the activity on page 84 of the Student Book. Encourage them to interpret the activity for themselves. Avoid leading groups too much. Encourage them to re-read the instructions so that they understand what they need to do. Make sure that you leave plenty of time for students to plan their presentations (the Stretch zone activity). It is important that the presentations are well prepared. When the groups have completed the activity and prepared their presentation, ask three groups, in turn, to present their solution.

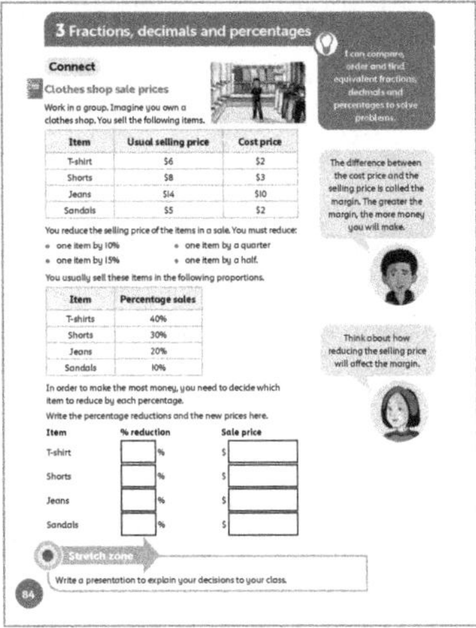

## Differentiation

All students should work in mixed-attainment groups and so engage with the group task.

**Stretch zone:** *Write a presentation to explain your decisions to your class.*

Ensure that the presentations include information on which item of clothing will be reduced by which percentage amount and include reasons for the decision. Students should look at the sales margin on each item and the percentage of sales of each item.

 **Reflection time**

At the end of the presentations, ask the class to imagine that they are the shopkeeper and to decide which solution to the problem is the best. Encourage students to give a clear explanation. *Why do you think this solution is best?*

| Differentiated outcomes | |
|---|---|
| **All students** | should contribute to the group discussion and presentation. |
| **Most students** | will support the preparation of the presentation by calculating percentages. |
| **Some students** | may explain to the class how the amounts were calculated and argue for the best solutuion. |

## Student Book page 84

Answers will vary because students make their own choices about how to reduce the prices. Check that their calculations of percentages are correct.

---

# 3 Fractions, decimals and percentages

### Review  Student Book page 85 • Practice Book page 77

### Global skills

- **Real-world skills:** interpreting information
- **Self-development skills:** reflecting on learning

### Student Book

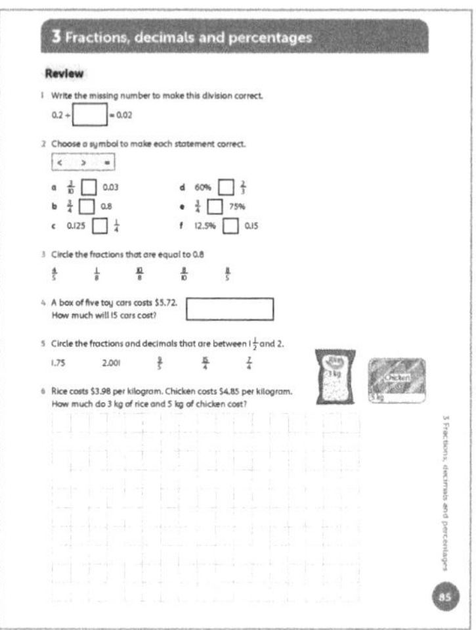

With young students, assessment activities are most effective when carried out as an everyday classroom activity. Students should have 100-squares available to support them. Watch as students represent and convert between fractions, decimals and percentages. Listen as they say the value of each decimal digit and say the numbers aloud. Listen to check that they understand how the place of a digit affects its value. If there are errors, help them to partition decimal numbers so they can understand the value of different digits.

When ordering fractions, decimals and percentages, it may help to model how to place numbers on number lines, by first converting each number to the same form.

The Review activity is an individual summative assessment and so should be completed by students working on their own.

**Answers**

## Student Book page 85

**1** 10

**2 a** >     **b** <     **c** <     **d** <     **e** =     **f** <

**3** $\frac{4}{5}$ and $\frac{8}{10}$

**4** $85.80

**5** 1.75, $\frac{9}{5}$, $\frac{7}{4}$

**6** $36.19

### Practice Book

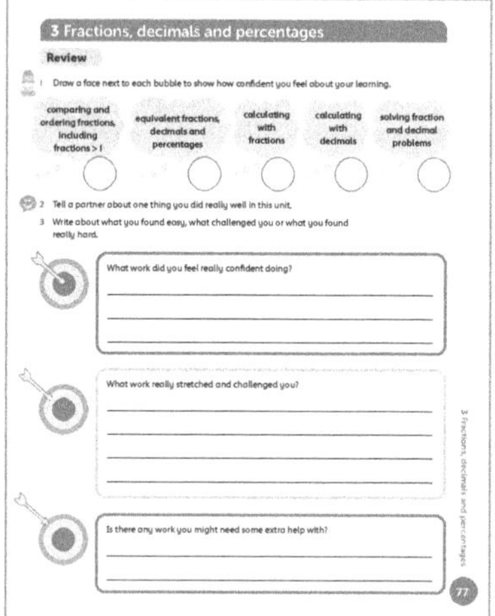

It is appropriate to complete this Practice Book Review as a whole-class discussion. You may choose to keep a record of the class discussion or a copy of the Review page for your own records. The Review provides an opportunity for students to reflect on their learning from the unit, to discuss any areas of mathematics that they feel went particularly well, and any areas that they feel less confident about. Ensure that all students have a copy of the Student Book as a reminder of the areas of mathematics that they have worked on in this unit.

Allow students plenty of time for discussion before asking them to complete the Practice Book page individually, and then, if appropriate, to share their responses with the rest of the class. If students complete this self-assessment at home, encourage them to discuss this with adults. Make a note of any areas that students still feel unsure about.

### Additional material

There are additional end-of-unit assessments available on the *Oxford Owl for School* website.

# 4 Ratio and proportion

## Overview

### Big idea

In this unit, students will develop their understanding about the multiplicative reasoning structures of ratio and proportion. It is important that students move beyond seeing multiplication as repeated addition and, by looking at ratio, proportion and scaling, they will develop a deeper conceptual sense of the nature of multiplication as a scaling or stretching process. They will explore the 'x times as many' or 'x times the size of' relationship that exists in different contexts and relate this to their knowledge of fractional parts.

The essential point that the unit will address is that proportion looks at one part being a part of a whole amount, whereas a ratio situation considers the relationship of one part to another part. A ratio cannot be expressed as a single number, and when two quantities are related and one changes, the other will change in a precise way. The quantities may change but the relationship between them remains constant.

### Look out for

- **Students who think that ratio problems can be solved by using additive structures.** Help them to understand that, for example in a ratio of 1 : 3, if the 1 increases 5 times to 5, the 3 will increase 5 times to 15. They should not see the increase from 1 to 5 as adding 4.
- **Students who think that scaling a length will scale areas and volumes by the same number.** For example, give students opportunities to see that doubling the sides of a rectangle will cause the area to increase to four times its original size.

### Possible misconceptions

- **Students think that all ratios must be whole numbers.** Support them to compare quantities in a multiplicative sense. For example, if two heights are 145 cm and 125 cm, encourage them to calculate the ratio as $\frac{145}{125} = 1.16$, and describe one as 1.16 times the other.
- **Students represent ratios as though they are proportions.** Help them to avoid thinking of, for example, 1 : 4 as $\frac{1}{4}$ by emphasising that 1 : 4 is a relationship between 5 parts, so 1 part is $\frac{1}{5}$ of the other.

### Key vocabulary

- ratio, one to four (1 : 4), proportion, 'for every', 'in every'
- fraction, percentage, bar model, percentage increase, percentage decrease
- scale, scale factor, enlarge, enlargement, scale drawing

### Coverage in lessons

| Learning objective | E | 4A | 4B | 4C | 4D | C | R |
|---|---|---|---|---|---|---|---|
| Solve problems involving the relative sizes of two quantities where missing values can be found by using integer multiplication and division facts. | ✓ | ✓ | ✓ | ✓ | ✓ | ✓ | ✓ |
| Solve problems involving the calculation of percentages (for example of measures, and such as 15% of 360) and the use of percentages for comparison. | ✓ | | ✓ | ✓ | | | |
| Solve problems involving similar shapes where the scale factor is known or can be found. | | | | | ✓ | ✓ | ✓ |

# 4 Ratio and proportion



**Engage** Student Book page 86

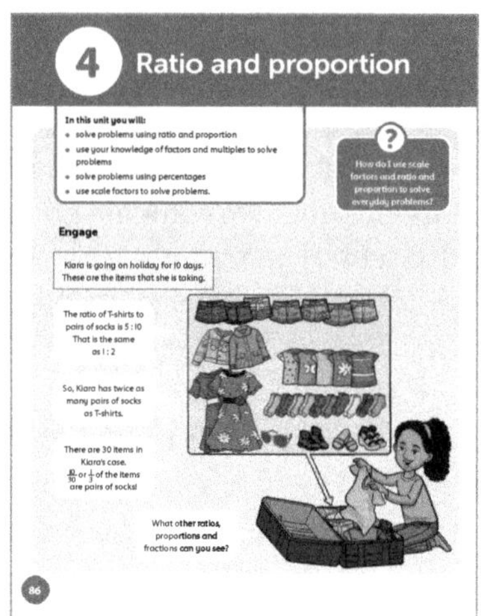

## Big question

- How do I use scale factors and ratio and proportion to solve everyday problems?

## Global skills

- **Creative skills:** problem solving
- **Self-development skills:** reflecting on learning

## Key vocabulary

- ratio, proportion, fraction

## Resources

- cubes or counters in two colours (for each pair)
- mini whiteboards and markers
- large sheets of paper

## Language support

Encourage students to use the language of fractions, proportions and ratios to describe sets of items, for example:

- There are twice as many pairs of socks as T-shirts.
- The ratio of pairs of socks to T-shirts is 2 : 1.
- 2 pairs of shorts out of 7 are blue.
- $\frac{2}{7}$ of the pairs of shorts are blue.

When we talk about proportion, we use the language 'in every' or 'out of'. When we talk about ratio, we use the language 'for every'.

 Introductory activity

Give each pair of students a set of cubes (or counters), in two colours, for example 4 blue, 8 purple. Ask a series of questions to help students describe the set of cubes in relation to their colours, for example:

- *How many cubes are there altogether? How many are purple?*
- *What fraction of the cubes are purple? What fraction are blue?*
- *How many purple cubes are there for every blue cube?*
- *What is the ratio of purple : blue cubes?*
- *What is the ratio of blue : purple cubes?*
- *If you had 12 blue cubes, how many purple cubes would you need to keep the same ratio?*

 Main activity

Look together at page 86 of the Student Book. Display on the IWB, if possible. Students can see the clothes Kiara is packing for her holiday. Ask them to work in pairs to write down as many statements as they can about the clothes and any fractions or ratios they can see. Ask them to simplify their fractions to the simplest terms where possible. For example, 4 of the 20 socks are blue, so $\frac{4}{20}$ or $\frac{1}{5}$ of the socks are blue.

Students can display their facts about the clothes on a poster for discussion in Reflection time.

### Differentiation

**Supporting:** Model for students the 'times as many' relationship using their cubes or the clothes in the images of packing for a holiday.

**Consolidating:** Ask students to write each proportion they found as a ratio and vice versa.

**Extending:** Challenge students to express ratios and proportions about the colours of the clothes as well as which item they are.

 Reflection time

Ask pairs of students to share their posters of ratios and proportions about the clothes. Ask them to explain some of their statements and describe how they found the relationship between the quantities. They should use the phrases 'in every', 'for every' and 'times as many' in their explanations.

## Discover  Student Book page 87 • Practice Book page 78

### Specific learning focus
- Solve simple problems involving ratio.

### Global skills
- **Creative skills:** problem solving

### Key vocabulary
- ratio, one to four (1 : 4), for every

### Resources
- blue, yellow and red counters or cubes
- squares of blue, yellow and red paper, reusable adhesive
- mini whiteboards and markers

### Language support
Model key words and phrases in your discussions with students, for example:
- ratio
- for every . . .

 **Introductory activity**

Stick three blue squares and one yellow square to the board. Explain that this shows a **ratio** of 1 : 3. Say 'one blue square to three yellow squares'. Explain that this means that **for every** one part of yellow, there are three parts of blue. Write this as 1 : 3 on the board.

Then write 3 : 1 on the board and ask students if they can give an example using the blue and yellow squares with the ratio 3 : 1. They should be able to describe this as being 'for every three parts of blue, there is one part of yellow'.

Write on the board the ratios 1 : 4, 1 : 5 and 1 : 6. Ask students to come to the front and stick on the board pieces of coloured paper to illustrate these ratios. Ask students to say the ratios ('one to four', 'one to five' and 'one to six'). Ask them to describe what the ratios 4 : 1, 5 : 1 and 6 : 1 would be, using the coloured squares.

 **Main activity**

Illustrate the idea of adding two different ratios. Ask one student to use red and blue coloured squares to represent a ratio of their choice (for example 2 : 3) and another student to use blue and red squares to represent a ratio of their choice (for example 1 : 5). Ask, *What is the overall ratio of blue to red squares?* (3 : 8)

Look together at page 87 of the Student Book. Display on the IWB, if possible. Discuss the worked example, making sure that all students are clear about what happens to the overall ratio when adding different ratios together. Ask students to work in pairs on the investigation on page 87. Give each pair a selection of counters or cubes. They can use these to model the problem. Ask students to represent the problem on their whiteboards. Encourage them to use sketches to illustrate their ideas (and not to try immediately to represent the solution using fractions). You may choose to group less-confident students together so that you can model the problem to the whole group.

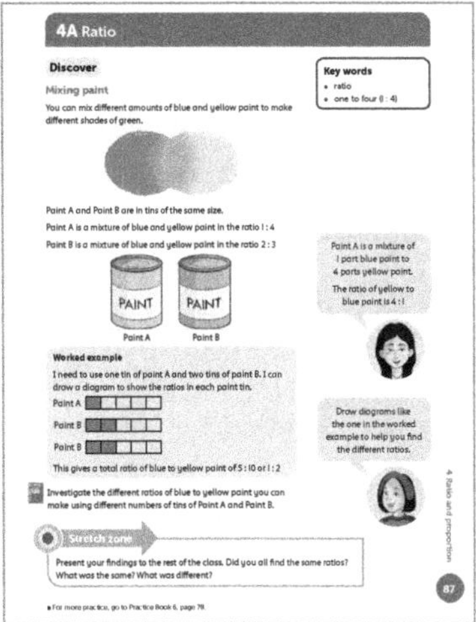

### Differentiation
**Supporting:** Model the problem for students using cubes and counters.

**Consolidating:** Ask students to model the problem for you using cubes and counters, and to use sketches to help them solve the problem.

**Extending:** Ask students to attempt the problem using sketches or written methods.

**Stretch zone:** *Present your findings to the rest of the class. Did you all find the same ratios? What was the same? What was different?*

As a follow-up activity, students can explore the characteristics of the class or school using ratios. For example, explore languages spoken, types of travel to school and ways of spending free time.

 **Reflection time**

Ask students to share ratios about the class that they may have found from the Stretch zone activity. For example, what is the ratio of boys to girls in the class? What is the ratio of girls to boys? What is the ratio of those who travel to school by car to those who travel by bus? What is the ratio of those with older siblings to those with younger siblings?

**Practice Book:** Students complete Practice Book page 78. They can do this directly after the Main activity, as homework, or as the focus of a separate mathematics session to help students consolidate their learning and build fluency.

Students write ratios of the number of cracked eggs to uncracked eggs in different-sized boxes. They draw diagrams to represent each problem. Remind students that ratios are written as 'part : part'.

| Differentiated outcomes | |
|---|---|
| **All students** | should attempt the investigation using counters and cubes to support them and solve the problems with support. |
| **Most students** | will solve the problems using counters and cubes, or sketches, for support. |
| **Some students** | may solve the problems using sketches and written methods. |

# 4A Ratio

## Explore   Student Book page 88 · Practice Book page 79

### Specific learning focus

- Solve simple problems involving ratio and direct proportion.

### Global skills

- **Creative skills:** problem solving

### Key vocabulary

- ratio, for every

### Resources

- blue, yellow and red counters or cubes
- squares of blue, yellow and red paper
- mini whiteboards and markers

### Language support

Ask students to write their own definition of 'ratio' in the glossary at the back of the Student Book. This will help you to monitor what students understand by this term. Support students by giving them model examples (as in Reflection time).



### Student Book page 87

Students will explore their own ratios for the paint mixtures, so check that the ratios are correct.

### Practice Book page 78

| | Eggs in box | Cracked eggs | Ratio (cracked : uncracked) | Ratio (simplest terms) |
|---|---|---|---|---|
| **1** | 12 | 4 | 4 : 8 | 1 : 2 |
| **2** | 16 | 2 | 2 : 14 | 1 : 7 |
| **3** | 24 | 6 | 6 : 18 | 1 : 3 |
| **4** | 15 | 3 | 3 : 12 | 1 : 4 |
| **5** | 20 | 15 | 15 : 5 | 3 : 1 |
| **6** | 15 | 12 | 12 : 3 | 4 : 1 |

Stretch zone: Students could draw a box of eggs with 7 cracked and 2 uncracked, or 14 cracked and 4 uncracked and so on.

 **Introductory activity**

Revisit the Introductory activity from 4A Discover, using different-coloured squares of paper. *What are the most important things to remember when solving a problem involving ratio?* Encourage students to draw pictures. This helps them to understand and solve problems. Reinforce the language of 'for every' in ratios. For example:

- There are 5 blue squares for every 2 red squares, so the ratio is 5 : 2.
- There are 4 pairs of socks for every T-shirt (4 : 1).

 **Main activity**

Draw this tile pattern on the board:

Explain to students that this is the pattern on one floor tile. Ask them to say how many black squares there are and how many white. *What is the ratio of black squares to white squares?* (5 : 4)

*If I need to tile a floor that is 8 tiles long by 5 tiles wide, how many tiles do I need in total?* (40) *How many black squares will there be on all the tiles?* (200) *How many white squares?* (160) *What is the ratio of the total number of black squares to white squares?* (200 : 160 or 5 : 4)

Ask students to work on the problems on page 88 of the Student Book in pairs for support and to encourage discussion of strategies. Give each pair a selection of counters or cubes. They can use these to model the problems. Ask students to use their whiteboards for sketches to illustrate their ideas (and not to try immediately to represent the solution using fractions). You may choose to group less-confident students together so that you can model the problems with the whole group.

*How do you know that your answers are correct? Can you show me a sketch to convince me? Or can you show me using cubes or counters?*

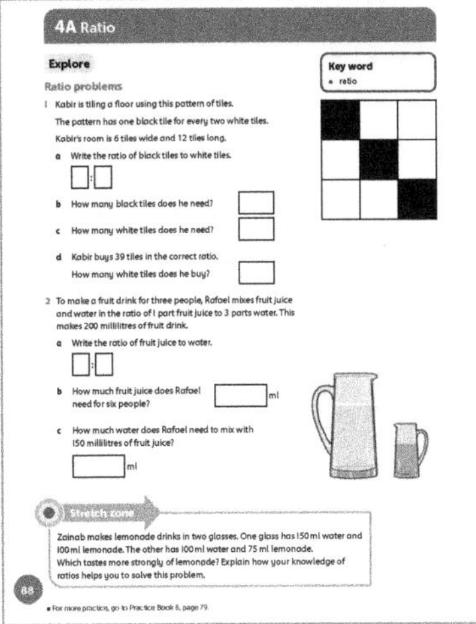

## Differentiation

**Supporting:** Model the problems using cubes or counters.

**Consolidating:** Ask students to model the problems for you using cubes or counters.

**Extending:** Ask students to attempt the problems using written methods.

**Stretch zone:** *Zainab makes lemonade drinks in two glasses. One glass has 150 ml water and 100 ml lemonade. The other has 100 ml water and 75 ml lemonade. Which tastes more strongly of lemonade? Explain how your knowledge of ratios helps you to solve this problem.*

Work through a solution to the problem with the class. The ratios of water to lemonade in the glasses are 150 : 100 and 100 : 75. Convert this to ratios that have one part the same number in each. For example, doubling the first and multiplying the second by 3 would give 300 : 200 and 300 : 225. You could then divide by 25 to get 12 : 8 and 12 : 9. The second ratio has more lemonade for the same amount of water, so will taste more strongly of lemonade.

Students could explore other practical activities that involve ratio, adapting recipes for larger or smaller groups of people, for example.

 **Reflection time**

Ask each pair to write a sentence to answer the question 'What does ratio mean?' Share some of the sentences. Make sure that students understand the sentences by asking them to give examples.

**Practice Book:** Students complete Practice Book page 79. They can do this directly after the Main activity, as homework, or as the focus of a separate mathematics session to help students consolidate their learning and build fluency.

| Differentiated outcomes | |
| --- | --- |
| **All students** | should attempt the problems using counters or cubes to support them and solve the problems with support. |
| **Most students** | will solve the problems using counters or cubes for support. |
| **Some students** | may solve the problems using written methods. |

## Answers

### Student Book page 88

**1 a** 1 : 2

   **b** 24 black tiles

   **c** 48 white tiles

   **d** 26 white tiles

**2 a** 1 : 3

   **b** 100 millilitres of juice

   **c** 450 millilitres water

### Practice Book page 79

**1** 8 : 7

**2** 1 : 1

**3** 30

**4** 1 : 1

**5** Check that students have drawn a pattern with the correct ratio.

Stretch zone: False, the next row will have more reds than blues, bringing the number of each tile to 10, so the ratio will change from 8 : 7 to 1 : 1. If I add another row, there will be more blues than reds so the ratio will become 13 : 12, another row will change the ratio back to 1 : 1 and so on.

# 4B Proportion

## Discover  Student Book page 89 • Practice Book page 80

### Specific learning focus

- Solve problems using proportion.

### Global skills

- **Creative skills:** problem solving
- **Interpersonal skills:** communication

### Key vocabulary

- proportion, ratio, fraction

### Resources

- counters and cubes in two colours
- mini whiteboards and markers

### Language support

Model the language of proportion and ratio from both quantities. For example, in a collection of 15 books, if 10 are non-fiction and 5 are fiction, then we can make statements such as:

- 10 of the 15 books are non-fiction, so the proportion of non-fiction books is 10 out of 15 or $\frac{10}{15}$, which is the same as $\frac{2}{3}$
- 5 books are fiction, so the ratio of fiction : non-fiction is 5 : 10 or 1 : 2
- 10 books are non-fiction, so the ratio of non-fiction : fiction is 10 : 5 or 2 : 1.

###  Introductory activity

Draw on the board 3 squares and 2 triangles. Ask students to tell you how many shapes there are altogether. *What **proportion** of the shapes are triangles? What proportion of the shapes are squares?* Students should be encouraged to use the 'in every' phrase to answer: '3 in every 5 shapes is a square' and '2 in every 5 shapes is a triangle'. Show them that they can write these proportions as fractions: squares make up $\frac{3}{5}$ and triangles make up $\frac{2}{5}$ of the whole set.

Add another 2 squares to the set and repeat.

### Main activity

Give pairs of students a set of coloured counters or cubes in two colours. Ask them to use their counters to make a rectangular array in which the proportion of one colour is $\frac{1}{3}$. Ask them to tell you what proportion the other colour will be. Say that the number of counters they use does not matter, as long as the proportions are correct.

When they have finished, ask different student pairs to share with the class their numbers of counters. For example, they might have 4 blue and 8 red, or 6 white and 12 black. *What is the ratio of the smaller number to the larger number?* Students should identify that whatever their numbers of counters, the ratio is 1 : 2.

Students can now complete the activities on page 89 of the Student Book. Ask them to draw diagrams to help them represent their solution.

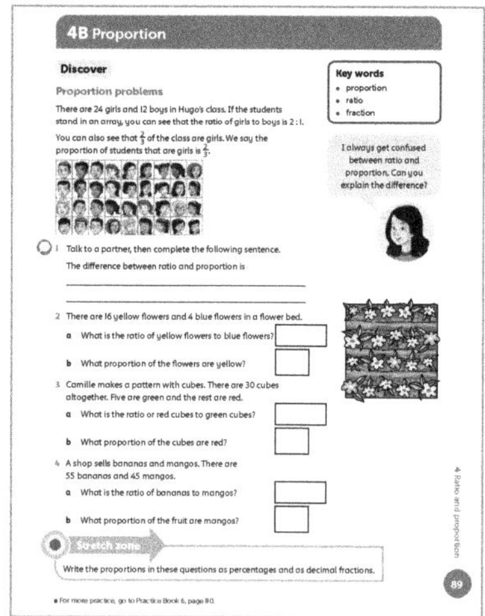

### Differentiation

**Supporting:** Model the proportions with cubes and help students with phrasing the proportions.

**Consolidating:** Ask students to explain how they found the fractions to represent the proportions.

**Extending:** Challenge students to work out the new proportions if some of one quantity is removed from the set.

**Stretch zone:** *Write the proportions in these questions as percentages and as decimal fractions.*

For example, in question 2, there are 20 flowers in total and 16 are yellow so $\frac{16}{20}$ are yellow, which is $\frac{8}{10}$ or 80% or 0.8.

###  Reflection time

Share solutions for each question and ask one pair to model their solution each time. Share the images that students have used to solve question 3. Model the problem practically using cubes. Repeat this for other proportions including:

- 25 cubes, 40% are red
- 20 cubes, 75% are red.

**Practice Book:** Students complete Practice Book page 80. They can do this directly after the Main activity, as homework, or as the focus of a separate mathematics session to help students consolidate their learning and build fluency.

Students shade diagrams to represent a given proportion. They then write the proportion that is shaded as a percentage and as a decimal fraction.

| Differentiated outcomes | |
|---|---|
| **All students** | should model simple proportions with counters or cubes, with support. |
| **Most students** | will solve the problems using drawings to help them. |
| **Some students** | may use mental methods to solve the problems relating proportion to knowledge of equivalent fractions. |

## Answers

### Student Book page 89

**1** Students should describe ratio in terms of the relationship between two parts, and proportion as one part compared to the whole.

**2 a** 16 : 4 or 4 : 1    **b** $\frac{4}{5}$

**3 a** 25 : 5 or 5 : 1    **b** $\frac{5}{6}$

**4 a** 55 : 45 or 11 : 9    **b** $\frac{9}{20}$

## Practice Book page 80

| | Proportion that is shaded | | |
|---|---|---|---|
| | **Fraction** | **Decimal** | **Percentage** |
| **1** | $\frac{6}{10} = \frac{3}{5}$ | 0.6 | 60 |
| **2** | $\frac{9}{10}$ | 0.9 | 90 |
| **3** | $\frac{1}{10}$ | 0.1 | 10 |
| **4** | $\frac{5}{10} = \frac{1}{2}$ | 0.5 | 50 |
| **5** | $\frac{8}{10} = \frac{4}{5}$ | 0.8 | 80 |

**6**  9 shaded

**7**  2 shaded

Stretch zone: Students should draw a rectangle of 8 equal sections with 5 shaded.

# 4B Proportion

## Explore  Student Book page 90 · Practice Book page 81

### Specific learning focus

- Calculate proportions as fractions and percentages.

### Global skills

- **Creative skills:** problem solving

### Key vocabulary

- proportion, ratio, percentage, bar model

### Resources

- mini whiteboards and markers
- coloured counters
- newspapers

### Language support

Model the language used, and support students by giving example problems, for example:

*One fifth of a pile of counters are blue. The other counters are red. What proportion are red? What **percentage** are blue?*

##  Introductory activity

Ask students to draw a square grid that is 8 squares wide and 5 squares high. They can choose three colours. Ask them to shade $\frac{3}{5}$ of the grid in one colour, 15% in a second colour and the rest in a third colour.

As students work, ask, *How many squares make $\frac{3}{5}$?* (24) *How many squares make 15%?* (6) *How many squares are in the third colour?* (10) *What fraction and percentage is this?* $\left(\frac{1}{4}\text{ or }25\%\right)$

## Main activity

Give pairs of students a pile of three different-coloured counters. Ask them to work out the proportions of each colour in their pile, and to write on their whiteboards the proportions as fractions and then as percentages. Encourage them to check their answers by adding the fractions to make sure that they total 1, and the percentages to make sure that they total 100%.

Set this problem:

*I have a drawer full of socks. $\frac{1}{2}$ of them are red, $\frac{1}{6}$ are green and $\frac{1}{3}$ are yellow. There are more than 30 and fewer than 40 socks altogether. How many socks of each colour do I have?*

Ask students to work with a partner to solve this problem. Encourage them to use drawings on their whiteboards to help them. Take feedback. Establish that for these proportions there must be a number that can be divided equally between 2, 3 and 6. 36 ÷ 2 = 18, 36 ÷ 3 = 12 and 36 ÷ 6 = 6, so there must be 36 socks: 18 red, 6 green and 12 yellow. 36 is a multiple of 2, 3 and 6.

Students now complete the activities on page 90 of the Student Book in their pairs. Encourage them to use counters or drawings to solve the problems.

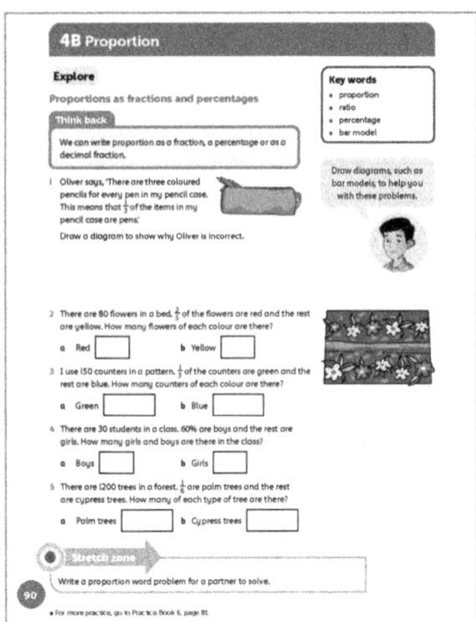

## Differentiation

**Supporting:** Help students to use counters to model proportions.

**Consolidating:** Encourage students to use drawings to model proportions.

**Extending:** Challenge students to use mental methods and jottings to solve problems.

**Stretch zone:** *Write a proportion word problem for a partner to solve.*

Students should check that they can solve the problem themselves.

## Reflection time

Invite pairs to share how they worked out the problems and to give their possible solutions. Students should share the drawings or diagrams they used to help them and explain why these helped.

### Practice Book page 81

| Content | World news (Provided) | National news | Local news | Politics | Business | Sport | Other |
|---|---|---|---|---|---|---|---|
| Number of pages | 10 | 15 | 8 | 5 | 4 | 6 | 2 |
| Proportion (fraction) | $\frac{1}{5}$ | $\frac{3}{10}$ | $\frac{4}{25}$ | $\frac{1}{10}$ | $\frac{2}{25}$ | $\frac{3}{25}$ | $\frac{1}{25}$ |
| Proportion (%) | 20% | 30% | 16% | 10% | 8% | 12% | 4% |

Check that students have completed a similar table for an actual newspaper.

Stretch zone: Check that students have drawn up a suitable table for their newspaper and that the fractions total 1 and the percentages total 100%.

**Practice Book:** Students complete Practice Book page 81. They can do this directly after the Main activity, as homework, or as the focus of a separate mathematics session to help students consolidate their learning and build fluency.

Students calculate the total number of pages in a fictitious newspaper and then work out the proportion of pages allocated to different types of content. They write the proportion as a fraction and as a percentage of the total number of pages. They do the same activity with a real newspaper.

| **Differentiated outcomes** | |
|---|---|
| **All students** | should model simple proportions with cubes or counters, with support. |
| **Most students** | will solve the problems using drawings to help them. |
| **Some students** | may use jottings and mental methods to solve the problems, relating proportion to their knowledge of equivalent fractions. |

## Answers

### Student Book page 90

**1** 3 coloured pencils for each pen is a ratio of 3 : 1 but the proportion of pens is $\frac{1}{4}$.

**2 a** 32 red     **b** 48 yellow

**3 a** 50 green     **b** 100 blue

**4 a** 18 boys     **b** 12 girls

**5 a** 200 palm trees     **b** 1000 cypress trees

# 4C Percentage problems



## Specific learning focus

- Solve problems involving percentages.

## Global skills

- **Creative skills:** investigating
- **Real-world skills:** interpreting information

## Key vocabulary

- percentage increase, percentage decrease

## Resources

- mini whiteboards and markers
- calculators

## Language support

Remind students about '**increase**' and '**decrease**' and how we often talk about price changes or population changes as percentages rather than numbers to make fair comparisons.

 ## Introductory activity

Split the class into small groups. Write $250 on the board. Ask students to work in their groups to write down as many fractions and percentages as they can of this amount of money, for example: $\frac{1}{2}$ of $250 = $125, 50% of £$50 = $125. Share all the different fractions that groups find. Encourage students to use percentages they know to find new percentages. For example, if they know that 10% is $25, then 20% is $50 (double) and 5% is $12.50 (half).

Explain that first they are going to increase $250 by 10%. How much will this be? They need to find 10% of $250 and add it to the original amount: $250 + $25 = $275.

*Now can you calculate a 20% increase on $250? How did you do it?*

Ask questions to encourage students to explain their answers, for example:

- *Which calculation will increase an amount by 25%?*
- *Which of these decreases is the largest: 10% of $50 or 20% of $40? How do you know?*

 ## Main activity

Look together at page 91 of the Student Book. Display on the IWB, if possible. Discuss with students how an election works. Explain that people vote for their preferred candidate and the candidate with the most votes is the winner. Some election candidates stand for the next election, and their percentage of the votes might increase or decrease (go up or down in comparison to the previous election).

Point out the first speech bubble on page 91. It explains how to find a number of votes as a percentage of the total number of votes. Work this out together to find the percentage for candidate A in the table. Refer students to the second speech bubble, which tells them to use a calculator and to round each percentage to the nearest whole number. So, for candidate A they would calculate:

$1525 \div 7032 = 0.216 \ldots$

$\times 100 = 21.6 \ldots$

$= 22\%$ to the nearest whole number.

Students complete the rest of the table and the remaining questions on page 91. As they work, ask them to explain how they worked out the answers, and make a note of students who can give clear explanations for discussion in Reflection time.

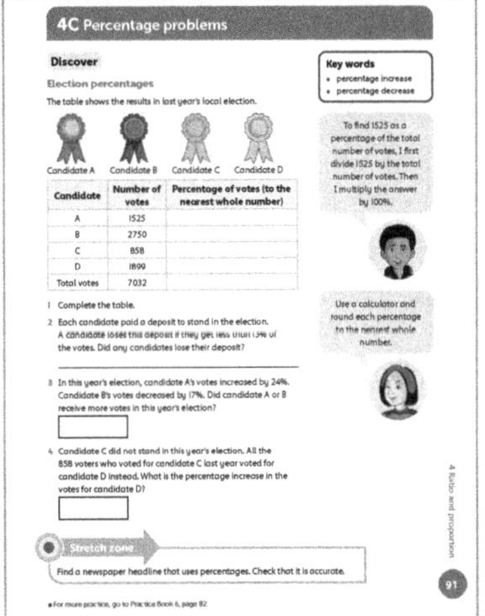

## Differentiation

**Supporting:** Model for students how to calculate percentages from numbers of votes in an election. Ask them to recall your strategies.

**Consolidating:** Ask students to find increases and decreases in percentages.

**Extending:** Challenge students to explain how to find changes in percentages.

**Stretch zone:** *Find a newspaper headline that uses percentages. Check that it is accurate.*

Ask students to explain how they checked the accuracy of the percentage in the headlines.

 ## Reflection time

Discuss together the answers to the questions from Student Book page 91. Choose students you identified earlier to share their clear explanations of the methods they used to find changes in percentages.

**Practice Book:** Students complete Practice Book page 82. They can do this directly after the Main activity, as homework, or as the focus of a separate mathematics session to help students consolidate their learning and build fluency.

Students work out profit margins by calculating a percentage of a given price for various items of fruit. They then use this information to help them explain what they would choose to sell each item of fruit for if they were a shopkeeper. This puts the percentage problem in a real-life context for students.

| Differentiated outcomes | |
| --- | --- |
| **All students** | should find changes in percentages with support. |
| **Most students** | will find changes in percentages independently. |
| **Some students** | may use changes in percentages to work out new amounts, totals or prices. |

## Answers

### Student Book page 91

| Candidate | Number of votes | Percentage of votes (to the nearest whole number) |
| --- | --- | --- |
| A | 1525 | 22% |
| B | 2750 | 39% |
| C | 858 | 12% |
| D | 1899 | 27% |
| Total votes | 7032 | 100% |

1 Check that students have completed the table correctly.

2 Candidate C

3 Candidate B

4 45%

### Practice Book page 82

| Item | Mango | Banana | Bag of nuts | Oat bar | Juice carton |
| --- | --- | --- | --- | --- | --- |
| Cost price | (Provided) $1.05 | $0.33 | $1.50 | $0.60 | $0.75 |
| 10% profit margin | $0.11 | $0.03 | $0.15 | $0.06 | $0.08 |
| 15% profit margin | $0.16 | $0.05 | $0.23 | $0.09 | $0.11 |
| $33\frac{1}{3}$% profit margin | $0.35 | $0.11 | $0.50 | $0.20 | $0.25 |

Students will make their own choices about how much to sell each item for. Check that their reasoning is sound. They may select to make the greatest profit margin for each item of fruit, or they may decide that, by doing that, the cost price is too high so they would not sell any fruit.

Stretch zone: False. For example, reducing an item that costs $1 by 50% will make it cost $0.50, but reducing this by a further 50% will make the cost $0.25, because you are not using 50% of the same amount in the second calculation.

# 4C Percentage problems

## Explore
Student Book page 92 • Practice Book page 83

### Specific learning focus

- Solve problems about tax using percentages.

### Global skills

- **Creative skills:** problem solving
- **Real-world skills:** financial literacy

### Key vocabulary

- percentage increase, percentage decrease

### Resources

- large sheets of paper

## Language support

Support students by explaining what tax is. Say that it is an amount added by the government to the price of many items. We pay this to provide the government with money, which it spends on services such as schools and hospitals.

 ## Introductory activity

Write on the board:

$4.80          $12.60          $34.50

Ask students to work in pairs and increase each amount by 15%. Prompt them by asking how they might find 15% of an amount. They might recognise that they can make 15% by finding 10%, halving it to get 5% and adding the two amounts together for 15%.

Ask students to work out 10%, 20% and 5% of $48.

Now ask students, in pairs, to use these percentages to make new ones. For example, they could use these percentages to find 15% of $48, or 35%, or $12\frac{1}{2}$%. They should record their calculations on a large piece of paper.

Ask questions to encourage students to explain their answers, for example:

- *Which calculation will increase an amount by 25%?*
- *Which of these decreases is the largest: 10% of $50 or 20% of $40? How do you know?*

Look together at page 92 of the Student Book. Display on the IWB, if possible. Explain that most countries add a sales tax to prices. This tax can vary from one country to another, but it is usually an amount that can easily be calculated from 10%.

Students should work in pairs to complete the table of prices on page 92.

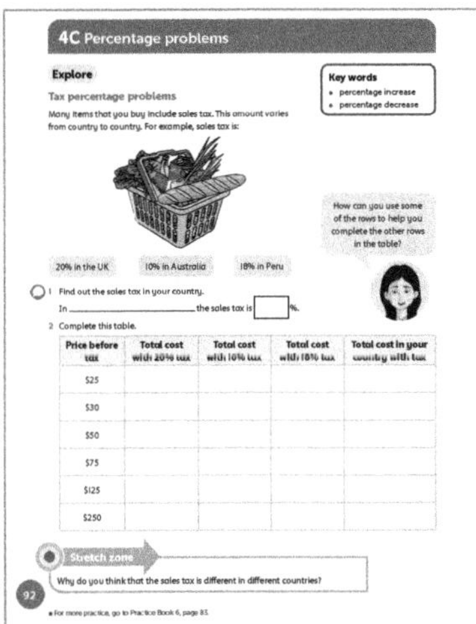

## Differentiation

**Supporting:** Help students to calculate simple percentages such as 20% and 10% of amounts of money. Ask them to explain their strategies.

**Consolidating:** Ask students to use tax percentages to find the new total cost of items.

**Extending:** Challenge students to write their own tax percentage problems for a partner to solve.

**Stretch zone:** *Why do you think that the sales tax is different in different countries?*

Students research the rates of sales tax using the internet and discuss possible reasons for why these are different in different countries. For example, some countries may have different ways of taxing people, such as income tax, so the sales tax might be lower. Students may mention other types of tax such as wealth tax (from people who have lots of money) or service tax (for using services such as hairdressing or beauty treatments). Add your own ideas to anything students raise and discuss how useful these taxes are and whether they are 'fair'. You could mention the historical use of 'window tax', which was charged to people wealthy enough to afford glass windows.

 ## Reflection time

Invite students to share their answers and strategies for solving the tax percentage problems. They can explain how they used known percentages to find costs.

**Practice Book:** Students complete Practice Book page 83. They can do this directly after the Main activity, as homework, or as the focus of a separate mathematics session to help students consolidate their learning and build fluency.

Students solve a series of problems about costs at a fitness club. Special offers are advertised with different percentage decreases so students calculate the cost including the percentage decrease. *What calculation would you do to work out a 'half price' cost? What calculation would you do to work out a reduction of 10% in the cost?*

| Differentiated outcomes | |
|---|---|
| **All students** | should calculate simple tax percentages with support. |
| **Most students** | will calculate a range of tax percentages. |
| **Some students** | may write their own tax percentage word problems. |

## Answers

### Student Book page 92

**1** Answers will vary.

**2**

| Price before tax | Total cost with 20% tax | Total cost with 10% tax | Total cost with 18% tax | Total cost in your country with tax |
|---|---|---|---|---|
| $25 | $30 | $27.50 | $29.50 | |
| $30 | $36 | $33 | $35.40 | |
| $50 | $60 | $55 | $59 | |
| $75 | $90 | $82.50 | $88.50 | |
| $125 | $150 | $137.50 | $147.50 | |
| $250 | $300 | $275 | $295 | |

Check the local rate of sales tax then, particularly if it is different from the other rates listed, verify the final column.

Discounts this month:

monthly fee = $25 – 25% = $18.75, swim = $8.50 – 10% = $7.65, gym = $7.50 – 10% = $6.75, personal training = $35 – 50% = $17.50

**1** $18.75 + 3 × ($7.65 + $30) = $131.70

**2** $18.75 + 3 × ($7.65 + $13.50) + $17.50 + $15 = $114.70

**3** $18.75 + $7.65 + $6.75 + $15 + $17.50 + $13.50 = $79.15

**4** $18.75 + 4 × ($7.65 + $6.75+ $30) = $196.35

Stretch zone: Check that students have kept within their $100 budget.

---

# 4D Scaling problems

## Discover   Student Book page 93 • Practice Book page 84

### Specific learning focus

- Enlarge shapes by a scale factor.

### Global skills

- **Creative skills:** investigating

### Key vocabulary

- scale factor, ratio, enlarge, enlargement

### Resources

- squared paper with 1 cm squares
- rulers

### Language support

Use phrases connected with enlargement to support students with scale factor problems, for example:

- *This shape is double the size of that one.*
- *Enlarge this shape by a scale factor 4.*

 **Introductory activity**

Ask pairs of students to discuss where they can find examples of **enlargements** in everyday life, for example on scale diagrams and models. Give each student a piece of squared paper. Ask them to draw a square measuring 4 cm along each side. Now ask whether they can draw a new square with sides that are double the length. They should draw a square with sides measuring 8 cm. Explain that doubling the lengths means increasing the size of the square by a **scale factor** of 2.

Ask students to draw two more shapes of their choice using straight sides, and then draw the same shapes again with sides that are 3 times larger, or a scale factor of 3.

 **Main activity**

Look together at page 93 of the Student Book. Display on the IWB, if possible. Ask students to look at the shapes in question 1. In the instructions, they are asked to **enlarge** the shapes by different scale factors. By referring to the lengths of the sides of the shapes, discuss how large the sides of each shape will be. For example, the yellow rectangle is 4 squares tall so, when enlarged by a scale factor of 3, its height will be 12 squares. Ask students questions like this about all three shapes.

Students should draw the enlarged shapes on squared paper. Check that everyone has drawn them the correct size. Students then complete the remaining activities on page 93.

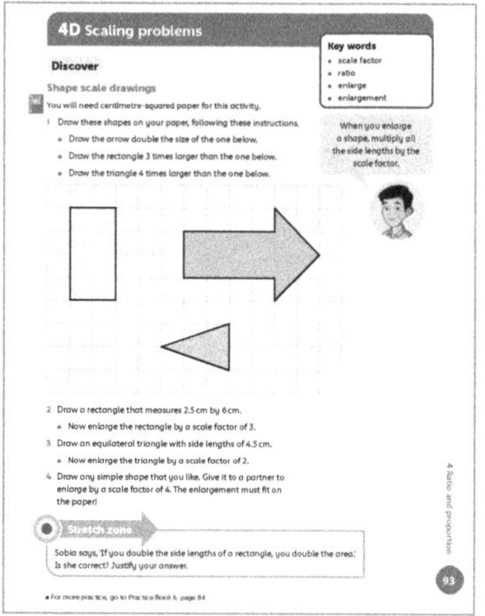

### Differentiation

**Supporting:** Support students to enlarge shapes by a given scale factor.

**Consolidating:** Ask students how they enlarge shapes by a given scale factor.

**Extending:** Encourage students to explore enlargements using different scale factors.

**Stretch zone:** *Sobia says, 'If you double the side lengths of a rectangle, you double the area.' Is she correct? Justify your answer.*

Students describe the relationship between the lengths and the areas of an enlarged shape to prove that Sobia is incorrect. For example, they may use an example of a rectangle measuring 3 cm × 2 cm, with an area of 6 cm$^2$, which is enlarged by a scale factor of 2. This would produce a rectangle with sides of 6 cm × 4 cm, which gives an area of 24 cm$^2$, which is 4 times bigger than the original area, not 2 times bigger.

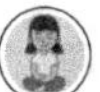 ## Reflection time

Ask students to draw their enlargements on a large grid on the board. Ask them to describe how they drew the enlargements and discuss the connection between the scale factor for lengths and the effect on the areas. *Is there a connection between the area increase and the scale factor?*

**Practice Book:** Students complete Practice Book page 84. They can do this directly after the Main activity, as homework, or as the focus of a separate mathematics session to help students consolidate their learning and build fluency.

Students draw shapes on a square grid and then increase or decrease the shape by a given scale factor. They draw the new shape on the same grid. *Use a ruler to ensure that you measure the side lengths and draw the lines accurately.*

| Differentiated outcomes | |
|---|---|
| **All students** | should enlarge shapes using a scale factor with support. |
| **Most students** | will enlarge shapes using a scale factor. |
| **Some students** | may explore enlargements using a range of scale factors and notice the effect on areas. |

## Answers

### Student Book page 93

Check that students have drawn each of the shapes the correct size according to the scale factors.

### Practice Book page 84

For each shape, check that students have drawn it the right size to begin with, then enlarged it, or reduced it, correctly by the scale factor given.

Stretch zone: Check that students have used the scale factor 1.5 correctly and ask them to explain their strategy.

# 4D Scaling problems

## Explore — Student Book page 94 • Practice Book page 85

### Specific learning focus

- Use scaling to solve recipe problems.

### Global skills

- **Creative skills:** problem solving
- **Real-world skills:** interpreting information

### Key vocabulary

- scale factor, ratio

### Resources

- mini whiteboards and markers
- recipes in books, magazines or online
- different-coloured counters
- sheets of A3 paper

### Language support

Remind students of the language of ratio, for example:

- Ratio is …
- What does a ratio of 1 : 5 mean?
- The ratio of onions to tins of tomatoes is 6 : 3.

 ## Introductory activity

Introduce students to this problem:

*Sue wants to make a cake for her son, Conor. The recipe includes 4 eggs. However, Conor cannot eat eggs, so Sue buys a packet of egg substitute made from soya.*
*On the packet, it states that the whole packet contains the equivalent of 8 eggs and weighs 490 g. How many grams of egg substitute should Sue use in the recipe?*

Ask students to discuss in pairs how to work out how much egg substitute is needed, noting ideas and answers on their whiteboards. Ask them to share their reasoning and answers. Write on the board the ratios 8 : 4 = 490 : ?

Students should be able to see that the number of eggs needed is half of the packet, so half of 490 g = 245 g.

 ## Main activity

Now extend the problem from the Introductory activity by changing the number of eggs. Suppose an omelette needs 3 eggs. How much egg substitute is needed?

Explain that finding the grams equivalent for 1 egg will help. If 490 g = 8 eggs, we can scale down to 1 egg, which will be $490 \div 8 = 61.25$ g. Multiply by 3 to scale up to the grams equivalent for 3 eggs: $61.25$ g $\times 3 = 183.75$ g.

Look together at page 94 of the Student Book. Display on the IWB, if possible. Introduce the recipe scaling problem. Explain that students should use ratios and scale factors to calculate the amounts needed for different numbers of people. For example, if the recipe for 12 people needs 6 onions, the ratio of people : onions is 12 : 6 or 2 : 1, so, for example, a recipe for 16 people would need 8 onions, or a recipe for 6 people would need 3 onions.

**Unit 4 Ratio and proportion** 123

Students should now complete the activities on page 94 using the recipe and their knowledge of ratios and scale factors.

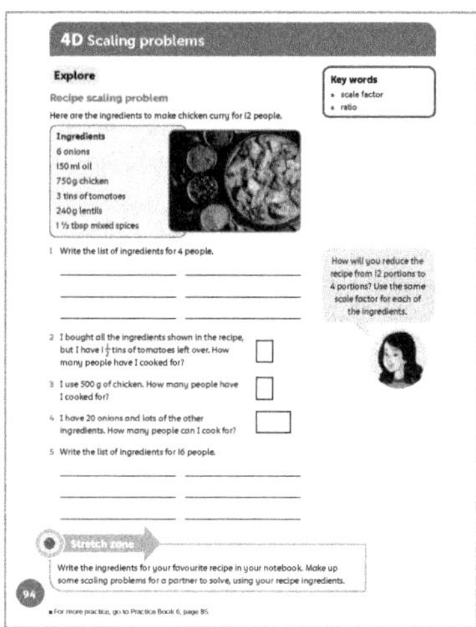

## Differentiation

**Supporting:** Help students to use different-coloured counters to model the ratios needed in the recipe.

**Consolidating:** Encourage students to use drawings to model ratios.

**Extending:** Challenge students to use mental methods to solve scaling problems.

**Stretch zone:** *Write the ingredients for your favourite recipe in your notebook. Make up some scaling problems for a partner to solve, using your recipe ingredients.*

Students find a recipe in a book, magazine or online and then ask their partner to scale it up or down for a new number of people. They use their knowledge of ratio to work out the amount of each ingredient needed.

 ## Reflection time

Ask different students to share their answers and methods to the recipe problems. Ask them to explain what scale factor they used in each problem and how they used this to calculate the amounts of ingredients.

Set this problem:

*A recipe needs 3 eggs for 8 people. How many eggs are needed for 4 people? How might this be solved practically?*

**Practice Book:** Students complete Practice Book page 85. They can do this directly after the Main activity, as homework, or as the focus of a separate mathematics session to help students consolidate their learning and build fluency.

Students draw a scale drawing of the solar system. The are given the distances from the Sun of each planet and they have to choose their own scale factor by which to reduce the distances.

| Differentiated outcomes | |
| --- | --- |
| **All students** | should solve recipe scaling problems with support. |
| **Most students** | will solve recipe scaling problems, using their knowledge of ratio. |
| **Some students** | may make their own recipe scaling problems that use ratio. |

## Answers

### Student Book page 94

**1**  2 onions

50 ml oil

250 g chicken

1 tin of tomatoes

80 g lentils

$\frac{1}{2}$ tbsp mixed spices

**2**  6 people

**3**  8 people

**4**  40 people

**5**  8 onions

200 ml oil

1 kg chicken

4 tins of tomatoes

320 g lentils

2 tbsp mixed spices

### Practice Book page 85

Students draw a scale diagram of the solar system. Check that they have chosen a suitable scale and drawn the solar system accurately.

Stretch zone: Check that students have chosen a suitable scale for the drawing of their home and drawn accurately to that scale.

# 4 Ratio and proportion

## Connect  Student Book page 95

### Big idea

I can solve everyday problems using scale factors, ratio and proportion. I often use my knowledge of fractions and of multiplication and division.

### Global skills

- **Creative skills:** problem solving
- **Interpersonal skills:** teamwork
- **Self-development skills:** reflecting on learning

### Key vocabulary

- scale drawing, scale factor, ratio

### Resources

- squared paper
- tape measures, rulers

### Language support

Ask questions to check that students understand the concept of proportion and scale factors, for example:

- *How are ratio and scale factors the same? How are they different?*
- *What does a ratio of 1 : 5 mean? What does a scale factor of 1 : 5 mean?*

 **Introductory activity**

Write 'Scale factor' in the middle of the board and draw five arrows coming from it. Ask students what they have been learning about that relates to scale factors (fractions, ratios, proportions). Write these at the end of the arrows. *What do you remember about each of these?* Note their feedback in the appropriate places on the diagram. Ask them to give examples. These might include: 'A ratio of 1 : 3 means that for every 1 part, there are 3 corresponding'.

 **Main activity**

Arrange the class into mixed-attainment groups. Explain that students are going to draw a scale drawing of the classroom. They will need to measure the length and width of objects in the room such as desks and cupboards, then draw a reduced version of them on squared paper.

Look together at page 95 of the Student Book. Display on the IWB, if possible. They need to measure the room and decide on the best scale factor so their drawing fits

on the grid on page 95. As they work, ask them how they chose their scale and how they are working out the size of each item on the drawing.

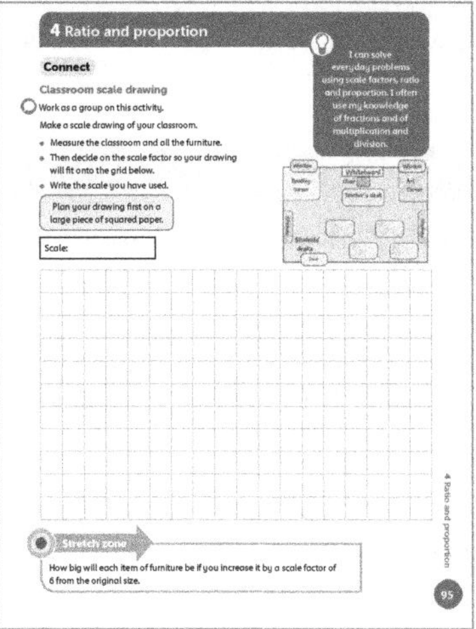

## Differentiation

Students work in mixed-attainment groups, supporting each other. They might distribute tasks by focusing on skills, for example some group members might be better at measuring, others better at drawing.

**Stretch zone:** *How big will each item of furniture be if you increase it by a scale factor of 6 from the original size?*

Students could work out what size sheet of paper they would need for a **scale drawing** of the classroom with everything increased by a scale factor of 6.

 **Reflection time**

Ask each group to give feedback on how they worked to complete their scale drawing. They should describe how they measured and calculated the scale. Compare drawings from different groups. Did they all use the same scale factor or were there differences?

Ask questions about the ratio of sizes of different items. For example, the teacher's desk might be twice as long as a student's desk, so the ratio of lengths is 2 : 1.

| Differentiated outcomes | |
| --- | --- |
| **All students** | should use a scale factor to construct a scale drawing of the classroom using peer support. |
| **Most students** | will use a scale factor to construct a scale drawing of the classroom in a group. |
| **Some students** | may use a scale factor to construct a scale drawing of the classroom, supporting peers in a group. |

## Answers

**Student Book page 95**

Check that students have measured accurately and used the scale factor to draw the items in the classroom correctly.

# 4 Ratio and proportion

## Review  Student Book page 96 · Practice Book page 86

### Global skills

- **Real-world skills:** presenting information/interpreting information
- **Interpersonal skills:** communication
- **Self-development skills:** reflecting on learning

## Student Book

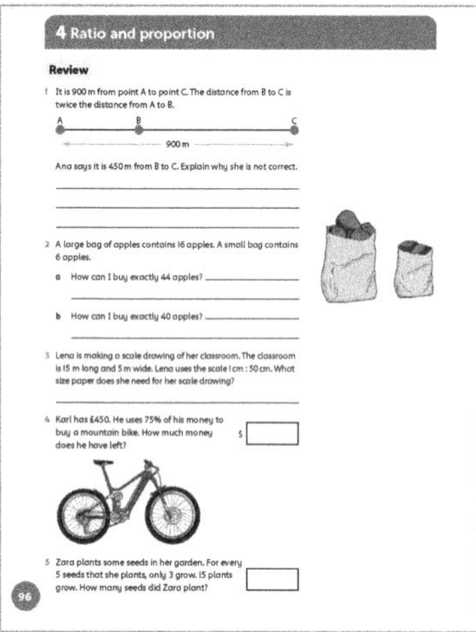

With young students, assessment activities are most effective when carried out as an everyday classroom activity. Students should have cubes and counters available to support them. Watch as students use their understanding of multiplication and division with whole numbers and decimals to calculate ratios and proportions. Support them to understand the difference between a ratio that is 'part to part', or a proportion that compares a part to a whole.

Students should be able to solve problems involving ratio and proportion and draw on their knowledge of factors and multiples. Students should be able to calculate with percentages and scale factors and apply these in problem-solving contexts.

The Review activity is an individual summative assessment and so should be completed by students working on their own.

## Answers

### Student Book page 96

1  If B to C was 450 km, then A to B would be 225, but 450 + 225 = 675, not 900.

2  **a**  2 large bags and 2 small bags

   **b**  1 large bag and 4 small bags

3  At least 30 cm × 10 cm

4  £112.50

5  25 seeds

## Practice Book

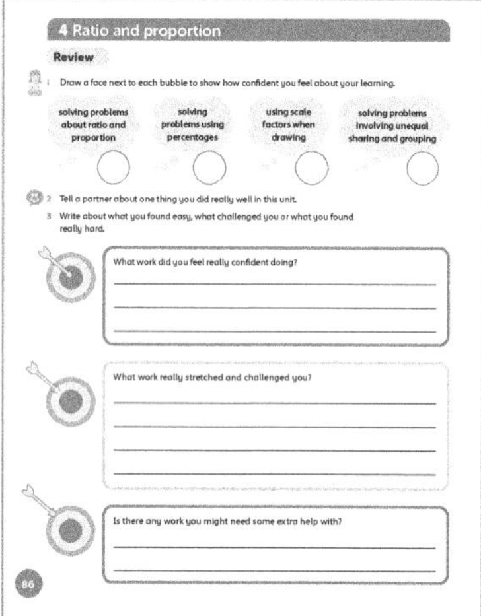

It is appropriate to complete this Practice Book Review as a whole-class discussion. You may choose to keep a record of the class discussion or a copy of the Review page for your own records. The Review provides an opportunity for students to reflect on their learning from the unit, to discuss any areas of mathematics that they feel went particularly well, and any areas that they feel less confident about. Ensure that all students have a copy of the Student Book as a reminder of the areas of mathematics that they have worked on in this unit.

Allow students plenty of time for discussion before asking them to complete the Practice Book page individually, and then, if appropriate, to share their responses with the rest of the class. If students complete this self-assessment at home, encourage them to discuss this with adults. Make a note of areas that students still feel unsure about.

## Additional material

There are additional end-of-unit assessments available on the *Oxford Owl for School* website.

# 5 Algebra

## Overview

### Big idea

In this unit, number concepts begin to be formalised by discussing number patterns and sequences and describing the rules that connect the numbers in a sequence. The rules are described using a combination of numbers and letters. Linear sequences arise from building patterns with shapes or by applying a rule to a number and repeating that rule to make further numbers. These numbers in a sequence are called 'terms'. Students will experience sequences, which they can use to predict later terms or find a rule that generates the terms.

Exploration of sequences of perimeters or areas, for example, will lead to early recording of algebraic formulae and the notion of variables. Once students are confident in reading and using formulae, these can be used to solve a range of problems where missing values are required.

### Look out for

- **Students who do not have the necessary language to describe situations algebraically.** Set problems in a 'story' context to allow students to use available language before introducing more formal algebraic terms such as variables and constants. For example, a cup of coffee and a cake together cost \$4.50 and two cups of coffee cost \$6. Write this as $f + k = 4.5$ and $2f = 6$ where $f$ is the price of a coffee and $k$ is the price of a cake.

- **Students who cannot continue a number sequence using the rule that created it.** Relate number sequences to real-life scenarios, perhaps increasing amounts of money, or sizes of a growing pattern of shapes, so that the sequences are within students' experience and not abstract.

### Possible misconceptions

- **Students identify a general expression, but are not precise when writing it down.** For example, in finding an expression for the perimeter of a rectangle of sides $p$ and $q$, students may write it as $p + q + p + q$ and not recognise that this can be simplified as $2(p + q)$. Remind students of the laws of arithmetic to help them write expressions more concisely.

- **Students confuse variables with objects.** For example, if a school has 20 times more students than teachers, this could be represented by $S = 20T$. Students may think that $S$ represents Students, rather than the number of students. Help students to understand that variables represent an amount of items, not the item itself.

### Key vocabulary

- formula/formulae, sequence, linear number sequence
- area, volume, base, height, distance, speed, time
- equation, variable, constant, unknown, patterns
- rule, term, position-to-term rule, term-to-term rule
- predict, generate, algebra, missing number, substitute, solve
- kilometres per hour (kph)

### Coverage in lessons

| Learning objective | E | 5A | 5B | 5C | 5D | 5E | C | R |
|---|---|---|---|---|---|---|---|---|
| Use simple formulae. | ✓ | | ✓ | ✓ | | ✓ | ✓ | ✓ |
| Generate and describe linear number sequences. | | ✓ | ✓ | | | | ✓ | ✓ |
| Express missing number problems algebraically. | ✓ | | | ✓ | ✓ | ✓ | ✓ | ✓ |
| Find pairs of numbers that satisfy an equation with two unknowns. | | | | | ✓ | | | ✓ |
| Enumerate possibilities of combinations of two variables. | | | | | ✓ | | | ✓ |

# 5 Algebra

## Engage  Student Book page 97

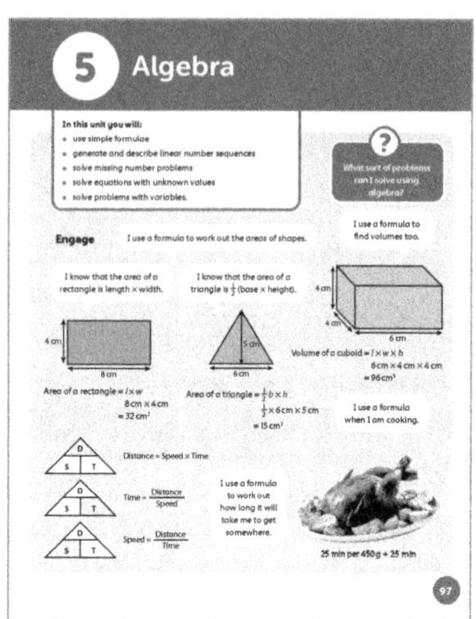

## Big question

- What sort of problems can I solve using algebra?

## Global skills

- **Creative skills:** exploring
- **Real-world skills:** research
- **Interpersonal skills:** communication
- **Self-development skills:** reflecting on learning

## Key vocabulary

- formula, area, volume, base, height, distance, speed, time, variable, constant

## Resources

- mini whiteboards and markers
- sets of squares and equilateral triangles
- large sheets of paper

## Language support

Model for students how to 'say' a formula to help with understanding how to use it when calculating missing numbers or amounts. For example:

Area of a triangle $= \frac{1}{2} b \times h$, say this as 'half the base times the height'.

Speed $= d \div t$, say this as 'speed is distance divided by time'.

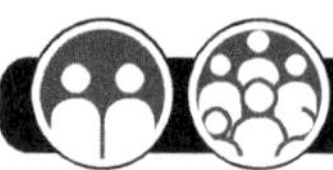 Introductory activity

Students work in pairs. Give pairs a small pile of squares and equilateral triangles. Ask them to take a square and say how many sides it has. Now take another square. *How many sides do two squares have altogether?* Repeat

for three squares, then four up to ten. *What is the relationship between the number of squares and the total number of sides?* Agree that for any number of squares, the number of sides is 4 times as many.

Now measure one side of a square (perhaps it is 5 cm). *What is the perimeter of the square? (20 cm). What would be the total perimeter of two squares? What would it be for three squares? What about ten squares?* Draw a table to represent the relationship between the number of squares and the total perimeter:

| Number of squares | Total perimeter |
|---|---|
| 1 | 20 cm |
| 2 | 40 cm |
| 3 | 60 cm |
| 4 | 80 cm |
| 5 | 100 cm |
| 10 | 200 cm |
| 20 | 400 cm |

Repeat with the equilateral triangles. Agree that for any number of triangles, the number of sides is 3 times as many. Say each side of the triangle is 4 cm. Ask students, using their whiteboards, to find a relationship for the total perimeter of any number of triangles. Ask a student to come to the front and complete a table to show this.

| Number of triangles ($t$) | Total perimeter ($p$) |
|---|---|
| 1 | 12 cm |
| 2 | 24 cm |
| 3 | 36 cm |
| 4 | 48 cm |
| 5 | 60 cm |
| 10 | 120 cm |
| 20 | 240 cm |

Explain that we can use algebra to generalise the rule. For example, if $p =$ perimeter and $t =$ the number of triangles, we can generalise the rule for finding the perimeter of any number of triangles with side lengths of 4 cm as: $p = 12t$.

 Main activity

Look together at page 97 of the Student Book. Display on the IWB, if possible. Ask students to discuss in pairs what the page is showing, then have a class discussion about what a formula is and when we use formulae. Point out how we can use algebra in each of the formulae to represent numbers we don't know. For example, if we don't know the side lengths of the rectangle, we can just say that the perimeter is $2(l + w)$ or that the area is $l \times w$.

Introduce students to the language of formulae, including 'variable' and 'constant'. Ask them whether they can think of any formulae they have seen in different situations. In small groups, students can make posters showing what they know about formulae, including anything they have learned from page 97.

## Differentiation

**Supporting:** Model the area relationship for rectangles using square tiles.

**Consolidating:** Ask students to change some of the measures in the formulae for shapes or letters and calculate the new answers.

**Extending:** Challenge students to write problems that use one of the formulae and swap with a partner to solve them.

 **Reflection time**

Ask groups to share their posters of formulae. Ask them to explain some of their formulae and how the formulae describe the relationship between the quantities. Before explaining their formulae, they should write on their whiteboards examples to use for each one they share.

# 5A Number sequences

## Discover   Student Book page 98 • Practice Book page 87

### Specific learning focus

- Recognise and extend number sequences.

### Global skills

- **Creative skills:** investigating

### Key vocabulary

- linear number sequence, pattern, terms, term-to-term rule

### Resources

- none needed

### Language support

Support students' understanding of the key language of sequences by asking questions and emphasising the key vocabulary, for example:

- *Can you extend the sequence?*
- *What number pattern can you see in the sequence?*
- *What do you think the next term is?*
- *Can you predict what the tenth term will be?*

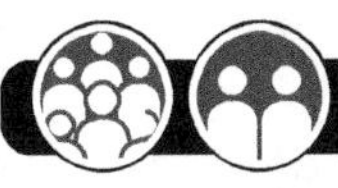 **Introductory activity**

Write on the board the first few numbers of a **linear number sequence**. Explain that numbers (or items) in a sequence are called **terms** and that we refer to them by their position in the sequence. For example, in this sequence, the first term is 1, the second term is 3, the fifth term is 9 and so on:

1, 3, 5, 7, 9, …

Ask students to work in pairs. *Can you tell me the next number in this sequence?*

Use the following sequences. Write the first two terms, then add the next terms one at a time until students can predict the next term in the sequence, and explain the

pattern of numbers, or the **term-to-term rule** for the sequence, for example +2 or +3.

- 1, 3, 5, 7, 9, …
- 3, 6, 9, 12, 15, …
- 2, 5, 8, 11, 14,
- 17, 11, 5, −1, −7, …

 **Main activity**

Ask students to write down their own number sequences. Ask one student to come to the front of the class and share their sequence. Whoever gives the correct term-to-term rule for the sequence can come to the front of the class and share their sequence.

Look together at page 98 of the Student Book. Display on the IWB, if possible. Refer students to the Think back definition. As students recall the meaning of 'perimeter', revise the names and side properties of some regular polygons, for example a pentagon has 5 equal sides.

Ask students to work in pairs on the activities on page 98 of the Student Book. Students will benefit from working together. Ask pairs to explain their methods to you. Look for good explanations that can be shared later. This helps you to check which pairs will be able to share their ideas with the whole class during Reflection time.

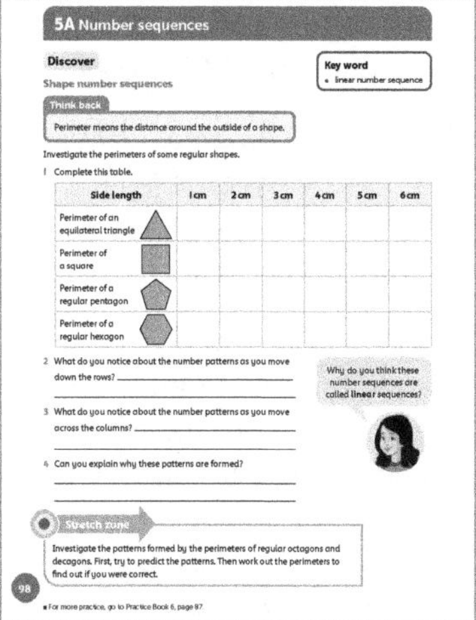

## Differentiation

**Supporting:** Support students in finding the patterns and rules for the sequences.

**Consolidating:** Ask students to tell you their thinking when they are finding the terms in a sequence.

**Extending:** Encourage students to generate their own sequences.

**Stretch zone:** *Investigate the patterns formed by the perimeters of regular octagons and decagons. First, try to predict the patterns. Then work out the perimeters to find out if you were correct.*

Encourage students to describe the patterns in the sequences of perimeters using algebra, if they can. For example, for regular decagons encourage them to say that, $p = 10d$.

### Reflection time

Ask students to share their findings from the shape sequences. *What patterns did you notice in each sequence? Could you predict the next term in each sequence? Can you describe the term-to-term rule for each sequence?*

Look together at the pattern of numbers in the sequence for the perimeter of a regular pentagon. *What was the sequence of perimeters?* Draw a table to show this:

| Side length | Perimeter |
|---|---|
| 1 cm | 5 cm |
| 2 cm | 10 cm |
| 3 cm | 15 cm |
| 4 cm | 20 cm |
| 5 cm | 25 cm |
| 6 cm | 30 cm |
| $s$ cm | $5s$ cm |

Explain to students how to use the table to relate the side length to the perimeter. For example, the second pentagon has perimeter 2 cm × 5, or the sixth pentagon has perimeter 6 cm × 5. For any pentagon in the sequence, its side length can be used to calculate the perimeter by multiplying the side length by 5. We can describe this using algebra by giving the side length a letter (let's use 's'). Any regular pentagon, with any side length, will have a perimeter of $5s$, so we can say $p = 5s$.

**Practice Book:** Students complete Practice Book page 87. They can do this directly after the Main activity, as homework, or as the focus of a separate mathematics session to help students consolidate their learning and build fluency.

Students create and tabulate number sequences based on the perimeter of rectangles with different side lengths, which increase to form a linear number sequence.

| Differentiated outcomes | |
|---|---|
| **All students** | should find the rules for the sequences with support. |
| **Most students** | will find the rules for the sequences independently. |
| **Some students** | may generate their own sequences. |





### Student Book page 98

1

| Side length | 1 cm | 2 cm | 3 cm | 4 cm | 5 cm | 6 cm |
|---|---|---|---|---|---|---|
| Perimeter of an equilateral triangle | 3 cm | 6 cm | 9 cm | 12 cm | 15 cm | 18 cm |
| Perimeter of a square | 4 cm | 8 cm | 12 cm | 16 cm | 20 cm | 24 cm |
| Perimeter of a regular pentagon | 5 cm | 10 cm | 15 cm | 20 cm | 25 cm | 30 cm |
| Perimeter of a regular hexagon | 6 cm | 12 cm | 18 cm | 24 cm | 30 cm | 36 cm |

2  The sequences match the times tables for 3, 4, 5 and 6. The times table matches the number of sides that each shape has.

3  The sequences increase by the same amount from one term to the next.

4  The side lengths and the number of sides are increasing by one in each direction.

### Practice Book page 87

1

| Dimensions of rectangle | Perimeter of rectangle |
|---|---|
| 1 × 2 | 6 |
| 1 × 3 | 8 |
| 1 × 4 | 10 |
| 1 × 5 | 12 |
| 1 × 6 | 14 |

2

| Dimensions of rectangle | Perimeter of rectangle |
|---|---|
| 2 × 2 | 8 |
| 2 × 3 | 10 |
| 2 × 4 | 12 |
| 2 × 5 | 14 |
| 2 × 6 | 16 |

3

| Dimensions of rectangle | Perimeter of rectangle |
|---|---|
| 3 × 2 | 10 |
| 3 × 3 | 12 |
| 3 × 4 | 14 |
| 3 × 5 | 16 |
| 3 × 6 | 18 |

4

| Dimensions of rectangle | Perimeter of rectangle |
|---|---|
| 4 × 2 | 12 |
| 4 × 3 | 14 |
| 4 × 4 | 16 |
| 4 × 5 | 18 |
| 4 × 6 | 20 |

**5**

| Dimensions of rectangle | Perimeter of rectangle |
|---|---|
| 5 × 2 | 14 |
| 5 × 3 | 16 |
| 5 × 4 | 18 |
| 5 × 5 | 20 |
| 5 × 6 | 22 |

Stretch zone: The numbers increase by 2 each time in the previous tables, so the perimeters will be:

| Dimensions of rectangle | Perimeter of rectangle |
|---|---|
| 6 × 2 | 16 |
| 6 × 3 | 18 |
| 6 × 4 | 20 |
| 6 × 5 | 22 |
| 6 × 6 | 24 |

# 5A Number sequences

## Explore
Student Book pages 99–100 • Practice Book page 88

### Specific learning focus

- Make general statements about number sequences.

### Global skills

- **Creative skills:** investigating

### Key vocabulary

- linear number sequence, rule, term, position-to-term rule

### Resources

- counting stick
- mini whiteboards and markers

### Language support

- Support students' understanding of the key language of number sequences by asking questions and emphasising the key vocabulary, for example: *What is the rule? What is the difference between each term in the sequence?*

 **Introductory activity**

Hold up the counting stick and tell students that one end is the number 3. Say that each step along the counting stick is going to add 7. *The term-to-term rule is add 7.* Count together: 3, 10, 17, 24, 31, 38, …

Now change the starting number to 5 but still add 7 each time: 5, 12, 19, 26, 33, …

This time, change the start number to 60 but subtract 7: 60, 53, 46, 39, 32, …

Explain that a sequence where the same number is added or subtracted each time is called a linear sequence. It has a constant term-to-term **rule**.

Show how to use algebra to describe any term of the sequence. For example in the first sequence 3, 10, 17, 24 …, if we write these terms below their position in the sequence in a table:

| Position (n) | 1 | 2 | 3 | 4 | 5 | 6 | 7 |
|---|---|---|---|---|---|---|---|
| Term (t) | 3 | 10 | 17 | 24 | 31 | 38 | 45 |

the rule can be written as $t = 7n - 4$, where $t$ is the term in the sequence and $n$ is the position of the term in the sequence. We call this the **position-to-term rule**.

Note: a good way to work out the position-to-term rule is first to find the term-to-term rule, and then see how much is added or subtracted from the term-to-term rule to reach the first term answer. For example, using the sequence 5, 12, 19, 26 …, the term-to-term rule is +7 and the first term is 5, which is two less than 7, so the position-to-term rule is $7n - 2$.

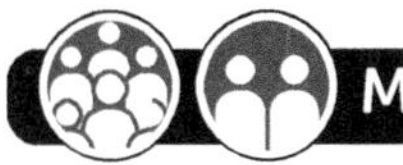 **Main activity**

Write the following sequence on the board:

1.1, 2.2, 3.3, 4.4, 5.5, …

Ask students whether they can tell you the rule for the sequence. *How much is added each time?* (1.1) *Is this a linear sequence?* (yes) *Can you describe the 'position-to-term' rule for this sequence?* ($t = 1.1n$)

Now write: 12, $8\frac{1}{2}$, 5, $1\frac{1}{2}$, $-2$, $-5\frac{1}{2}$, … . Again, ask students whether they can describe the term-to-term rule ($-3\frac{1}{2}$), whether this is a linear sequence (yes) and the position-to-term rule ($t = -3.5n + 15.5$).

Ask students to work in pairs. Each student makes up a linear sequence, then they swap sequences and try to work out their partner's term-to-term rule and then the position-to-term rule.

Look together at pages 99–100 of the Student Book. Display on the IWB, if possible. Ask students to work in pairs on the activities. They write the term-to-term rule for each sequence, as shown in the worked example on page 99. Challenge them to also write the position-to-term rule for each sequence.

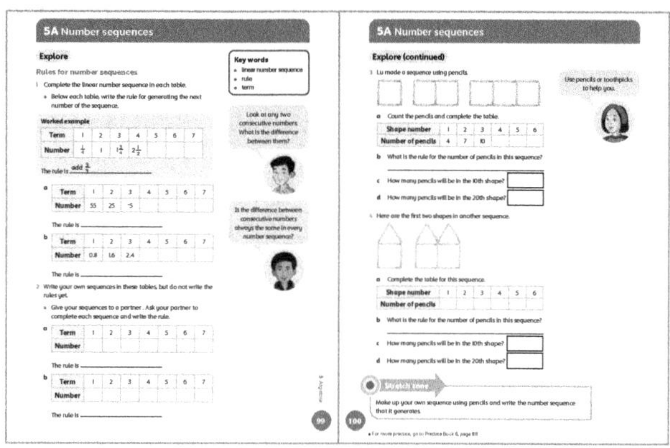

## Differentiation

**Supporting:** Help students to identify the rules for linear sequences.

**Consolidating:** Ask students to explain how they found the rules for linear sequences.

**Extending:** Challenge students to create an easy linear sequence and a hard linear sequence for a partner to find the rules.

**Stretch zone:** *Make up your own sequence using pencils and write the number sequence that it generates.*

Students can draw pictures on their whiteboards to try to prove some of the generalisations they have made.

 **Reflection time**

Write the following sequence on the board and ask students to say whether it is a linear sequence and if so, what the term-to-term rule is, and what the position-to-term rule is:

4, 25, 46, 67, 88, 109, …

Share some of the sequences made for the Stretch zone activity. Students can either show the sequences as drawings on their whiteboards or give the numbers. The rest of the class can try to work out the rule.

**Practice Book:** Students complete Practice Book page 88. They can do this directly after the Main activity, as homework, or as the focus of a separate mathematics session to help students consolidate their learning and build fluency.

Students write their own number sequences by writing a start number and a rule and then completing the sequence. Encourage them to use both positive and negative numbers in their sequences. *Can you tell me the term-to-term rule for this sequence? What about the position-to-term rule?*

## Differentiated outcomes

| Differentiated outcomes | |
| --- | --- |
| **All students** | should work out the rules for a linear sequence with support. |
| **Most students** | will work out the rules for a linear sequence. |
| **Some students** | may create linear sequences using shape patterns or rules. |

## Answers

### Student Book pages 99–100

**1 a** 55, 25, ⁻5, ⁻35, ⁻65, ⁻95, ⁻125, the term-to-term rule is subtract 30

The position-to-term rule is $^-30n + 85$

**b** 0.8, 1.6, 2.4, 3.2, 4, 4.8, 5.6, the term-to-term rule is add 0.8

The position-to-term rule is $0.8n$

**2** Students make their own sequence for a partner to work out. Check that the sequence is linear.

**3 a** 4, 7, 10, 13, 16, 19

**b** Add 3

**c** 31 ($p = 3n + 1$)

**d** 61

**4 a** 6, 11, 16, 21, 26, 31

**b** Add 5

**c** 51 ($p = 5n + 1$)

**d** 101

### Practice Book page 88

Students choose their own start numbers and rules. Check that their sequences are completed correctly and that they have made appropriate comments about what they notice.

Stretch zone: 2, 9, 16, 23, 30

Rule: +7 and $7n - 5$

# 5B Using a formula

## Discover
Student Book page 101 • Practice Book page 89

### Specific learning focus

- Explore sequences using the formulae for perimeter and area of shapes.

### Global skills

- **Creative skills:** investigating

### Key vocabulary

- formula/formulae, unknown, variable, constant

### Resources

- mini whiteboards and markers

### Language support

Support students' understanding of the key language of formulae by asking questions and emphasising the key vocabulary, for example:

- *What are the variables in the formula?*
- *Are there any constants in the formula?*
- *Can you use the formula to predict what the hundredth term will be?*

 **Introductory activity**

Ask students to draw on their whiteboards three different rectangles and label the lengths of their sides. Now ask them to recall how to find the area and perimeter of their rectangles. They should remember that the perimeter is the sum of the side lengths, and that the area is found by multiplying the length by the width.

Ask them to work out the perimeter and area of their three rectangles. *Did any of the perimeters or areas work out to be equal? Does the rectangle with the largest area have the largest perimeter? What about the rectangle with the smallest area? Is there a relationship between the area and perimeter of a rectangle?*

 **Main activity**

Look together at page 101 of the Student Book. Display on the IWB, if possible. Ask students to look at the Think back example at the top of the page.

Ask students to notice that for any rectangle where we don't know the length of the sides, we can label the length as $l$ and the width as $w$. Point out that the calculation for the perimeter (as shown in the Think back) becomes:

$w + l + w + l$, which is the same as $w + w + l + l$.

Ask students why these are the same; they might recall the commutative rule for addition.

Talk through and explain the idea of **variables** being numbers (in this case, side lengths) that can change, and that the length and width of a rectangle is something that can vary because not all rectangles are the same size (refer them to the Introductory activity). Look in the Think back example at how the variables ($l$ and $w$) have been simplified to $2w + 2l$, which is the same as $2(w + l)$, giving the perimeter formula as $P = 2(w + l)$. Explain that the 2 is a **constant** as it never changes in this formula, as there are always 2 sides for the length and 2 sides for the width of a rectangle.

Now ask students to complete the activities on Student Book page 101.

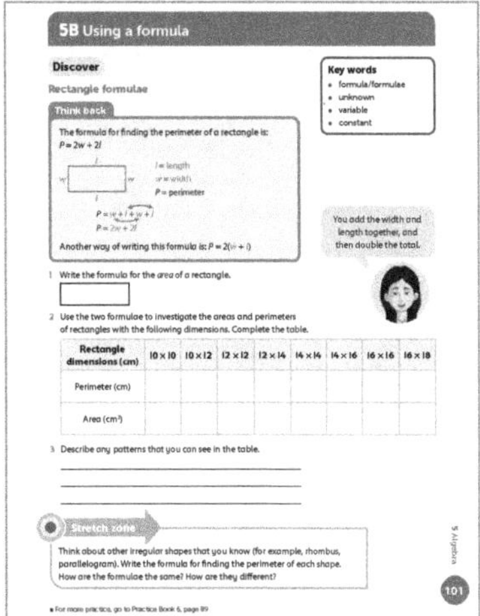

## Differentiation

**Supporting:** Help students to identify the variables in the formula by showing that they can be different values in different rectangles.

**Consolidating:** Ask students to explain how they used the formula to calculate the perimeter of different rectangles.

**Extending:** Challenge students to use the formula for rectangles with sides that are very large, or fractions or decimals.

**Stretch zone:** *Think about other irregular shapes that you know (for example rhombus, parallelogram). Write the formula for finding the perimeter of each shape. How are the formulae the same? How are they different?*

Students might need to research the formulae for some shapes using the internet.

## Reflection time

Ask students to share the results from the activities. *Can you describe the formula for the area of a rectangle? Does this formula contain variables or constants?* Ask students what patterns they noticed in question 2. *Did you notice any linear sequences?*

**Practice Book:** Students complete Practice Book page 89. They can do this directly after the Main activity, as homework, or as the focus of a separate mathematics session to help students consolidate their learning and build fluency.

Students are introduced to the formula for finding the area of a triangle. They then use this formula to work out the area of triangles that have areas that form linear sequences. They write about what they notice.

| Differentiated outcomes | |
|---|---|
| **All students** | should use the formulae to calculate rectangle perimeter and area with support. |
| **Most students** | will use the formulae to calculate rectangle perimeter and area. |
| **Some students** | may find formulae to calculate areas and perimeters of other shapes. |



### Student Book page 101

**1** Area of a triangle = length $\times$ width = $l \times w$.

**2**

| Rectangle dimensions | Perimeter (cm) | Area (cm²) |
|---|---|---|
| **10 × 10** | 40 | 100 |
| **10 × 12** | 44 | 120 |
| **12 × 12** | 48 | 144 |
| **12 × 14** | 52 | 168 |
| **14 × 14** | 56 | 196 |
| **14 × 16** | 60 | 224 |
| **16 × 16** | 64 | 256 |
| **16 × 18** | 68 | 288 |

One the sides of the rectangle increases by 2 cm each time, and the perimeters increase by 4 cm each time.

### Practice Book page 89

**1** Areas: 2 cm², 3 cm², 4 cm², 5 cm², 6 cm² – they increase by 1 cm² each time.

**2** Areas: 3 cm², 4.5 cm², 6 cm², 7.5 cm², 9 cm² – they increase by 1.5 cm² each time.

**3** Areas: 4 cm², 6 cm², 8 cm², 10 cm², 12 cm² – they increase by 2 cm² each time.

Stretch zone: Check that students have created a spreadsheet that will calculate triangle areas when the base and height are entered.

# 5B Using a formula

## Explore
Student Book page 102 • Practice Book page 90

### Specific learning focus

- Explore sequences using the formulae for distance, time and speed.

### Global skills

- **Creative skills:** investigating

### Key vocabulary

- formula/formulae, unknown, variable, constant, kilometres per hour (kph)

### Resources

- mini whiteboards and markers



Remind students of the key components in a formula, by asking, for example:

- *What are the variables in the formula?*
- *Are there any constants in the formula?*

Set students this problem:

*A bus travels for 3 hours and covers 90 kilometres. If it travelled at the same speed for the whole 3 hours, at what speed was it travelling in kilometres per hour?*

Students might refer to their knowledge on ratios to help them with this; the ratio of distance : time = 90 : 3 = 30 : 1, or 30 **kilometres per hour** (30 kph).

Now ask, *If the bus travelled at that speed for 5 hours, how far would it travel?* Students should discuss with a partner and note their ideas and answer on their whiteboards. Ask pairs to share to their answers and reasoning. They should work out that travelling for 5 hours at 30 kph will cover 150 kilometres.

# Main activity

Look together at page 102 of the Student Book. Display on the IWB, if possible. Point out the diagram at the top of the page. Relate the triangles to the problems discussed in the Introductory activity. Talk through each triangle in turn and use an example for each. The first triangle is used for the bus distance question from the Introductory activity, and the third triangle is the speed per hour from the first bus problem.

For the middle triangle, use this problem:

*How long will it take a train to travel 280 kilometres at a speed of 70 kph?*

Explain that the triangles are used to help students to remember which operation to use. When finding a distance, multiply the speed and time (they are beside each other in the triangle). When finding a speed, the triangle shows distance is over time, so divide distance by time. When finding a time, the triangle shows distance is over speed, so divide distance by speed, as shown in the speech bubble. *Which formula will you use to work out how long it will take a train to travel 280 kilometres at a speed of 70 kph?* Agree on the middle formula shown in the Student Book and work out the answer:

$280 \div 70 = 4$, so it will take the train 4 hours.

Ask students to work on the activities on page 102. They should use the formulae triangles to help them.

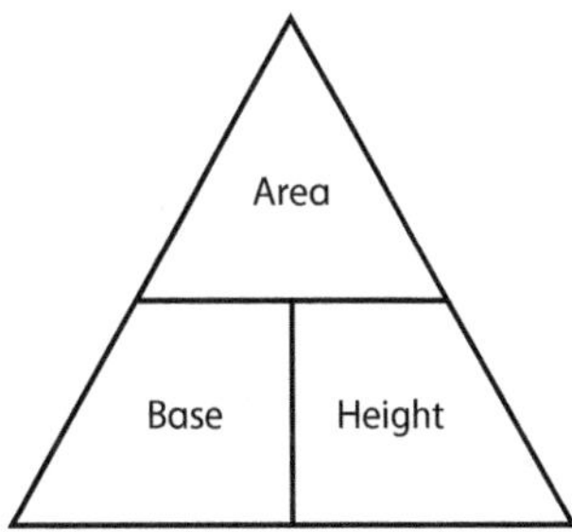

## Differentiation

**Supporting:** Help students to identify which version of the formulae triangles to use for each problem.

**Consolidating:** Ask students to explain to you how they used the formulae triangles to solve the problems.

**Extending:** Challenge students to explain how to use the formulae to other students.

**Stretch zone:** *What do you notice about the patterns in the answers?*

Students comment on any patterns they recognise from using the formulae triangles. For example, the faster the speed, the further the distance is covered in the same amount of time; the faster the speed, the less time it takes to cover the same distance.

##  Reflection time

Ask students to share any of the questions they found particularly challenging and discuss them as a class. Ask different students to share their explanations to help each other understand the formulae and the triangles.

Share this problem with the class, which uses a similar formula triangle:

*A parallelogram has an area of 84 m². If it has a base of 12 m, what is the height of the parallelogram?*

Share solutions and ask students to explain the method. They should be able to see from the triangle that the height = area ÷ base.

**Practice Book:** Students complete Practice Book page 90. They can do this directly after the Main activity, as homework, or as the focus of a separate mathematics session to help students consolidate their learning and build fluency.

Students write the output of various function machines after they are given the input number and the formula (function) of the machines.

| Differentiated outcomes | |
| --- | --- |
| **All students** | should use the distance–speed–time formula with support. |
| **Most students** | will use the distance–speed–time formula. |
| **Some students** | may use the distance–speed–time formula and other similar formulae to solve a range of problems. |

# Answers

## Student Book page 102

**1** 10 km

**2** 30 km

**3** 90 km

**4** 2 hours

**5** $\frac{2}{3}$ hour or 40 minutes

**6** $\frac{1}{6}$ hour or 10 minutes

**7** 4 kph

**8** 20 kph

**9** 50 kph

## Practice Book page 90

**1** Outputs: 3, 5, 7, 9, 11, 13

**2** Outputs: 6, 9, 12, 15, 18, 21

**3** Outputs: 9, 13, 17, 21, 25, 29

**4** Outputs: 12, 17, 22, 27, 32, 37

Stretch zone: The outputs for machine 4 will increase by one more each time than machine 3, for example, 9 to 12 is 3, 13 to 17 is 4 and so on.

---

# 5C Missing number problems

## Discover   Student Book page 103 · Practice Book page 91

### Specific learning focus

- Use algebra to find unknown numbers.

### Global skills

- **Creative skills:** problem solving

### Key vocabulary

- equation, missing number, unknown

### Resources

- mini whiteboards and markers

### Language support

Model the language associated with equations and missing numbers. For example, point out that saying $4c = 12$ means that 4 times the *value* of c makes 12, so the value of c must be $12 \div 4 = 3$.

## Introductory activity

Explain that you are thinking of a number and you will give students clues to help them work out what your number is. *4 times my number is 32. What is my number?* Students should quickly work out that your number is 8. You can write this on the board in the form $4 \times \square = 32$

Now, explain that you are thinking of a new number. *This time, if I multiply my number by 3 and then add 2, I get 20. What is my number?* (6) Write this as : $\square \times 3 + 2 = 20$

Then choose another number. *This time, multiplying my number by 3 and adding 2 gives 50. What is my number?* (16) Write this as: $\square \times 3 + 2 = 50$

Finally, say that we can replace the missing number box with a letter (choose any letter), for example $a \times 3 + 2 = 50$. What is value of *a*?

## Main activity

Draw on the board a row of three triangles. *Each triangle is worth the same number, but we don't know what that number is yet.* Now write next to the triangles on the board: = 15. *If the total value of the three triangles is 15, then how much each triangle is worth?* Students should be able to divide 15 by 3 and say that each triangle is worth 5.

Now explain that the value of the triangles has changed and draw a fourth triangle. Change the total value of the triangles to 28 and ask students to work out the new value of each triangle.

Look together at page 103 of the Student Book. Display on the IWB, if possible. Ask students to look at question 1. Point out the first speech bubble and explain how the shapes can be represented by letters in an equation, just as we replaced the empty number box with a letter in the Introductory activity. Using the same approach as finding the **missing number** from the Introductory activity, the values of the shapes or letters can be calculated.

Students should work in pairs to complete the activities on page 103.

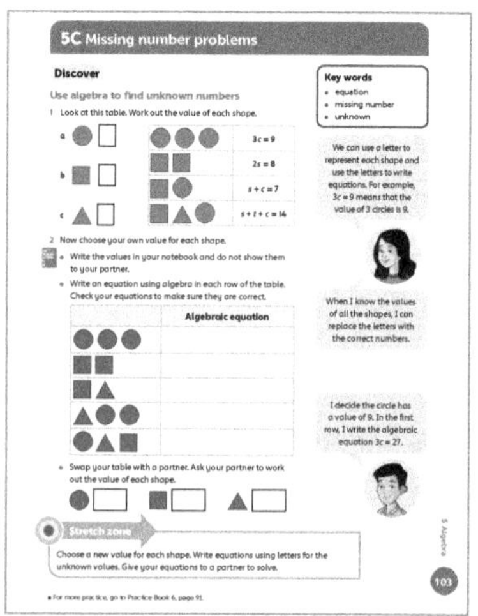

## Differentiation

**Supporting:** Help students relate the shapes and letters to help them find the values of the shapes.

**Consolidating:** Ask students to explain how they found the shape values.

**Extending:** Challenge students to make up an equation for a partner to solve using letters to represent missing numbers.

**Stretch zone:** *Choose a new value for each shape. Write equations using letters for the **unknown** values. Give your equations to a partner to solve.*

For extra challenge, encourage students to use numbers that are not whole numbers.

## Reflection time

Discuss the solutions to the equations as a class. Ask pairs of students to say something about how they found out the value of each shape or letter. Some students will be able to make statements comparing shapes to each other, for example the circle is worth one less than the square.

Students who made up their own equations in the Stretch zone activity can share them for the class to try.

**Practice Book:** Students complete Practice Book page 91. They can do this directly after the Main activity, as homework, or as the focus of a separate mathematics session to help students consolidate their learning and build fluency.

Students use the information provided to work out the value of individual shapes and then work out the value of a combination of different shapes.

Encourage students to write on their whiteboards the combinations of shapes using letters and symbols. This will help them to deduce the value of each individual shape. For example, encourage them to write $3s = 12$, $t + 2c = 11$, $c + t + s = 12$.

| Differentiated outcomes | |
| --- | --- |
| **All students** | should solve simple missing number equations with support. |
| **Most students** | will solve simple missing number equations. |
| **Some students** | may solve missing number equations and make up their own missing number problems for others to solve. |

## Answers

### Student Book page 103

**1 a** 3      **b** 4      **c** 7

**2** Students choose their own values. Check that their equations are correct.

### Practice Book page 91

| | | |
| --- | --- | --- |
| **1** 12 | **4** 4 | **7** 2 |
| **2** 9 | **5** 5 | **8** 3 |
| **3** 3 | **6** 8 | |

Stretch zone: Students make their own puzzles. Check that the equations and the values are correct.

---

# 5C Missing number problems

**Explore**   Student Book page 104 • Practice Book page 92

### Specific learning focus

- Use algebra to solve problems.

### Global skills

- **Creative skills:** problem solving
- **Real-world skills:** interpreting information

### Key vocabulary

- equation, unknown, solve

### Resources

- mini whiteboards and markers

## Language support

Use a range of reasoning language to help students understand the thinking.

For example, use 'if' and 'then' when working on problems: *If a magazine and book together cost $9 and the book costs $6.45, then the magazine must cost $2.55.*

Encourage students to use 'if' and 'then' when solving problems.

##  Introductory activity

Look together at page 104 of the Student Book. Display on the IWB, if possible. Ask students to look at the worked example. Discuss the steps in **solving** the problem about the magazine and book prices. Use the method in the Language support section of emphasising the 'if' and 'then'.

Write this problem on the board:

*A cup of coffee costs $3.20. A cup of coffee costs more than a cake. If you buy one of each, then the total is $4.05. How much more does a cup of coffee cost than a cake?*

Give students some time to think about how they might solve this problem. Ask whether anyone can write the problem as an equation on their whiteboard.

Agree that you can write it as coffee + cake = $4.05, so $3.20 + cake = $4.05. Subtracting $3.20 from $4.05 leaves the price of a cake as 85¢. Subtracting again shows that the coffee costs $2.35 more than the cake.

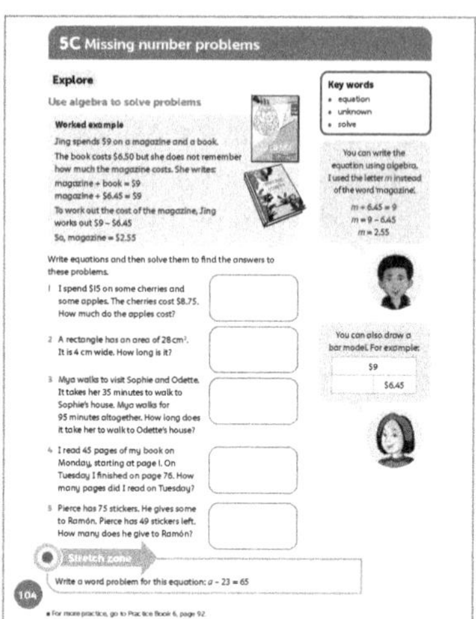


## Main activity

Give students this problem:

*Mr Williamson weighs 80 kg. He wants to find the weight of his wife and son. His scales are broken and won't weight anything less than 100 kg, so he weighs two people at a time.*

*Mr Williamson and his wife together weigh 155 kg. Mrs Williamson and the son together weight 147 kg. How much do they each weigh?*

Tell students that we can write a word equation for each weighing, and then replace each person's full name with a letter, for example Mr Williamson (*W*), Mrs Williamson (*M*) and the son (*S*).

Mr Williamson + Mrs Williamson = 155 kg (*W* + *M* = 155). We know that Mr Williamson is 80 kg, so we know that 80 kg + Mrs Williamson = 155 kg (80 + *M* = 155). We can find Mrs Williamson's weight by subtracting 80 from 155, so Mrs Williamson weighs 75 kg.

Mrs Williamson + the son = 147 kg (*M* + *S* = 147). Ask how we can calculate the son's weight. Students should be able to say that because Mrs Williamson weighs 75 kg, the son weighs 147 − 75 = 72 kg.

Ask students to complete the questions on page 104 using equations. Encourage them to choose a letter to represent each person or object in the problems. This will help them to write an equation to solve each problem.

### Differentiation

**Supporting:** Model how to set up the equation for each word problem.

**Consolidating:** Ask students to explain how they used the equations to solve the problems.

**Extending:** Challenge students to represent the equations using bar models.

**Stretch zone:** *Write a word problem for this equation: a − 23 = 65*

Encourage students to find more than one context for this equation.

### Reflection time

Ask students to share their strategies for solving the problems. Ask a student to write their equation on the board for each question and see whether the class agree. *Which operation is needed to solve the equation? Can you see a connection between the operation in the equation and the operation used to solve it?*

**Practice Book:** Students complete Practice Book page 92. They can do this directly after the Main activity, as homework, or as the focus of a separate mathematics session to help students consolidate their learning and build fluency.

Students solve word problems by writing equations using algebra and then substituting letters with numbers once they are known. The problems involve both constants and variables.

| Differentiated outcomes | |
| --- | --- |
| **All students** | should solve word problems using simple algebraic equations with support. |
| **Most students** | will solve word problems using simple algebraic equations. |
| **Some students** | may solve word problems using simple algebraic equations and make up problems to fit equations. |

### Answers

**Student Book page 104**

1 $6.25

2 7 cm

3 60 minutes

4 31 pages

5 26 stickers

**Practice Book page 92**

1 $175

2 $225

3 $275

4 Total = 45 + 15*w*

5 $60

6 $120

7 $195

8 Wall standing = 12 − 1.5*d*

9 10.5 m

10 7.5 m

11 3 m

12 8 days

Stretch zone: Jada has saved the same amount each day but started with a 'once only' birthday gift.

# 5D Problems with two unknowns



## Specific learning focus

- Solve problems with two unknowns.

## Global skills

- **Creative skills:** problem solving

## Key vocabulary

- unknown, variable

## Resources

- mini whiteboards and markers

## Language support

Support students to describe the strategies they are using. Ask, for example:

- *How did you work that out?*
- *What did you do next?*
- *Why did you decide to use that strategy?*

 **Introductory activity**

Explain that you are thinking of two different numbers. They must work out what the numbers are from the clues you will give them.

The two numbers add up to 12. The difference between the numbers is 4.

Ask students to write down an equation for the first clue. They can label the missing numbers as $s$ and $t$. Agree that the clue can be represented by $s + t = 12$. Ask 'what numbers could $s$ and $t$ be?' Students should suggest answers such as 7 and 5, 8 and 4, 9 and 3, 10 and 2, 11 and 1 of 12 and 0. Now look to see which of these pairs has a difference of 4. The correct answer is 8 and 4. Note that $s$ and $t$ are interchangeable, but if we had specified $s - t = 4$, then $s = 8$ and $t = 4$.

 **Main activity**

Ask students in pairs to work on the activity on page 105 of the Student Book so they can compare answers and discuss strategies. You may need to give some students support in setting up the equation for the problem. Refer to the speech bubbles on page 105.

As students work on the problem, ask the class to share some of their strategies. Ask individual students who successfully complete the problem to share their strategies with the rest of the group. Note that it is

possible to complete the problem by using mental methods by finding pairs of numbers that have the right numbers of heads and legs.

If the problem is solved quickly, ask students to make up their own animal problem and challenge a partner to work out how many of each animal there are.

## Differentiation

**Supporting:** Help students to set up equations to test different numbers of animals.

**Consolidating:** Ask students to explain their strategies for solving the problem.

**Extending:** Challenge students to form their own examples for a partner to solve.

**Stretch zone:** *Is there more than one possible answer? Justify your answer.*

There is only one answer to the problem: 15 emus and 9 zebras. There are many ways to make 24 heads and many ways to make 66 legs, but only one way to make both these numbers correct.

 **Reflection time**

Discuss the solutions to the problem as a class. Ask pairs to say something about what they found out about the numbers of animals needed to match the statement. Some students will be able to make statements about how they listed the numbers systematically until they found the correct answer.

**Practice Book:** Students complete Practice Book page 93. They can do this directly after the Main activity, as homework, or as the focus of a separate mathematics session to help students consolidate their learning and build fluency.

Students solve another animal problem, this time using spiders and beetles. Encourage them to take what they learned from the Main activity and apply the same systematic working to solve this problem. *What do you think would be a good equation to write, to start with?*

<table>
<tr><th colspan="2">Differentiated outcomes</th></tr>
<tr><td>All students</td><td>should solve the problem for two unknowns with support.</td></tr>
<tr><td>Most students</td><td>will solve the problem for two unknowns.</td></tr>
<tr><td>Some students</td><td>may create similar problems and explain the strategy to others.</td></tr>
</table>

# 5D Problems with two unknowns

## Explore  Student Book page 106 · Practice Book page 94

### Specific learning focus

- Solve number problems that have more than possibility.

### Global skills

- **Creative skills:** problem solving

### Key vocabulary

- unknown, variable

### Resources

- mini whiteboards and markers

### Language support

Help students to understand the term 'variable'. Ensure that they understand that it means 'something that can be changed' and can represent different numbers in an expression. For example, in the expression $p = 5t$, $p$ and $t$ are variables. If we change the value of $t$, then the value of $p$ will change.

### Introductory activity

Write on the board:

$2a + b = 30$

Ask students to think of a pair of whole numbers for $a$ and $b$ to fit the equation. Take an answer from one student, perhaps $a = 7$ and $b = 16$. Now ask whether anyone has a different answer. Students will probably have a variety of possible answers, so collect these and list them:

$a = 0, b = 30; a = 1, b = 28; a = 2, b = 26, a = 3, b = 24$ and so on.



### Student Book page 105

15 emus and 9 zebras.

### Practice Book page 93

8 spiders and 6 beetles.

Stretch zone: Students make up their own problems about starfishes and octopuses. Check that they have formed a suitable problem and can show the equation to represent it.

Agree that, in this case, $a$ and $b$ are variables because they can be changed to produce many different answers. All the answers are correct, as long as they fit the equation.

### Main activity

Discuss with students the concept that a problem can have one solution or many solutions, as the example in the Introductory activity showed. Now give students this problem to work through on their whiteboards:

*Find pairs of whole numbers* a *and* b *that fit this equation:*

$3a - b = 20$

Give students time to find at least two pairs of numbers, then ask for answers and ask students to explain how they found them. Possible answers include:

$a = 10, b = 10; a = 7, b = 1; a = 12, b = 16.$

Now ask students to complete the activities on page 106 of the Student Book. Allow them to work in pairs to exchange ideas and test each other's solutions.

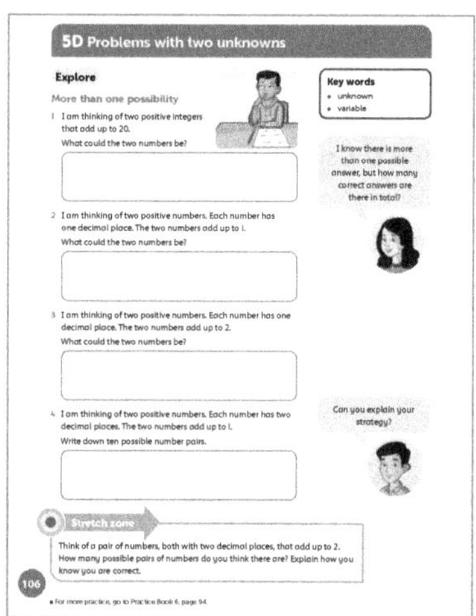

### Differentiation

**Supporting:** Help students to find a range of possible answers to a problem.

**Consolidating:** Ask students to explain how they solved each problem and found multiple answers.

**Extending:** Challenge students to create their own problems to share with a partner or with the rest of the class in the Main activity.

**Stretch zone:** *Think of a pair of numbers, both with two decimal places, that add up to 2. How many possible pairs of numbers do you think there are? Explain how you know you are correct.*

Students should work systematically to find all the possible pairs. There are 99 pairs:
0.01 + 1.99, 0.02 + 1.98 … 0.99 + 1.01

## Reflection time

Discuss the answers to the Student Book questions as a class. Ask pairs of students to say something about what they found out about the possible pairs of numbers. Some students will be able to describe the range of possible numbers for each unknown. Ask for the largest value of *a* that anyone in the class has found. Then see whether students can find a pair with larger values for *a*.

**Practice Book:** Students complete Practice Book page 94. They can do this directly after the Main activity, as homework, or as the focus of a separate mathematics session to help students consolidate their learning and build fluency.

Students solve number puzzles from given clues. For each puzzle, they use algebra to write an equation that contains variables. They then write all the possible solutions that fit the equation. *How do you know you have found all the possible solutions?*

| Differentiated outcomes | |
| --- | --- |
| **All students** | should solve problems with multiple answers with support. |
| **Most students** | will solve problems with multiple answers. |
| **Some students** | may create and solve problems with multiple answers. |

## Answers

### Student Book page 106

**1** 1 and 19, 2 and 18, 3 and 17, 4 and 16, 5 and 15, 6 and 14, 7 and 13, 8 and 12, 9 and 11, 10 and 10

**2** 0.1 and 0.9, 0.2 and 0.8, 0.3 and 0.7, 0.4 and 0.6, 0.5 and 0.5

**3** 0.1 and 1.9, 0.2 and 1.8, 0.3 and 1.7, 0.4 and 1.6, 0.5 and 1.5, 0.6 and 1.4, 0.7 and 1.3, 0.8 and 1.2, 0.9 and 1.1

**4** There are many answers, for example: 0.01 and 0.99, 0.06 and 0.94, 0.29 and 0.71, 0.11 and 0.89, 0.16 and 0.84, 0.25 and 0.75, 0.31 and 0.69, 0.34 and 0.66, 0.42 and 0.58, 0.43 and 0.57, 0.49 and 0.51

### Practice Book page 94

**1** $2x + y = 15$; 0 and 15, 1 and 13, 2 and 11, 3 and 9, 4 and 7, 5 and 5, 6 and 3, 7 and 1

**2** $\frac{1}{2}x + y = 15$; 0 and 15, 2 and 14, 4 and 13, 6 and 12, 8 and 11, 10 and 10, 12 and 9, 14 and 8, 16 + 7, 18 and 6, 20 and 5, 22 and 4, 24 and 3, 26 and 2, 28 and 1, 30 and 0

**3** $2x + 2y = 16$; 0 and 8, 1 and 7, 2 and 6, 3 and 5, 4 and 4, 5 and 3, 6 and 2, 7 and 1, 8 and 0

**4** $2(x + y) = 16$; 0 and 8, 1 and 7, 2 and 6, 3 and 5, 4 and 4, 5 and 3, 6 and 2, 7 and 1, 8 and 0

Stretch zone: The solutions to 3 and 4 are the same, even though the equations look different because $2(x + y) = 2x + 2y$.

### Specific learning focus

- Use variable expressions to calculate costs.

### Global skills

- **Creative skills:** investigating
- **Real-world skills:** financial literacy

### Key vocabulary

- unknown, variable, constant, equation

### Resources

- mini whiteboards and markers

### Language support

Reinforce for students the notion of variables standing for an amount of something that can change. For example, if there are 25 students ($s$) to every teacher ($t$), then the correct expression for the relationship between teachers and students is $25t = s$. When saying this, reinforce it as '25 times the number of teachers equals the number of students'.

 **Introductory activity**

Introduce students to the idea that some costs are made up of a fixed amount and then a variable amount. Say, for example, a builder needs to work out the cost of bathroom tiles. The tiles cost $5 per square metre, plus a $10 fixed charge for delivery. If the area of the bathroom floor in square metres is represented by the variable ($a$), ask students whether they can write on their whiteboards a formula for the total cost of the tiles and delivery.

Agree that the formula can written as $5a + 10$ and then ask students to use the formula to work out the cost of tiling an area of $12\,\text{m}^2$.

$5 \times 12 + 10 = 70$, so it will cost $70.

 **Main activity**

Write this problem on the board:

*An electricity bill is made up of a 'standing charge' of $15, plus a variable charge of $0.75 per unit of electricity (e) used. If I use 24 units of electricity what will my bill be?*

Students work in pairs to agree a formula for the total electricity bill, then use it to calculate the bill for 24 units on their whiteboards. Ask some pairs to share their strategy and formula, then explain how they calculated the bill. Look out for students who may have written the formula the wrong way round, as $15e + 0.75$.

Agree that the formula should be $15 + 0.75e$, and that for 24 units, this is $15 + 0.75 \times 24 = 15 + 18 = \$33$. Ask students to calculate the bill for 1 unit, 10 units and 100 units of electricity on their whiteboards.

Students can then complete the activities on page 107 of the Student Book.

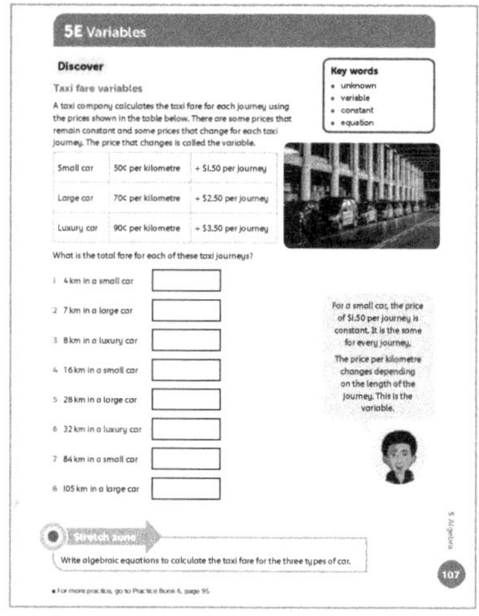

### Differentiation

**Supporting:** Model how to calculate costs using the fixed and variable charges.

**Consolidating:** Ask students to explain how they solved each problem.

**Extending:** Challenge students to create their own problems to share with a partner.

**Stretch zone:** *Write algebraic equations to calculate the taxi fare for the three types of car.*

Students should use the speech bubble to help them.

 **Reflection time**

Ask pairs of students to give feedback on how they worked out the total cost for different distances in different cars. They should describe how they calculated the cost from the fixed charge and the variable charge per kilometre. As a class, compare answers given by different groups.

Ask students to work out which vehicle will take them the furthest for $40. How far will the other sizes of car take them for that price?

**Practice Book:** Students complete Practice Book page 95. They can do this directly after the Main activity, as homework, or as the focus of a separate mathematics session to help students consolidate their learning and build fluency.

Students complete tables for the values of two variables – $x$ and $y$ – where $x$ is known and they are given a formula to work out the value of $y$.

| Differentiated outcomes | |
| --- | --- |
| **All students** | should calculate costs using fixed and variable charges with support. |
| **Most students** | will calculate costs using fixed and variable charges. |
| **Some students** | may use the formula for fixed and variable charges to compare journeys costing the same amount in different cars. |

## Answers

### Student Book page 107

**1** $3.50

**2** $7.40

**3** $10.70

**4** $9.50

**5** $22.10

**6** $32.30

**7** $43.50

**8** $76

Stretch zone: Small car: $0.50a + 1.50$

Large car: $0.70b + 2.50$

Luxury car: $0.90c + 3.50$

### Practice Book page 95

**1** $y = ^-2, 1, 4, 7, 10$

**2** $y = ^-3, 0, 3, 6, 9$

**3** $y = ^-2, 0, 2, 4, 6$

**4** $y = ^-3, ^-1, 1, 3, 5$

**5** $y = 0, 4, 8, 12, 16$

Stretch zone: Differences are 3, 3, 2, 2, 4. The difference is the same amount as the multiple of $x$ in the equation.

# 5E Variables

**Explore**  Student Book page 108 • Practice Book page 96

### Specific learning focus

- Write equations using variables.

### Global skills

- **Creative skills:** investigating

### Key vocabulary

- unknown, variable, substitute, equation

### Resources

- set of digit cards 1, 2, 3, 4, 6, 10, 12, 25, $n$, +, −, =
- mini whiteboards and markers

### Language support

Help students by reinforcing the language associated with solving problems that use equations. Ask them to describe their method using the correct vocabulary, for example:

- My equation for this problem is …
- The variables are …
- The constants are …
- The unknown is …
- The solution to my equation is …

 **Introductory activity**

Write on the board: $3m = 21$. Ask students whether they can say what the value of $m$ will be. Using their knowledge of multiplication and division, they should be able to say that $m = 7$. Use other examples of this type, for example $4m = 32$, $8m = 40$ and $11m = 121$.

Now extend the equations by adding or subtracting a constant, for example $3m − 4 = 17$. Ask students what the value of $3m$ was before the 4 was subtracted and they will get that $3m = 21$, which they can solve as above.

Offer further examples of this type for them to solve on their whiteboards:  $4m + 3 = 35$, $8m − 11 = 29$ and $11m − 23 = 98$.

 **Main activity**

Look together at page 108 of the Student Book. Display on the IWB, if possible. Give each pair of students a set of cards as listed in Resources. Model how to play the 'equation game'. One student should choose two constants, an operation and a total to make an equation, for example $4n − 2 = 10$ or $6n + 1 = 25$. Their partner solves the equation to find the unknown, $n$. For this practice activity, students should write on their whiteboards.

Reinforce the thinking so, for example, to solve $4n − 2 = 10$ say that if $4n$ subtract 2 makes 10, then $4n$ must make 12, then $n$ must be 3. Point out that students are using inverse operations to simplify the equation, adding 2 to the 10 and then dividing by 4 to find $n$.

Students then play the game in pairs and take turns to make an equation for their partner to solve. They write their equations in the table on page 108 of the Student Book. Point out that, depending on the numbers they choose, the value of the unknown could be a fraction and not always a whole number.

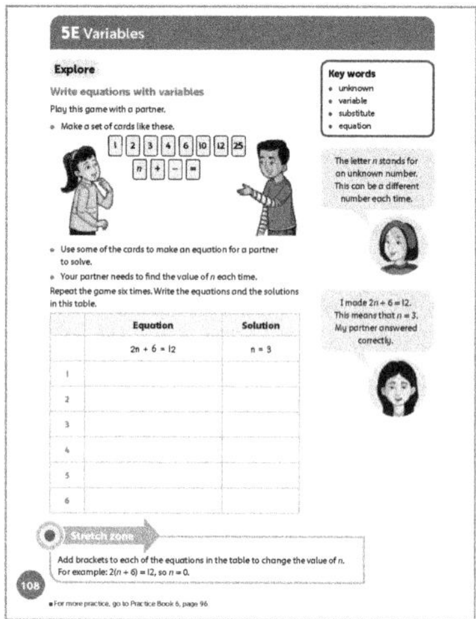

## Differentiation

**Supporting:** Model how to write and solve equations.

**Consolidating:** Ask students to explain how they solved each equation.

**Extending:** Challenge students to create equations with the unknown in a different position, for example $25 - 3n = 10$.

**Stretch zone:** *Add brackets to each of the equations in the table to change the value of* n. *For example: 2*(n + 6) = 12, *so* n = 0.

Students should be encouraged to make as varied equations as possible.

 **Reflection time**

Discuss as a class some of the equations that were formed and how they were solved. First, ask students for examples where the value of $n$ was a whole number, then move on to look at examples where it was a fractional answer. Ask students to share an easy example and a hard example.

**Practice Book:** Students complete Practice Book page 96. They can do this directly after the Main activity, as homework, or as the focus of a separate mathematics session to help students consolidate their learning and build fluency.

Students solve equations where an unknown is represented by a letter. They write out their workings, then check their answer by writing out the equation again, this time **substituting** the unknown (the letter) with their answer. *Is the answer to the equation correct? If not, can you see where you have gone wrong?*

| Differentiated outcomes | |
|---|---|
| **All students** | should solve simple linear equations with support. |
| **Most students** | will solve simple linear equations. |
| **Some students** | may solve linear equations in different arrangements. |

## Answers

### Student Book page 108

Students make equations from digit cards. Check that their equations are solved correctly.

### Practice Book page 96

**1** a = 9      **4** d = 7

**2** b = 4      **5** e = 3

**3** c = 4      **6** f = 4

# 5 Algebra

## Connect Student Book page 109

### Big idea

I can solve real-life problems to find out missing or unknown information. I can find relationships in patterns of numbers.

### Global skills

- **Creative skills:** investigating
- **Interpersonal skills:** teamwork
- **Self-development skills:** reflecting on learning

### Key vocabulary

- relationship, formulae, pattern

### Resources

- tape measures
- mini whiteboards and markers
- circular objects (tins, discs, wheels)
- calculators
- string

### Language support

Model and encourage the use of phrases that explain the relationship between sets of numbers, for example:

- *In the pairs of numbers (1, 4), (2, 8), (3, 12) and so on, one number is 4 times as many as the other.*
- *As the first number in each pair increases by 1, the other increases by 4.*

 **Introductory activity**

Put students into small groups and give each group a tape measure and a small collection of circular items. Tell groups to measure the diameter of each item (across the centre) and its circumference (around the outside). To measure the circumference, they use string to surround the circumference and then measure the length of string needed. They record measurements for each item on their whiteboards.

On the board, draw up a table of results for each item, recording its circumference (C) and diameter (d). Then calculate $C \div d$ for each object to two decimal places.

Ask students to look at the results and share what they notice. Ask them to describe whether they think there is a **relationship** between the circumference and diameter of circles.

 **Main activity**

Look together at page 109 of the Student Book. Display on the IWB, if possible. Explain that students are going to work in groups to investigate whether there is a relationship between their head circumference and their height. Introduce to them the idea of a hypothesis (a statement that needs to be proved). Write on the board:

'The taller a person, the larger their head.'

Explain to students that they will measure the head size and height of everyone in their group and then calculate the ratio of height to head size to see whether there is a relationship. Ask students to think about what the data will show if there is a relationship, and if there is not.

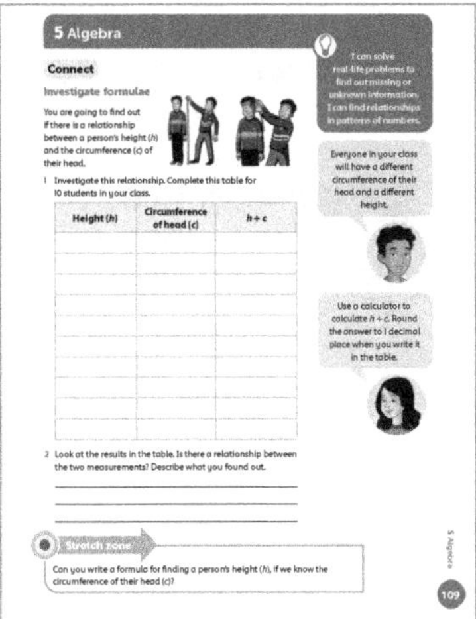

### Differentiation

Students work in mixed-attainment groups, supporting each other. They might choose to distribute the tasks by focusing on the different skills of the group. For example, some might be better at measuring and others might be better at recording and calculating.

**Stretch zone:** *Can you write a formula for finding a person's height (h), if we know the circumference of their head (c)?*

Students should try to write a formula using the results of their measurements.

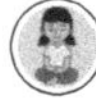 **Reflection time**

Ask each group to give feedback on how they worked to complete their investigation into head sizes and heights. They should describe how they measured and calculated the ratio. Compare results from the different groups. *Did you all find similar outcomes or were there differences?*

| Differentiated outcomes | |
| --- | --- |
| **All students** | should measure heights and head circumferences and look for a relationship between the values with support. |
| **Most students** | will measure heights and head circumferences and look for a relationship between the values. |
| **Some students** | may describe the relationship between height and head circumferences. |

## Answers

### Student Book page 109

Students will measure the heights and head circumferences of 10 students in the class and then divide the height by the circumference to look for a common relationship. Check that they have calculated accurately from their measurements and recorded what they notice about the relationship.

# 5 Algebra

**Review**  Student Book page 110 · Practice Book page 97

## Global skills

- **Self-development skills:** reflecting on learning

## Student Book

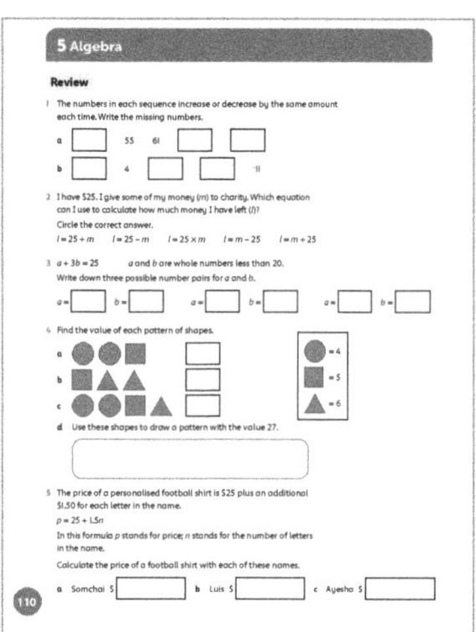

With young students, assessment activities are most effective when carried out as an everyday classroom activity. Watch as students use their understanding of algebra in exploring number sequences and using equations to solve problems. Support them to understand that some problems can have multiple answers.

Students should be able to discover the rule for linear number sequences and use it to extend the sequence. They should solve missing number problems by formulating algebraic expressions and they should be able to work out costs or times based on equations containing variables and constants.

The Review activity is an individual summative assessment and so should be completed by students working on their own.

## Answers

### Student Book page 110

**1 a** 49, 55, 61, 67, 73    **b** 9, 4, ⁻1, ⁻6, ⁻11

**2** $l = 25 - m$

**3** three responses out of the following: $a = 1, b = 8$; $a = 4, b = 7$; $a = 7, b = 6$; $a = 10, b = 5$, $a = 13, b = 4$; $a = 16, b = 3$; $a = 19, b = 2$; $a = 22, b = 1$

**4 a** 13    **b** 17
   **c** 19    **d** Various possible answers, for example:

**5 a** $35.50    **b** $31    **c** $34

## Practice Book

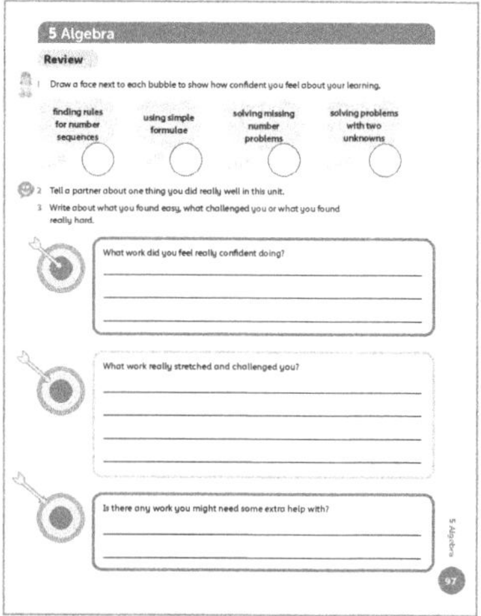

It is appropriate to complete this Practice Book Review as a whole-class discussion. You may choose to keep a record of the class discussion or a copy of the Review page for your own records. The Review provides an opportunity for students to reflect on their learning from the unit, to discuss any areas of mathematics that they feel went particularly well, and any areas that they feel less confident about. Ensure that all students have a copy of the Student Book as a reminder of the areas of mathematics that they have worked on in this unit.

Allow students plenty of time for discussion before asking them to complete the Practice Book page individually, and then, if appropriate, to share their responses with the rest of the class. If students complete this self-assessment at home, encourage them to discuss this with adults. Make a note of areas that students still feel unsure about.



There are additional end-of-unit assessments available on the *Oxford Owl for School* website.

# 6 Length, mass and capacity

## Overview

### Big idea

Students need to use their understanding of place value and their knowledge of decimals to help them convert between units of measure, such as kilometres to metres or kilograms to grams. They also begin to look at converting imperial units to metric units and vice versa, for example converting miles and kilometres.

In this unit, students also develop their knowledge of scales. This will help them to use measures to an appropriate degree of accuracy. It is important to introduce students to the idea that measure is always only accurate to a chosen degree of accuracy. We can never measure a line exactly. The measurement is always to the nearest mm, cm, m and so on. Students therefore need to decide an appropriate degree of accuracy to use.

### Look out for

- **Students who forget equivalences between units (for example how many millilitres make one litre).** Measure is a topic that is not frequently taught. Rehearse these facts regularly and refer to these when covering fractions, decimals and mental and written calculations.
- **Students who struggle to read scales on measuring devices accurately.** Use a range of different scales on different rulers and containers so that students become familiar with a range of different scales for measuring length, mass and capacity.
- **Students who may not think they know any imperial units.** These are introduced in this unit. You can use a video or visit a setting in which imperial units are used to show these units in context.

### Possible misconceptions

- **Students think that a measure such as 230 cm is different from 2.3 m. They may think that these measurements are different because 230 is a larger number.** You can use measuring equipment that is calibrated in both cm and m, or l and ml, in practical activities to show students how to convert between units.
- **Students think that metric and imperials units are accurately connected.** For example, they may think that 25 miles is the same as 2500 kilometres or similar. Provide opportunities for students to see and use the approximate equivalents that exist for converting between metric and imperial units.

### Key vocabulary

- units of measure, standard unit, non-standard unit, metric unit, imperial unit, convert, conversion graph
- milli-, centi-, kilo-, kilometre (km), metre (m), centimetre (cm), millimetre (mm); mile, yard, feet, foot, inches, inch; ruler, metre stick, tape measure, trundle wheel
- mass, weight; tonne, kilogram (kg), gram (g), milligram (mg), pounds, ounces; measuring cylinder, measuring jug
- capacity; litre (l), centilitre (cl), millilitre (ml), pint, gallon
- scale drawing, degree of accuracy, to the nearest …

## Coverage in lessons

| Learning objective | E | 6A | 6B | 6C | 6D | C | R |
|---|---|---|---|---|---|---|---|
| Solve problems involving the calculation and conversion of units of measure, using decimal notation up to three decimal places where appropriate. | ✓ | ✓ | ✓ | ✓ | ✓ | ✓ | ✓ |
| Use, read, write and convert between standard units, converting measurements of length, mass, volume and time from a smaller unit of measure to a larger unit, and vice versa, using decimal notation to up to three decimal places. | ✓ | ✓ | ✓ | ✓ | ✓ | ✓ | ✓ |
| Convert between miles and kilometres. | | | | | ✓ | | |

# 6 Length, mass and capacity

## Engage  Student Book page 111

### Big question

- Why do we need to convert between units?

### Global skills

- **Creative skills:** exploring
- **Real-world skills:** research
- **Interpersonal skills:** communication
- **Self-development skills:** reflecting on learning

### Key vocabulary

kilometre (km), metre (m), centimetre (cm), millimetre (mm) mile, yard, feet, foot, inches, inch, tonne, kilogram (kg), gram (g), milligram (mg), pounds, ounces, litre (l), centilitre (cl), millilitre (ml), pint, gallon

### Resources

- mini whiteboards and markers

### Language support

During the presentations, listen carefully to students' pronunciation of the units of measurement. Model the correct pronunciation. Make a display of all the units. Students can refer to this during the unit.

 **Introductory activity**

Ask students, in pairs, to talk about this question: *When have you used measurement this week?* They can note examples on their whiteboards. After five minutes, take feedback. List on the board all the units of measure they mention. Students can refer to this list when working on the Main activity. If they do not mention units, prompt them, asking: *What units did you measure in?* Refer to the vocabulary list above. For any of these units that students haven't used, ask, *Has anyone measured something in … this week?*

 **Main activity**

Organise the class into mixed-attainment groups of four students. *Which occupations use measuring? What sort of measuring do they need to use?* Students discuss ideas and write them on their whiteboards. After ten minutes, list all the suggested occupations on the board along with what the people in these occupations measure and what units they use.

Look together at page 111 of the Student Book. Display on the IWB, if possible. The Student Book gives some examples of people in occupations who use measuring on a daily basis. Ask each group to select a different occupation, either from the Student Book or from the class list, and to plan a short role-play that shows them carrying out measurements at work. Students can present these role-plays during Reflection time.

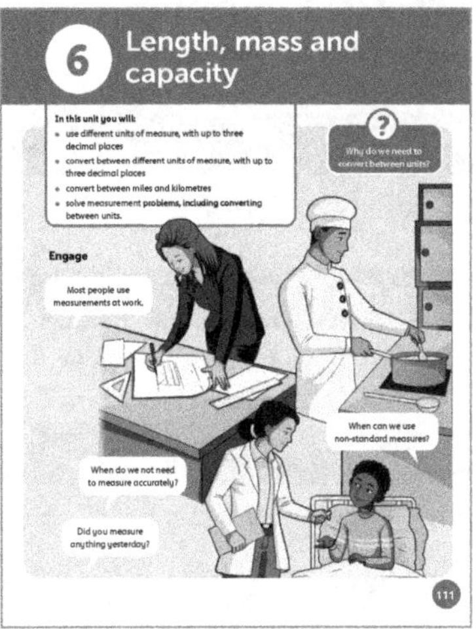

### Differentiation

**Supporting:** Help students to contribute to the discussion by suggesting some units of measurement and/or the occupations that use them.

**Consolidating:** Ask students to think of examples for units of measurement and occupations that have not yet been used.

**Extending:** Ask students to think of examples for all the units of measurement in the vocabulary list.

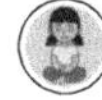 **Reflection time**

During the presentations, ask every group to make a list of all the units of measurement used in the presentations, along with how or when they are used. Discuss units that are not mentioned, including Imperial units such as mile, yard, foot, inch, pint, gallon, pound, ounce. *Why are some units used more commonly than others?* Many countries have adopted metric units as standard because they are easily converted using powers of ten. Some countries use imperial units for historical or cultural reasons, for example the USA. Some countries use a mix of both metric and imperial units, for example the UK, where metric units have been taught in schools for many years, but imperial measures are standard for some units, such as miles and pints.

# 6A Comparing units of measure

## Discover  Student Book page 112 • Practice Book page 98

### Specific learning focus

- Select and use standard units of measure.

### Global skills

- **Creative skills:** problem solving

### Key vocabulary

- kilometre (km), tonne, standard units

### Resources

- mini whiteboards and markers

### Language support

Refer back to the vocabulary list of units of measure that was started in the previous lesson. Encourage students to use the units in their responses throughout this lesson. Pick five students who are confident in their use of language to act as models for the use of language in Reflection time. For example, ask:

- *What unit can we use to measure the length of the classroom?*
- *What units can we use to measure the amount of fruit juice in a bottle?*
- *What units do we use to measure our own mass?*

 **Introductory activity**

Ask students to discuss in pairs and write on their whiteboards an object or measurement that fits each of the following descriptions (read them out to students):

- something that is 1.5 cm long
- something that weighs 2 kg
- a place that is 15 km away
- the width of the classroom
- the capacity of a can of drink.

 **Main activity**

Ask students which units were easy to find examples for in the Introductory activity and which were more difficult. To support students, prepare examples before the lesson, such as: a UK 5p coin is just over 1.5 cm (there may be coins this size in other countries), a bag of sugar could be 2 kg, name a town 15 km away from the school, measure the classroom before the lesson, drink cans could hold about 400 ml.

Now ask students to think about where larger units, for example **kilometres** and **tonnes**, are used. Can they think of examples of objects that weigh more than 1 tonne? Do they know anywhere that is more than 100 km away from where they live?

Students should work in pairs on the activities on page 112 of the Student Book to support one another and to aid discussion. As students work on the activities, ask:

- *How do you know that is correct?*
- *Can you show me how long that is?*
- *Did that surprise you?*

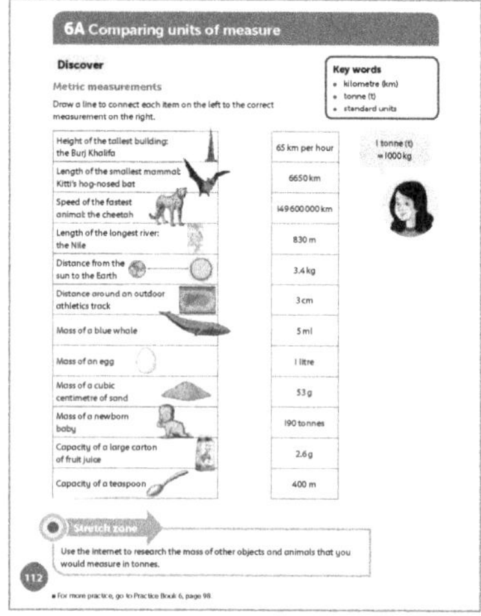

## Differentiation

**Supporting:** Support students in matching units by asking probing questions, such as *How many cm is the same length as 12 inches? 1 pound is roughly how many grams?*

**Consolidating:** Ask students to explain and justify their answers.

**Extending:** Encourage students to think of other objects or distances similar in measure to those given.

**Stretch zone:** *Use the internet to research the mass of other objects and animals that you would measure in tonnes.*

Students may find out some interesting facts such as the tongue of a blue whale weighs as much as an elephant or that a whale's heart weighs as much as a car. Ask students to share during Reflection time some of the interesting facts they find.

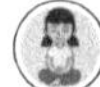 **Reflection time**

Talk through the list of items from the Main activity and let students suggest which measurement each one relates to. Ask whether there were any items that needed more consideration than others. *Can you think of a different object that would measure the same as that one?*

**Practice Book:** Students complete Practice Book page 98. They can do this directly after the Main activity, as homework, or as the focus of a separate mathematics session to help students consolidate their learning and build fluency.

Students find objects, measure them and then convert the measurements into different metric units.

| Differentiated outcomes | |
|---|---|
| **All students** | should match units to objects with support. |
| **Most students** | will match objects to units, supporting others. |
| **Some students** | may find additional items or measures approximately equal to those given. |

## Answers

### Student Book page 112

| | |
|---|---|
| Tallest building | 830 m high |
| Smallest animal | 3 cm long |
| Fastest animal | 65 km per hour |
| Longest river | 6650 km |
| Distance from Earth to Sun | 149 600 000 km |
| Distance around athletics track | 400 m |
| Mass of a blue whale | 190 tonnes |
| Mass of an egg | 53 g |
| Mass of a cubic centimetre of sand | 2.6 g |
| Mass of a newborn baby | 3.4 kg |
| Capacity of a large carton of fruit juice | 1 litre |
| Capacity of a teaspoon | 5 ml |

# 6A Comparing units of measure

**Explore** Student Book pages 113–114 • Practice Book page 99

### Specific learning focus

- Convert between units of measurement.

### Global skills

- **Creative skills:** investigating

### Key vocabulary

- kilometre (km), tonne (t), standard units

### Resources

- coloured pencils
- large sheets of paper

### Language support

Develop the language of unit conversion, for example:

- How many … is the same as 1 … ?
- One … is the same as …

## Practice Book page 98

Students will find examples of objects to measure and record using different units. Check that, for each item, they have given a reasonable measurement and converted into the different units.

Stretch zone: Check students' measurement and conversion between units for a light object.

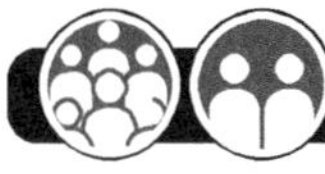 **Introductory activity**

Prepare examples to support students with comparing measures. For example, you could collect a 1 kg bag of sugar for students to use for comparisons of mass. Together, look back at page 112 of the Student Book. Display on the IWB, if possible. Ask students to look back at the list of items and their measurements as a reminder of some approximate measures. *Can you think of something that weighs close to 1 kg?* (A small melon would weigh approximately 1 kg.) *Can you think of something that is the same length as your arm?* (The height of their desk might be a similar length.)

Ask them to work in pairs to think of two objects of similar lengths, two objects of similar mass and two objects of similar capacity. Share their findings and record them on a poster to display in the class.

 **Main activity**

Look together at page 113 of the Student Book. Display on the IWB, if possible. Look at the worked example, which compares the height of the Burj Khalifa to twice around an athletics track. Ask students to write some similar comparisons using known measures. For example, using the kilogram mass, they might say that a desk weighs about the same as 15 bags of sugar. Using the table from Student Book page 112, they might say that the fastest animal would take a little over 100 hours to run the length of the longest river. Students record these comparisons on page 113 of the Student Book. Select five students who have interesting comparisons to give feedback in Reflection time.

Now ask students to complete the activities on page 114, where students will match equivalent measures displayed in different units.

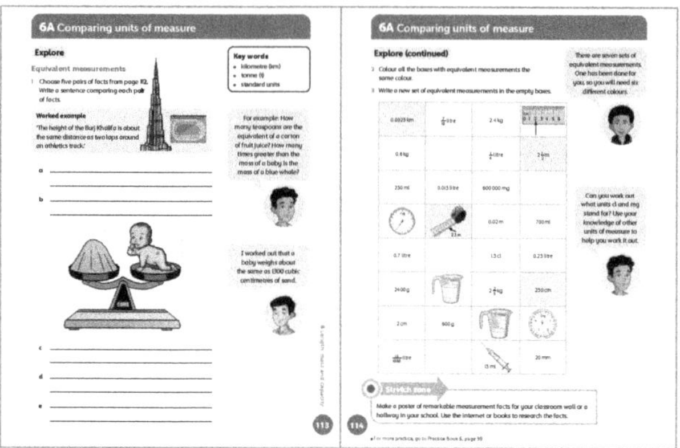

## Differentiation

**Supporting:** Help students to make the conversions between units, for example mm to m or g to kg.

**Consolidating:** Ask students to explain how they converted between units.

**Extending:** Challenge students to find equivalents to less common measures such as centilitres or decimetres.

**Stretch zone:** *Make a poster of remarkable measurement facts for your classroom wall or a hallway in your school. Use the internet or books to research the facts.*

Students can use some of the facts found earlier in the comparison activity.

 **Reflection time**

Ask students to talk about the different unit conversions they have been doing. *Which conversions did you find easiest? Which were most difficult? Why?*

Ask the students selected earlier to share their measurement comparisons with the class.

**Practice Book:** Students complete Practice Book page 99. They can do this directly after the Main activity, as homework, or as the focus of a separate mathematics session to help students consolidate their learning and build fluency.

Students solve word problems that involve units of measures. Some of the problems require conversion from one unit to another before calculating. *What do you need to do before you calculate? Will you convert the grams to kilograms first or grams to kilograms first?*

| Differentiated outcomes | |
|---|---|
| **All students** | should convert between different units with support. |
| **Most students** | will convert between different units. |
| **Some students** | may convert using less common units. |

## Answers

### Student Book pages 113–114

1 Check that students have made sensible comparisons between the measures of different items.

2 $\frac{7}{10}$ litre, 700 ml, 0.7 litre; 2.4 kg, $2\frac{2}{5}$ kg, 2400 g; 2 cm, 0.02 m, 20 mm; 0.6 kg, 600 000 mg, 600 g; $\frac{1}{4}$ litre, 250 ml, 0.25 litre; 0.015 litre, 1.5 cl, 15/1000 litre, 15 ml

3 Answers will vary, for example 4.5 cm, 45 mm, 0.045 m, 0.000045 km

### Practice Book page 99

1 10.05 litres

2 19.025 kg; 1.025 kg over

3 $15

# 6B Converting units of measure



**Discover**  Student Book pages 115–116 • Practice Book page 100

## Specific learning focus

- Convert between units of measurement, using decimals to three places.

## Global skills

- **Creative skills:** exploring
- **Self-development skills:** reflecting on learning

## Key vocabulary

- convert, milli-, centi-, kilo-

## Resources

- range of measuring equipment and items to measure
- calculators

## Language support

During the Main activity, ask students to explain their thinking about multiplying and dividing by 100 and 1000. You can use place-value grids to remind them that they are not 'adding zeros' but moving the digits across the decimal point. Ask, for example:

- *What is the value of the 3 in 5430?*
- *What is the value of the 7 in 98.7?*
- *When I multiply 89.7 by 100, what is the new value of the '8'?*

 **Introductory activity**

Prepare a series of measuring activities using everyday objects that involve measuring length, finding the capacity of containers and weighing the mass of objects. Ask each group to find the measurement in the most sensible unit then **convert** this to other units. For example:

'My hand span is 12 cm. This is equivalent to 120 mm, 0.12 m and 0.00012 km.'

You may need to remind students of the conversion factors and that **milli-** means 1000. For example: there are 1000 millilitres in 1 litre.

Use mixed-attainment groups so that students can discuss the measurements and more-confident students can model the vocabulary to less-confident students.

 **Main activity**

Look together at page 115 of the Student Book. Display on the IWB, if possible. Refer students to the Think back information as a reminder of how to convert from one metric unit to another. Ask students to discuss, as a group, the conversion of the units for the Student Book activity on pages 115–116 and then complete the tables individually. If they do not agree, they can check using a calculator. When they try to think of objects, ask them to start with objects they can visualise easily. They may not be able to complete them all. To support students, find objects they are familiar with before the lesson to demonstrate every measurement. As students complete the tables, select three or four objects to use in Reflection time.

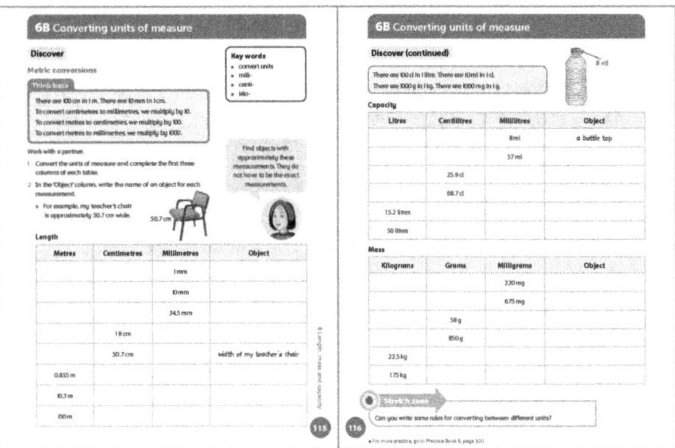

## Differentiation

**Supporting:** Support students in measuring objects accurately in the Introductory activity, and converting units.

**Consolidating:** Ask students to explain their thinking when they convert units.

**Extending:** Ask students to think of good examples for each of the measurements in the tables.

**Stretch zone:** *Can you write some rules for converting between different units?*

Remind students to use their knowledge of place value when writing their rules.

As a follow-up activity, students could carry out the same activity but start with objects. For example, say, *Find the capacity and height of a can of cola and write the units in different ways.*

 **Reflection time**

Have on display the three or four the objects you selected during the Main activity. State a measurement for one of the objects, from which it is difficult to recognise the object, for example: *This object has a capacity of 1000 ml.* (a large carton of fruit juice)

Students can decide in pairs which of the objects you are describing.

**Practice Book:** Students complete Practice Book page 100. They can do this directly after the Main activity, as homework, or as the focus of a separate mathematics session to help students consolidate their learning and build fluency.

Students complete a conversion table for metric units of length, mass and capacity. They then convert various measurements from one metric unit to another.

| Differentiated outcomes | |
| --- | --- |
| **All students** | should measure objects accurately and convert units with support. |
| **Most students** | will measure objects accurately and convert units. |
| **Some students** | may measure objects accurately and convert units and think of good examples. |

## Answers

### Student Book pages 115–116

**1**

| Metres | Centimetres | Millimetres |
| --- | --- | --- |
| 0.001 m | 0.1 cm | 1 mm |
| 0.01 m | 1 cm | 10 mm |
| 0.0345 m | 3.45 cm | 34.5 mm |
| 0.18 m | 18 cm | 180 mm |
| 0.507 m | 50.7 cm | 507 mm |
| 0.855 m | 85.5 cm | 855 mm |
| 10.3 m | 1030 cm | 10 300 mm |
| 150 m | 15 000 cm | 150 000 mm |

| Litres | Centilitres | Millilitres |
| --- | --- | --- |
| 0.008 l | 0.8 cl | 8 ml |
| 0.057 l | 5.7 cl | 57 ml |
| 0.259 l | 25.9 cl | 259 ml |
| 0.687 l | 68.7 cl | 687 ml |
| 15.2 l | 1520 cl | 15 200 ml |
| 50 l | 5000 cl | 50 000 ml |

| Kilograms | Grams | Milligrams |
| --- | --- | --- |
| 0.000 22 kg | 0.22 g | 220 mg |
| 0.000 675 kg | 0.675 g | 675 mg |
| 0.058 kg | 58 g | 58 000 mg |
| 0.850 kg | 850 g | 850 000 mg |
| 22.5 kg | 22 500 g | 22 500 000 mg |
| 175 kg | 175 000 g | 175 000 000 mg |

**2** Check that the objects students suggest for each measure are appropriate.

### Practice Book page 100

**1** 1 km = 1000 m  
**2** 1 m = 100 cm  
**3** 1 cm = 10 mm  
**4** 1 litre = 100 cl  
**5** 1 litre = 1000 ml  
**6** 1 cl = 10 ml  
**7** 1 kg = 1000 g  
**8** 1 g = 1000 mg  
**9** 8700 m  
**10** 875 cm  
**11** 895 mm  
**12** 55.5 cl  
**13** 845 ml  
**14** 8950 g  
**15** 9650 mg  
**16** 5.5 km  
**17** 9.85 litres  
**18** 6.75 kg  
**19** 1575 ml  
**20** 2.55 m  

Stretch zone: Students should explain how they remember the unit prefixes, for example **centi-** = 1 hundred.

 **Unit 6 Length, mass and capacity**

# 6B Converting units of measure

## Explore
Student Book pages 117–118 • Practice Book page 101

### Specific learning focus
- Convert between units of measurement, using decimals to three places.

### Global skills
- **Creative skills:** problem solving

### Key vocabulary
- convert units, milli-, centi-, kilo-

### Resources
- range of measuring equipment
- calculators

### Language support

Support students with the language of problem solving – problems involving conversions between units can be complex and can involve addition, subtraction, multiplication or division. Suggest that students make notes to break down the question into easier chunks so they can make sense of what is being asked.

As students work on the word problems, remind them to check their answers. *Does that answer make sense?*

 Introductory activity

Ask the class to imagine building a human tower by standing on one another's shoulders. *How tall do you think the tower would be?* Write some of the estimates on the board. Ask students to work in pairs to measure one another using a tape measure. List all the heights on the board.

Give four students calculators. Read out all the measurements. Ask the four students to total them using the calculators. Make sure that students agree on the total. Then write down the answer and compare it to the estimates. Try to think of something that is the same height as the total height of their human tower. Alternatively, you can measure out an actual distance with a trundle wheel. Carry out the initial calculation in cm using an estimate for a person's height in cm, and then carry out the conversion into m and km. This will involve three places of decimals. Briefly discuss rounding to the nearest m or km in this context.

 Main activity

Split the class into five mixed-attainment groups. Each group should start on a different problem from pages 117–118 of the Student Book, then move on to complete as many questions as possible during the time available.

Encourage groups to break down each problem into small steps. For example, for question 1 they can measure the desk and divide that measurement by 10. They can then find out how many currants fit into that measure and multiply this answer by 10 to find the answer.

For question 4, it may be easier to give students the mass of the highest value coin.

For question 5, students may need help in working out the distance between their home and the school.

Encourage students to estimate when they see three places of decimals. Encourage them to make sense of the decimals up to three decimal places in the context of these measurements. Encourage individual students to ask questions if they do not understand what someone else is saying.

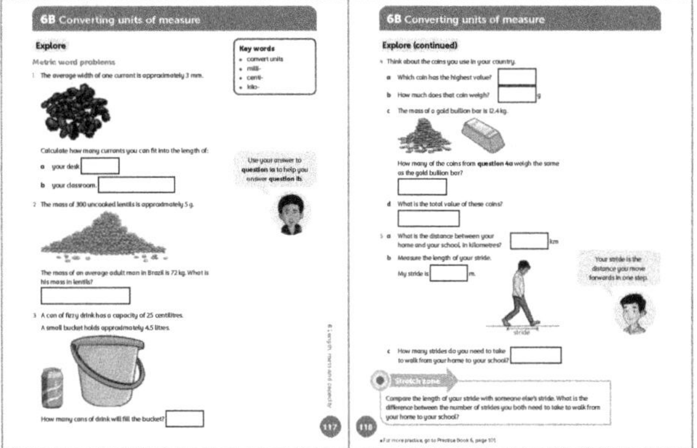

### Differentiation
Students will bring a range of experience to the task that they can share.

**Supporting:** Support students in decoding the word problems and deciding which conversions or calculations are needed.

**Consolidating:** Ask students to explain their thinking when they solve the problems and convert units.

**Extending:** Ask students to estimate the answer to each of the problems before calculating.

**Stretch zone:** *Compare the length of your stride with someone else's stride. What is the difference between the number of strides you both need to take to walk from your home to your school?*

Students can set their own word problems for one another based on measurements.

Ask each group to present their solution to one of the problems. Encourage other groups to ask questions to make sure that all students understand the solution. After each presentation, students in other groups can check their answers in their Student Book.

**Practice Book:** Students complete Practice Book page 101. They can do this directly after the Main activity, as homework, or as the focus of a separate mathematics session to help students consolidate their learning and build fluency.

Students solve word problems that involve converting between different units of measure. They may need to use a calculator for some of the calculations.

| Differentiated outcomes | |
|---|---|
| **All students** | All students will benefit from working in groups on these activities and from hearing the solutions presented. Less-confident students will hear vocabulary modelled by other students and gain problem solving skills. More-confident students benefit from having to explain their thinking and the problem-solving process. |
| **Most students** | |
| **Some students** | |

# 6C Reading scales and measuring accurately

## Discover  Student Book page 119 • Practice Book page 102

### Specific learning focus

- Measure and use scales to draw a map.

### Global skills

- **Creative skills:** problem solving
- **Real-world skills:** presenting information
- **Interpersonal skills:** teamwork

### Key vocabulary

- scale drawing, trundle wheel

### Resources

- tape measures, metre sticks
- trundle wheels
- calculators
- sheets of plain paper and squared paper

## Answers

### Student Book pages 117–118

1 Answers will vary as they depend on the size of students' desks and classroom. Check that the numbers are reasonable.

2 4 320 000 lentils

3 18 cans

4 Answers depend on the local currency.

5 Answers depend on the distance of each student's home from school and the student's stride length.

### Practice Book page 101

1 Answers will vary as the distance depends on the location of students' homes. Check that the destination is reasonable.

2 90.625 bottles

3 50 g

Stretch zone: Answers will vary as the amount depends on how many breaths students calculate they breathe in a week. Check that students' methods are reasonable. For example, if they take 12 breathes per minute they would breathe about 6000 ml of air in a minute and therefore 360 000 ml of air in an hour, 8 640 000 ml in a day and 60 480 000 ml in a week, or 60 480 litres.

## Language support

Take the opportunity to develop vocabulary relevant to maps and scale drawings by asking questions such as:

- *What features shall I include on my map?*
- *How accurately do I need to measure the features?*
- *What length on the map is a length of 12 m?*

 **Introductory activity**

Ask the class:

*Imagine we want to draw a map of the school grounds. What do we need to do to draw this map?*

Ask students to discuss, in pairs, how they might go about drawing a map of the school grounds. Their discussion may include reference to measuring the school and its grounds and choosing a scale for the map so it can be drawn on a sheet of paper.

Write on the board: Scale 1 : 1000.

Ask students what this means. Ask confident students to explain the scale and give examples. Ask questions such as: *How long on a map of this scale would a classroom with a length 10 m be drawn?* Since 10 m = 10 000 cm, then with a scale of 1 : 1000, the map of the classroom would be 10 cm. Repeat with other measurements in metres.

# Main activity

Arrange students into small mixed-attainment groups. Look together at page 119 of the Student Book. Display on the IWB, if possible. Ask the groups to decide what measurements they need to make and what scale they will use to draw their map.

Refer students to the first speech bubble on the page. *Where will we find a map of our school?* You could use an internet search engine to find a satellite image of your school and ask students to use this as the basis of their **scale drawing**. *What features will you include on your scale drawing?*

Refer students to the second and third speech bubbles. *What scale will you use to ensure that your scale drawing fits on a piece of paper?*

As they work, check that groups are calculating their scale measurements correctly and using the **trundle wheel** to measure as accurately as possible.

Encourage students to sketch their map first before completing an accurate scale version. They should include as many features as is practicable.

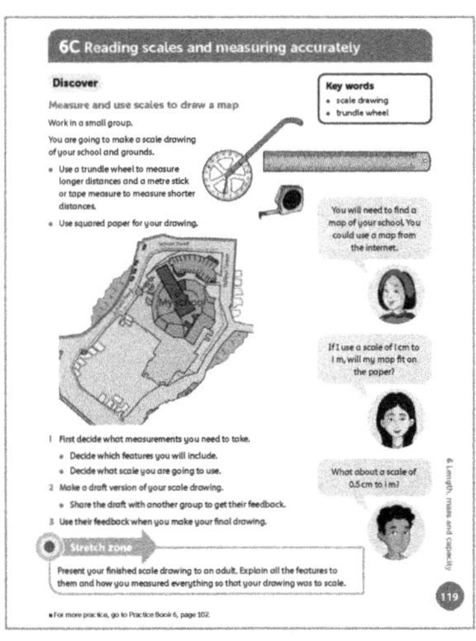

## Differentiation

Students will bring a range of experience to the task that they can share.

**Supporting:** Help students to measure correctly using a trundle wheel and convert measurements using the chosen scale.

**Consolidating:** Ask students to explain, to other students, the scale and how to use it.

**Extending:** Challenge students to include features that are not made up of straight lines.

**Stretch zone:** *Present your finished scale drawing to an adult. Explain all the features to them and how you measured everything so that your drawing was to scale.*

Students can devise questions about their map for others to answer. For example, if the school yard is 30 cm long on the map, how long is the real yard?

##  Reflection time

Ask each group to share their maps with the rest of the class, explaining how they chose their scale. Ask them to share some examples of the features they included on their map and how they worked out what size these features needed to be, using the scale.

**Practice Book:** Students complete Practice Book page 102. They can do this directly after the Main activity, as homework, or as the focus of a separate mathematics session to help students consolidate their learning and build fluency.

Students design a pencil case big enough to hold all the equipment they need for lessons. In this lesson they will decide on the contents and measure the length of each item.

| Differentiated outcomes | |
| --- | --- |
| **All students** | All students will benefit from working on this investigation and from hearing the solutions presented. Less-confident students will hear vocabulary modelled by other students and gain problem-solving skills. More-confident students benefit from having to explain their thinking and the findings. |
| **Most students** | |
| **Some students** | |

# Answers

## Student Book page 119

Check the finished maps of each group for accuracy and good use of the scale.

## Practice Book page 102

Answers will vary because students measure objects around them. Check that students' measurements are reasonable.

Stretch zone: Check that students' estimates are reasonable and that the measurements are accurate.

# 6C Reading scales and measuring accurately

## Explore
Student Book pages 120–121 • Practice Book page 103

### Specific learning focus
- Interpret readings on different scales.

### Global skills
- **Creative skills:** exploring

### Key vocabulary
- degree of accuracy

### Resources
- rulers or metre sticks
- measuring jugs
- weighing scales, spring balances
- range of objects to measure

### Language support
Support students by focusing on the language of units and accuracy. Ask, for example:

- *Are you measuring to the nearest centimetre or millimetre?*
- *Are you measuring to the nearest gram or milligram?*
- *Are you measuring to the nearest centilitre or millilitre?*

Encourage them to answer, for example:

- *I am measuring to the nearest …*

 **Introductory activity**

Set up a range of activities in the classroom. On one table, place a range of objects and some rulers or metre sticks to measure length; on another table, place a range of containers and some measuring jugs to measure capacity; and on a third table, place a range of objects and weighing scales to measure mass. Ask students to work in groups. Arrange for groups to move around the tables measuring the objects in turn and making a note of the longest item, the heaviest item and the container with the greatest capacity. *What degree of accuracy will you use when you measure?*

 **Main activity**

After all the groups have measured all the items, ask groups to feed back. Check their measurements. Discuss the degrees of accuracy they used. Now ask them to write their measurements to different **degrees of accuracy**, for example:

- to the nearest metre
- to the nearest litre
- to the nearest kilogram.
- Students then complete the activities on pages 120–121 of the Student Book. Check that they are reading the scales correctly as they mark the various measures.

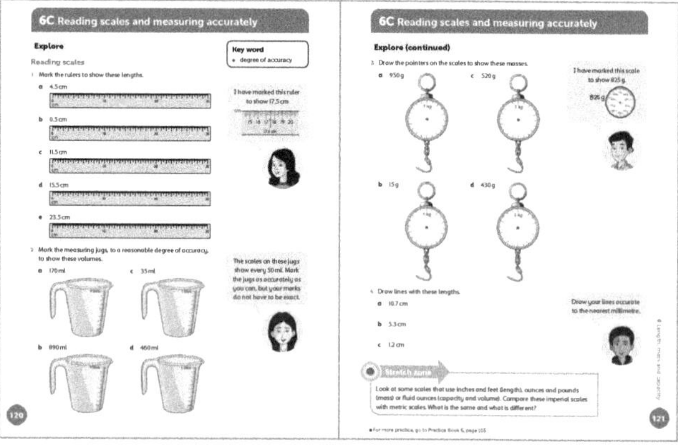

### Differentiation
**Supporting:** Support students to measure accurately, modelling the process.

**Consolidating:** Ask students to tell you the degree of accuracy they are using.

**Extending:** Challenge students to explain how to measure accurately to other students.

**Stretch zone:** *Look at some scales that use inches and feet (length), ounces and pounds (mass) or fluid ounces (capacity and volume). Compare these imperial scales with metric scales. What is the same and what is different?*

Students will notice that all scales are divided into large units and smaller units (often the smaller units are represented by a line only).

As an additional activity, ask students to find and measure the longest or widest object in their house. They can bring the measurements in to school to compare with those of other students. Similarly, they can find the largest container or the heaviest object, measured as accurately as possible.

**Reflection time**

Ask students to share the results of their measuring skills. Have a set of measuring tools available (tape measures, metre sticks, measuring jugs, weighing scales and spring balances) and ask students to demonstrate how they use them.

Ask students to decide on what degree of accuracy is appropriate for different items, for example does the mass of a car need to be to the nearest gram?

**Practice Book:** Students complete Practice Book page 103. They can do this directly after the Main activity, as homework, or as the focus of a separate mathematics session to help students consolidate their learning and build fluency.

Students continue with their pencil case design. In this lesson, they bundle all their equipment together and measure its dimensions. They use this information to decide what shape their pencil case will be and what the dimensions will need to be to fit all the equipment.

| Differentiated outcomes | |
| --- | --- |
| **All students** | should measure items accurately and mark some measurements on scales with support. |
| **Most students** | will measure items to different degrees of accuracy and mark measurements on scales. |
| **Some students** | may check others' measurements and scale markings and correct them. |

# 6D Imperial units

## Discover  Student Book page 122 • Practice Book page 104

### Specific learning focus

- Use standard units and convert between imperial and metric units.

### Global skills

- **Creative skills:** investigating
- **Real-world skills:** research

### Key vocabulary

- imperial units, metric units

### Resources

- scissors, glue or sticky tape
- large sheets of paper
- local newspapers or magazines that students have regular access to

### Language support

Encourage students to read their sentences aloud to you. Focus particularly on the units. You may need to model the language so that students hear the correct pronunciation.

## Answers

### Student Book pages 120–121

1  Check that students have marked the correct lengths on the rulers.

2  Check that students have marked the correct capacities on the measuring jugs.

3  Check that students have marked the correct weights on the scales.

4  Check that students have drawn the lines to the correct lengths.

### Practice Book page 103

Students measure objects around them. Check that students' measurements are reasonable.

Stretch zone: Students will have individual responses for their own pencil cases. Check that the sketch of the net is reasonably accurate.

 Introductory activity

Ask students to look through the magazines and newspapers and cut out articles or advertisements that use units of measure related to length, mass or capacity. Collate the articles on a large sheet of paper then find the conversion factor to turn the measurements into **imperial units** (or if the articles use imperial units, to convert these into **metric units**).

 Main activity

Look together at page 122 of the Student Book. Display on the IWB, if possible. Refer students to the worked example, which introduces imperial units and provides an example of a metric to imperial conversion: 1 km = approximately $\frac{5}{8}$ of a mile. Students should use the internet or other resources for research to find the equivalent measurements for other imperial units on the page. They should create a list of conversions for measures of length, mass and capacity. Use this list to complete the sentences in question 1 on page 122 of the Student Book.

Ask students to work in small groups to research interesting facts about an animal to answer question 2 and plan a poster presentation for the class. Remind them that their animal facts need to use imperial units.

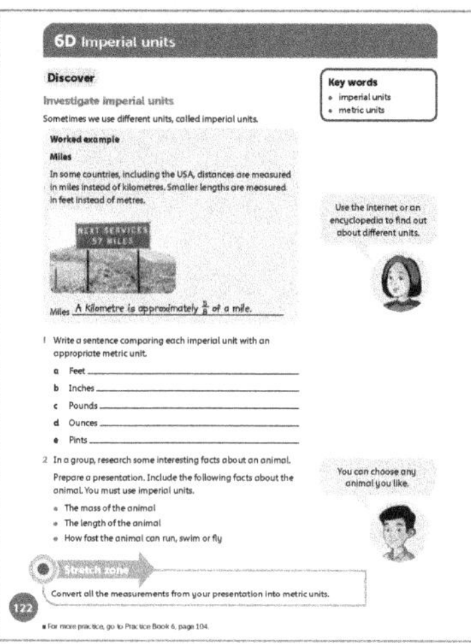

## Differentiation

**Supporting:** Help students to find examples of imperial units to add to the conversion list.

**Consolidating:** Ask students to think of examples of items that are measured in the imperial units they find.

**Extending:** Challenge students to write the calculation needed to convert from imperial to metric, and vice versa, for all the units.

**Stretch zone:** *Convert all the measurements from your presentation into metric units.*

Students can present the conversions as part of their presentations.

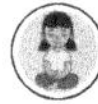 ## Reflection time

Ask each group to present their research into imperial units, sharing their examples of animals and their measures, and explain how they converted to metric units. Some students will be able to share the reverse conversions from metric to imperial units.

**Practice Book:** Students complete Practice Book page 104. They can do this directly after the Main activity, as homework, or as the focus of a separate mathematics session to help students consolidate their learning and build fluency.

Students complete imperial-to-metric conversion tables for units of length, mass and capacity. They are given the imperial measurements and the equivalent metric measurement for 1 imperial unit. They then use this to work out other measurements.

| Differentiated outcomes | |
|---|---|
| **All students** | should research imperial units and how to convert these into metric units, with support. |
| **Most students** | will research imperial units and how to convert these into metric units. |
| **Some students** | may research imperial units and be able to convert imperial measurements from and to metric units. |

## Answers

### Student Book page 122

1 Students' sentences comparing imperial and metric units will vary. They might include the following.

   a Three feet is approximately equal to one metre.

   b An inch is approximately equal to 2.5 centimetres.

   c Two pounds is a little less than 1 kilogram.

   d One ounce is just less than 30 grams.

   e One pint is just over half a litre.

2 Students will give their presentations in Reflection time.

### Practice Book page 104

| Inches | 1 | 10 | 20 | 50 | 100 |
|---|---|---|---|---|---|
| Centimetres | 2.54 | 25.4 | 50.8 | 127 | 254 |

| Yards | 1 | 10 | 20 | 50 | 100 |
|---|---|---|---|---|---|
| Metres | 0.91 | 9.1 | 18.2 | 45.5 | 91 |

| Pints | 1 | 10 | 20 | 50 | 100 |
|---|---|---|---|---|---|
| Litres | 0.57 | 5.7 | 11.4 | 28.5 | 57 |

| Pounds | 1 | 10 | 20 | 50 | 100 |
|---|---|---|---|---|---|
| Kilograms | 0.45 | 4.5 | 9 | 22.5 | 45 |

| Ounces | 1 | 10 | 20 | 50 | 100 |
|---|---|---|---|---|---|
| Grams | 28.35 | 283.5 | 567 | 1417.5 | 2835 |

Stretch zone: Check that students have measured their book accurately and converted the measurement from centimetres to inches.

# 6D Imperial units

## Explore  Student Book page 123 • Practice Book page 105

### Specific learning focus

- Convert between miles and kilometres.

### Global skills

- **Creative skills:** exploring
- **Real-world skills:** research

### Key vocabulary

- imperial units, metric units, mile, kilometre, conversion graph

### Resources

- mini whiteboards and markers
- atlases or internet access
- calculators
- counting stick

### Language support

Model the language for students relevant to graphs that they may need reminding of, for example:

- *The x- and y-axis are the horizontal and vertical axes.*
- *A conversion is used to change from imperial units to metric or vice versa.*
- *A linear graph is a straight-line graph.*

 **Introductory activity**

Write on the board the conversion 8 km = 5 **miles**. Ask students what calculation they would therefore need to do to find out how many kilometres there are to 1 mile. They should agree that it is 8 ÷ 5 = 1.6. *If we multiply the number of miles by 1.6 to find the number of kilometres, what calculation would we do to find the number of miles, if we know the number of kilometres?*

Use a counting stick and count up and back in multiples of 1.6 with students:

1.6, 3.2, 4.8, 6.4, 8, 9.6, 11.2, 12.8, 14.4, 16

Ask students to say how many kilometres is the same as each of these distances:

| | | |
|---|---|---|
| 10 miles | 50 miles | 25 miles |
| 100 miles | 12.5 miles | 200 miles |

*How can we work out each one? Which known facts can we use to work out new ones?*

 **Main activity**

Students should work in pairs. They use an atlas or the internet to find distances between different world cities. They should choose five different pairs of cities and record on their whiteboards the distance in miles between the cities in each pair. They can use the approximate conversion factor of 1 mile = 1.6 km to convert the distances into kilometres and record the distances.

Students now complete the questions on page 123 in the Student Book individually. As they work, ask how they are using the conversion rates, and check that they are completing the **conversion graph** correctly.

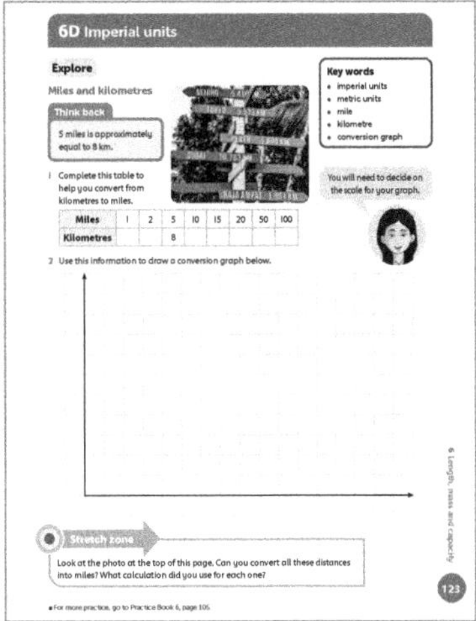

### Differentiation

**Supporting:** Help students to use multiplication to convert imperial units (miles) into metric units (kilometres).

**Consolidating:** Ask students to describe their strategy for converting between miles and kilometres.

**Extending:** Challenge students to convert from miles to kilometres and vice versa.

**Stretch zone:** *Look at the photo at the top of this page. Can you convert all these distances into miles? What calculation did you use for each one?*

Ask students whether they can identify each place on a globe or in an atlas.

 **Reflection time**

Students share their results with the class, presenting distances between one pair of cities. The class could estimate the distance between the cities before the pair reveal the true distance in miles and then explain their conversion calculation into kilometres.

**Practice Book:** Students complete Practice Book page 105. They can do this directly after the Main activity, as homework, or as the focus of a separate mathematics session to help students consolidate their learning and build fluency.

Students complete a conversion table for miles and kilometres. They are given the number of miles and they use the calculation from the Main activity to work out the equivalent number of kilometres. They then solve some word problems involving the conversion of kilometres to miles and vice versa.

| Differentiated outcomes | |
| --- | --- |
| **All students** | should convert miles to kilometres with support. |
| **Most students** | will convert miles to kilometres. |
| **Some students** | may convert kilometres to miles and miles to kilometres. |

## Answers

### Student Book page 123

**1**

| Miles | 1 | 2 | 5 | 10 | 15 | 20 | 50 | 100 |
| --- | --- | --- | --- | --- | --- | --- | --- | --- |
| Kilometres | 1.6 | 3.2 | 8 | 16 | 24 | 32 | 80 | 160 |

Check that students have plotted the conversion correctly.

# 6 Length, mass and capacity

## Connect  Student Book page 124

### Big idea

We can convert between units to solve everyday problems with different metric units and we can convert between metric and imperial units.

### Global skills

- **Creative skills:** problem solving
- **Interpersonal skills:** teamwork
- **Self-development skills:** reflecting on learning

### Key vocabulary

- metre (m), diameter, circular, rectangle, millilitres (ml)

### Resources

- rulers
- squared paper
- A4 paper
- pairs of compasses

### Language support

Develop the language of problem solving by asking, for example:

- *Are there other possible arrangements?*
- *How do you know that this is the best arrangement?*

### Practice Book page 105

**1**

| Miles | 1 | 2 | 5 | 10 | 25 | 100 |
| --- | --- | --- | --- | --- | --- | --- |
| Kilometres | 1.6 | 3.2 | 8 | 16 | 40 | 160 |

**2**  6.25 miles

**3**  62.5 miles

**4**  240 km

**5**  Hamad walks further, by 12 km.

**6**  2.5 miles

Stretch zone: Aidan is not correct. To convert miles to kilometres, you divide by 5 and multiply by 8.

 Introductory activity

Give each student a sheet of A4 paper. Check that students can measure in centimetres and draw circles of a given diameter using compasses. Model this if needed.

Now ask them to estimate how many circles they think they could draw on their sheet of A4 paper with a diameter of 5cm that do not overlap, and then check by drawing. Encourage students to make the circles close together, not with a lot of unused space between them.

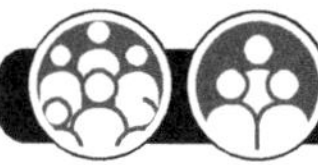 Main activity

Look together at page 124 of the Student Book. Display on the IWB, if possible. Draw students' attention to the pictures of pies and explain that a pie is made of a pastry case filled with meat, fish, fruit or other foods. The pie is usually circular and so the top and bottom need to be made of pastry circles. It is important that students, in groups, decide how to solve the problem without support.

Give each group squared paper, rulers and compasses. Give them time to think through the problem and work out their approach before you offer help. If a group is struggling to start, suggest that they speak to others who have found a way to start. The best way to approach the solution is to draw a section of the pastry (say 25 cm × 25 cm) and make cut-outs of pie bases and tops. The cut-outs can then be arranged on the pastry.

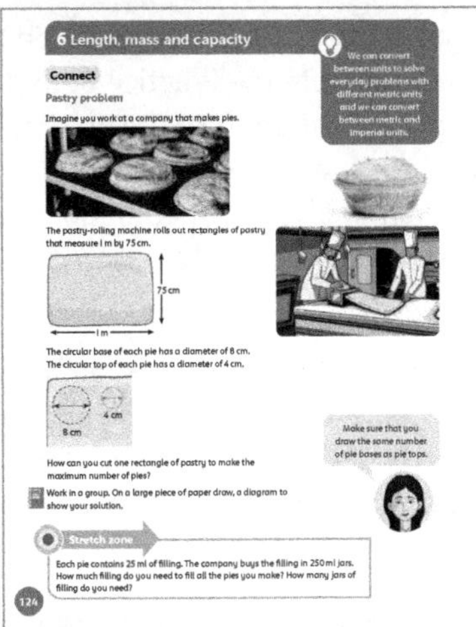

## Differentiation

Students should work in mixed-attainment groups to support one another in working through the investigation on page 124 of the Student Book.

**Supporting:** Model how to draw the pie tops and bases using compasses.

**Consolidating:** Ask students to support others in any groups who need it.

**Extending:** Challenge students to find the best possible arrangement of circles on the rectangle.

**Stretch zone:** *Each pie contains 25 ml of filling. The company buys the filling in 250 ml jars. How much filling do you need to fill all the pies you make? How many jars of filling do you need?*

## Reflection time

Ask groups to present their solutions. It is useful if alternative solutions are presented so that the class can discuss which solution and method they think is best.

The activity on page 124 of the Student Book is a good activity to carry out practically. Students can roll pastry and then re-roll the pastry when the first set of bases and tops have been made. The class can also sell the pies and think about budgeting and appropriate selling prices.

| Differentiated outcomes | |
|---|---|
| **All students** | should work in groups to solve the problem. |
| **Most students** | will calculate how many pastry circles can be cut from a sheet. |
| **Some students** | may extend the problem by changing the measures used. |

## Answers

### Student Book page 124

Students will have different solutions, some of which are a more efficient use of the pastry than others. Check that everyone's solution is reasonable.

# 6 Length, mass and capacity

## Review   Student Book page 125 • Practice Book page 106

### Global skills

- **Self-development skills:** reflecting on learning

### Student Book

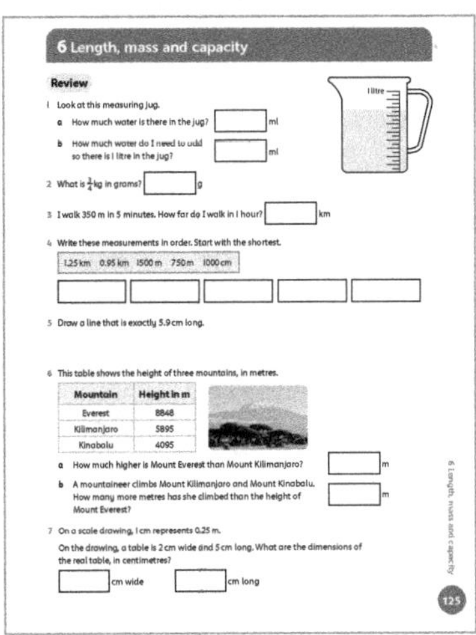

With young students, assessment activities are most effective when carried out as an everyday classroom activity. Students should be able to explore measures in different contexts and extend their previous knowledge to include conversions between metric units and from metric to imperial units. As they work on the activity on page 125 of the Student Book, they should be able to use their knowledge of measures and units. Point out how measures link to other units in mathematics, such as adding and subtracting, and multiplying and dividing.

## Answers

### Student Book page 125

**1 a** 775 ml   **b** 225 ml

**2** 750 g

**3** 4.2 km

**4** 1000 cm, 750 m, 0.95 km, 1.25 km, 1500 m

**5** Check that students have drawn a line of the correct length.

**6 a** 2953 m   **b** 1142 m

**7** 50 cm wide, 125 cm long

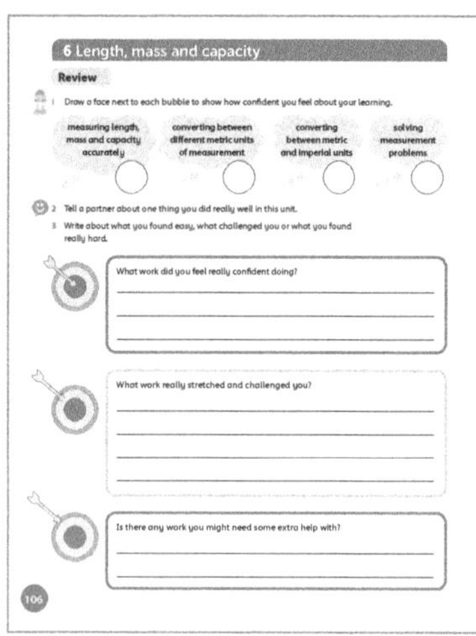

It is appropriate to complete this Practice Book Review as a whole-class discussion. You may choose to keep a record of the class discussion or a copy of the Review page for your own records. The Review provides an opportunity for students to reflect on their learning from the unit, to discuss any areas of mathematics that they feel went particularly well, and any areas that they feel less confident about. Ensure that all students have a copy of the Student Book as a reminder of the areas of mathematics that they have worked on in this unit.

Allow students plenty of time for discussion before asking them to complete the Practice Book page individually, and then, if appropriate, to share their responses with the rest of the class. If students complete this self-assessment at home, encourage them to discuss this with adults. Make a note of areas that students still feel unsure about.

As measures are found in many everyday contexts, revisit these ideas regularly. Build estimating and measuring into other aspects of everyday practice, for example when sharing things or finding a part of an amount or quantity.

## Additional material

There are additional end-of-unit assessments available on the *Oxford Owl for School* website.

# 7 Area, perimeter and volume

## Overview

### Big idea

The Big idea in this unit is the calculation of perimeters, areas and volumes using formulae. Three key concepts are covered. First is the notion that shapes with different perimeters can have different areas and vice versa. Second is the idea of using a formula linking the length and width of a rectangle to help calculate the areas of triangles and parallelograms. For a triangle, the area is half that of a rectangle with the same length and width, although these are called base and (perpendicular) height. For a parallelogram, the area will be the same as a rectangle of the same base and (perpendicular) height. This is illustrated in the following diagrams:

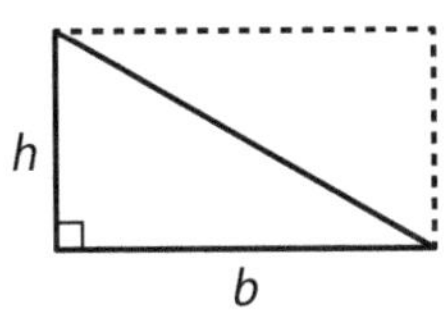 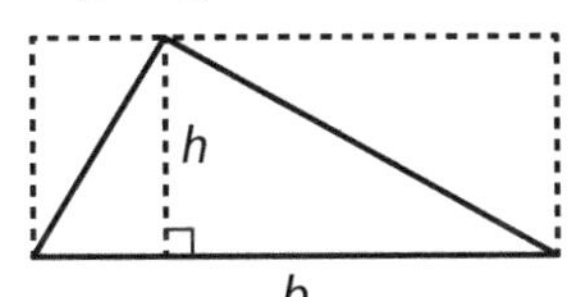

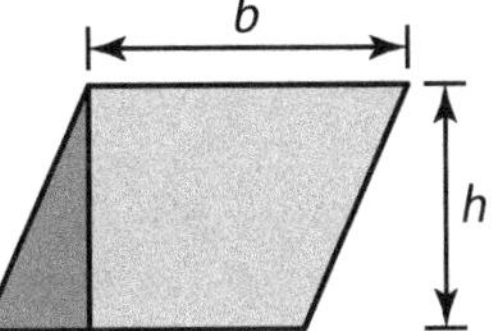 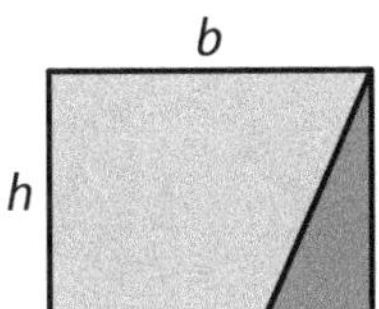

Third, the unit looks at finding the volume of shapes, using a formula. Students have met all the key ideas before. This unit expects them to apply their knowledge in a range of investigational situations.

### Look out for

- **Students who think that different shapes have different areas.** This is not necessarily so. You could give students pieces of paper that are the same size to cut up and rearrange so that they can appreciate that different shapes can have the same area.

- **Students who think that there is a relationship between the area and perimeter of shapes.** In fact, there is not. For any given perimeter, there can be a range of different areas. For any given area, there can be a range of perimeters. Students need to explore this early on in their experiences of these concepts. You could give students four small pieces of paper that are the same size. Tell them to place these side by side in different ways. This enables them to see that the perimeters are different simply by counting the edges of the pieces of paper (but that the areas are the same).

### Possible misconceptions

- **Students confuse area and perimeter.** This is an issue of language rather than a misconception within the mathematics. Thinking about the other meanings of the words can help. For example, students may understand area as a 'place'. For example, they may say, 'I live in the downtown area'. This suggests coverage, rather than something that surrounds, as in 'perimeter fence'.

- **Students do not understand the conservation of volume.** For example, they may not see that when they pour liquid from one container into a different-shaped container, the volume of liquid remains the same. Carrying out activities that involve pouring liquids into different-shaped containers or filling different-shaped containers with centimetre cubes is the best way to support students in overcoming this misconception.

### Key vocabulary

- length, width, height, perpendicular height, base, depth, breadth
- volume, dimension, edge, perimeter, area, surface, face, surface area
- formula, square centimetre (cm$^2$), square metre (m2), square millimetre (mm$^2$)
- rectilinear shape, composite shape, tangram

| Learning objective | E | 7A | 7B | 7C | 7D | C | R |
|---|---|---|---|---|---|---|---|
| Recognise that shapes with the same area can have different perimeters and vice versa. | ✓ | ✓ | | | | | |
| Recognise when it is possible to use formulae for area and volume of shapes. | | ✓ | | ✓ | ✓ | ✓ | ✓ |
| Calculate the area of parallelograms and triangles. | | | ✓ | ✓ | ✓ | | |
| Calculate, estimate and compare volume of cubes and cuboids using standard units, including cubic centimetres ($cm^3$) and cubic metres ($m^3$), and extending to other units (for example $mm^3$ and $km^3$). | | | | | ✓ | ✓ | ✓ |

# 7 Area, perimeter and volume

## Engage  Student Book page 126

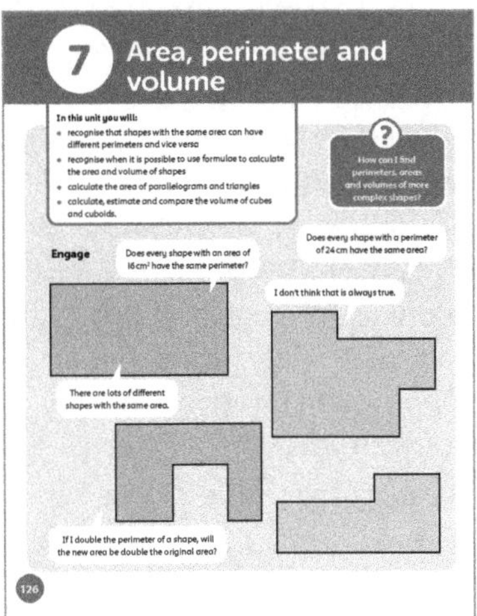

## Big question

- How can I find perimeters, areas and volumes of more complex shapes?

## Global skills

- **Creative skills:** exploring
- **Interpersonal skills:** communication
- **Self-development skills:** reflecting on learning

## Key vocabulary

- length, width, perimeter, area, surface, square centimetre ($cm^2$), square metre ($m^2$), square millimetre ($mm^2$), rectilinear

## Resources

- mini whiteboards and markers
- squared paper

## Language support

Model the use of the vocabulary and make a list of all the terms as they arise during the Main activity. As groups report back, model the correct use of the vocabulary, for example:

- *What is the width/length?*
- *How can I find the perimeter of this shape?*
- *How can I find the area of this rectilinear shape?*

Give pairs five minutes to consider this question: *What do you know about area and perimeter?* Ask them to discuss their ideas. After five minutes, take one piece of feedback from each pair.

During the feedback, list the key vocabulary on the board.

Look together at page 126 of the Student Book. Display on the IWB, if possible. Arrange the class into mixed-attainment groups of four or five students. Tell each group to explore one of the statements in the speech bubbles on page 126 of the Student Book. Different groups can explore the same question. Encourage students to try examples (using the squared paper and their whiteboards) to see whether the statements are always true, sometimes true or never true.

If there is time, students can then explore similar statements linked to volume. Ask them to consider questions such as:

- *If you double the dimensions of a cuboid, what happens to the volume?*
- *Cuboids are measured along their height, depth and breadth, so what happens if they are all doubled?*

### Differentiation

**Supporting:** Listen to the group discussion and support by clarifying any misconceptions.

**Consolidating:** Ask students to find examples to match the statements.

**Extending:** Challenge students to explain the statements to others.

 **Reflection time**

Two students from each group can report what they found out to the rest of the class. Encourage them to use the vocabulary you listed at the beginning of the lesson. Add to the word list any new vocabulary that students suggest.

# 7A Area and perimeter of rectilinear shapes

## Discover
Student Book page 127 • Practice Book page 107

### Specific learning focus
- Measure and calculate the perimeter and area of rectilinear shapes.

### Global skills
- **Creative skills:** exploring

### Key vocabulary
- perimeter, area, rectilinear shape

### Resources
- mini whiteboards and markers
- squared paper

### Language support
As you work with groups in the Main activity, model the use of the key vocabulary. For example:

- Perimeter: 'all the way around the outside', a measure of length
- Area: 'the amount of surface', a measure of 2D space
- The width of this land is …
- The length of this land is …
- The area of this land is …

During the reporting back in Reflection time, focus on the vocabulary developed in the activity.

 Introductory activity

If possible, use a large open space for this activity. Ask students to join hands and form a rectangle. Allow them to organise this for themselves. *What is the **perimeter** of the shape we have made?* (Take the answer in units of arm spans.) Now ask students to form a different rectangle. *What is the perimeter of this new shape?* (Allow time for students to realise and suggest that the perimeter remains the same whichever rectangle they form.)

After you have repeated this several times, ask pairs to sketch on squared paper the rectangles they formed and to predict and then find out what the **areas** of these rectangles are. They may be surprised that the areas are different.

 Main activity

Look together at page 127 of the Student Book. Display on the IWB, if possible. Refer students the Think back information to remind them that a **rectilinear shape** is a polygon made of straight lines that meet at right angles. Ask each student to draw on their whiteboard two rectilinear shapes with a perimeter of 24. Ask students, in pairs, to work out the areas of their four shapes initially, then to work on the activity on page 127 of the Student Book. The pairs should explore different shapes to try to find the maximum area. As students work on the activity, listen to their discussions and find a pair who are able to explain their solution as a good model to the rest of the class. Support them by asking them to practise their reporting back with you.

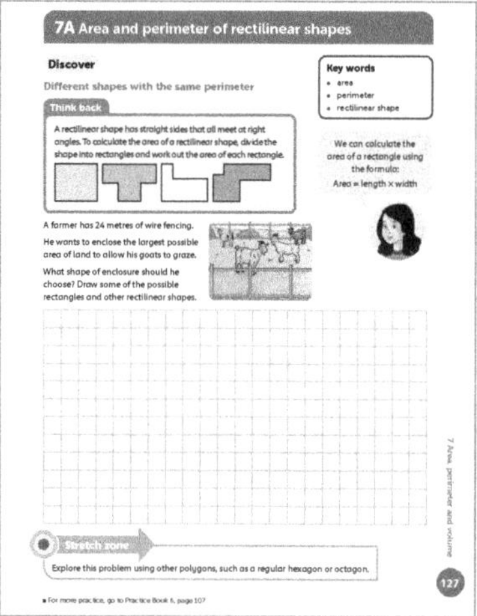

### Differentiation
**Supporting:** Support students to find the areas of the shapes they have drawn.

**Consolidating:** Ask students how they find the areas and which shapes give them the largest areas.

**Extending:** Ask students to explore a wider range of shapes, including compound shapes.

**Stretch zone:** *Explore this problem using other polygons, such as a regular hexagon or octagon.*

Students can extend the activity in the Student Book to look beyond rectilinear shapes. The maximum area is actually made by a circle of circumference 24 m.

 Reflection time

Ask the pair you selected to report back to the group. Encourage them to talk through the whole process and not just to give you their answer. Compare this solution with other solutions. As a class, agree on the best solution, which is a square with side lengths of 6 m. Explain that a circular area with a perimeter of 24 m would actually give a larger area than the square, but as circles are not rectilinear shapes, they were not included in the investigation.

**Practice Book:** Students complete Practice Book page 107. They can do this directly after the Main activity, as homework, or as the focus of a separate mathematics session to help students consolidate their learning and build fluency.

Students complete an investigation to design a school playground with an area of 40 m$^2$. They must come up with three different designs, all with the same area. *Do all your designs have the same perimeter? Which design do you think it best for a school playground?*

| Differentiated outcomes | |
| --- | --- |
| **All students** | should find areas of a range of rectilinear shapes with support. |
| **Most students** | will find an area that they can recommend as a solution. |
| **Some students** | may explore other shapes, including non-rectilinear shapes. |

# 7A Area and perimeter of rectilinear shapes

**Explore**   Student Book page 128 • Practice Book page 108

## Specific learning focus

- Measure and calculate the perimeter and area of rectilinear shapes.

## Global skills

- **Creative skills:** exploring

## Key vocabulary

- perimeter, area, rectilinear shape

## Resources

- mini whiteboards and markers
- squared paper

## Language support

As you work with students in the Main activity, model the use of key vocabulary, for example:

- *What are the dimensions of your shapes?*
- *What is the length of that side?*

During the reporting back in Reflection time, focus on the vocabulary developed in the activity.

## Answers

### Student Book page 127

Answers will vary because students draw their own rectangles and rectilinear shapes. Check that students have worked out the areas correctly.

### Practice Book page 107

Answers will vary because students draw plans of their own playgrounds that have areas of 40 square metres. Check that the areas shown are all equal.

Stretch zone: Check that students have calculated the perimeters of their designs correctly.

For their chosen design, check that the fencing cost has been calculated correctly.

 **Introductory activity**

Draw a rectangle on the board. Tell students, *The area of this rectangle is 48 cm$^2$.*

*What are possible dimensions for the length and width?* Ask pairs to write their answers on their whiteboards. Emphasise that the sketch is not to scale. Ensure that you get all possible answers ($1 \times 48$, $2 \times 24$, $3 \times 16$, $4 \times 12$, $6 \times 8$).

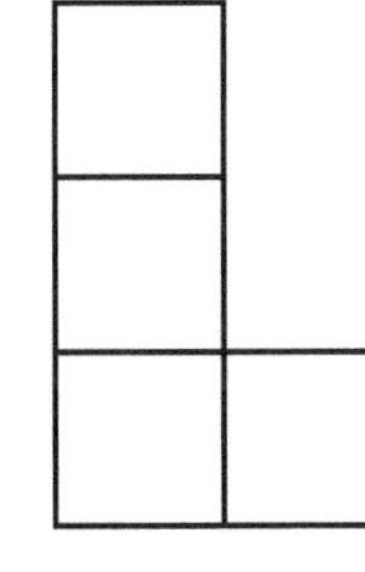

Repeat this with an L-shape, made up of four squares, with a total area of 100 cm$^2$. *What will the area of each square be? So, what will the dimensions of each square be?* Agree that each square will have side lengths of 5 cm and an area of 25 cm$^2$. And, in this case, the side lengths of the rectilinear shape are: 5 cm + 10 cm + 5 cm + 5 cm + 10 cm + 15 cm so the perimeter = 50 cm.

## Main activity

Ask students, in pairs, to explore question 1 on page 128 of the Student Book. First, they should create their own shapes, each with an area of 36 cm$^2$ and record the perimeter of each shape. Their partner should then check that their calculations are correct. Ask individuals to explain how they decided on the dimensions of the shapes. If students are drawing different-sized rectangles for their shapes, make the link to using factors of 36 to decide on the dimensions of the rectangles. If students are drawing more complex rectilinear shapes, they must first split their shape into a number of rectangles and then use their knowledge of factors for each rectangle, so that, in total, the areas add up to 36 cm$^2$.

Students go on to explore rectilinear shapes with larger areas in question 2.

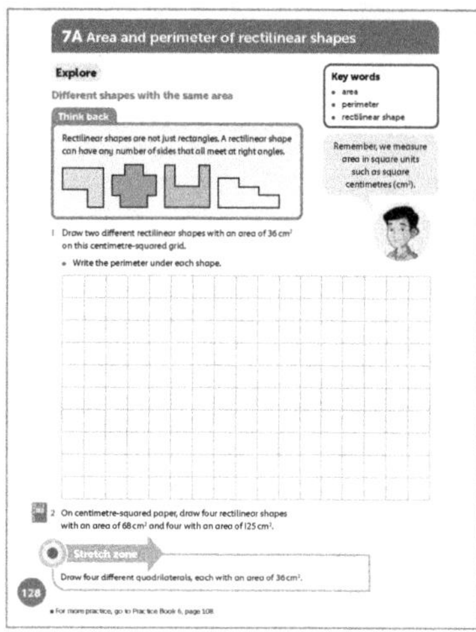

## Differentiation

**Supporting:** Support students to find the areas of their shapes.

**Consolidating:** Ask students how they find the areas and decide on the dimensions of the rectangles.

**Extending:** Ask students to explore a wider range of shapes with the given areas.

**Stretch zone:** *Draw four different quadrilaterals, each with an area of 36 cm².*

Ask students to explain how they know that their quadrilaterals all have an area of 36 cm².

You can extend the activity to work with other polygons, particularly with triangles.

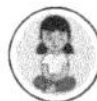 **Reflection time**

Take several shapes from students to share with the class.

Ask students whether there is a quick way to find the perimeter of a rectangle. Agree that the perimeter of a rectangle can be found using the formula $2(l \times w)$. *Can this formula be applied to the perimeter of their rectilinear shapes? Can you explain your answer?*

Remind them of the formula for rectangles:
area = length × width (or base × height)

**Practice Book:** Students complete Practice Book page 108. They can do this directly after the Main activity, as homework, or as the focus of a separate mathematics session to help students consolidate their learning and build fluency.

Students draw, on squared paper, a variety of rectilinear shapes with given areas or perimeters. *How do you know that shape has an area of 18 cm²? How do you know that shape has a perimeter of 32 cm?*

| Differentiated outcomes | |
| --- | --- |
| **All students** | should find areas and perimeters of rectilinear shapes with support. |
| **Most students** | will explain their strategies carefully. |
| **Some students** | may explore non-rectilinear shapes, such as triangles, with the given areas. |

## Answers

### Student Book page 128

Answers will vary because students draw their own rectilinear shapes for the given area. Check that the perimeters are correct for each shape.

### Practice Book page 108

Answers will vary because students draw their own rectilinear shapes for the given perimeters and areas. Check that students' shapes have the required properties.

Stretch zone: Check that students have drawn rectilinear shapes with perimeters 20 cm and found which has the largest area. This will be a square with side lengths of 5 cm.

# 7B Finding the area of triangles and parallelograms

**Discover** Student Book pages 129–130 · Practice Book page 109

## Specific learning focus

- Calculate the areas of triangles and parallelograms.

## Global skills

- **Creative skills:** investigating

## Key vocabulary

- length, perpendicular height, formula

## Resources

- squared paper

## Language support

Explain that when we talk about the formula of a rectangle, we usually use the terms 'length' and 'width', but that we could say 'base' and 'height'. When we talk about the sides of a triangle or a parallelogram, we always use the terms 'base' and 'height'. 'Height' is the perpendicular height of the shape. It is always perpendicular to the base.

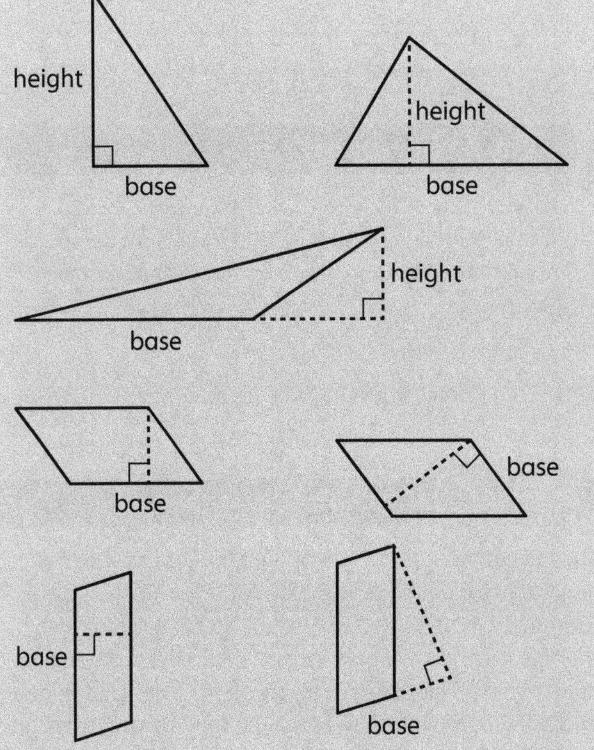




## Introductory activity

Draw on the board a rectangle labelled as having sides of 12 m by 7 m. Ask students to describe what it means to find the area of a rectangle. They should describe it as being an amount of 2D space that the shape takes up. Ask students to recall how the area of a rectangle is calculated. Agree that it is found by multiplying the **length** by the width, so in this case the area is $12 \times 7 = 84\,\text{m}^2$.

Ask students to calculate the areas of rectangles with the following dimensions, using the **formula** length × width:

5 cm × 3 cm        20 m × 4.5 m        6.4 m × 150 cm

## Main activity

Revisit the rectangle on the board with sides 12 m and 7 m. Ask students to suggest ways of drawing a single straight line across the rectangle to divide it into two equal pieces. Students may suggest drawing a vertical line down the middle, or a horizontal line across the middle. Some students might suggest drawing a diagonal line from one corner to the opposite corner. Illustrate each of these by drawing new rectangles on the board and marking in the lines as suggested.

Point to the rectangle divided vertically. Ask students whether they can say what the area of half the rectangle is. Agree that it will be half of the total area, or half of $84\,\text{m}^2 = 42\,\text{m}^2$. Now point to the rectangle divided horizontally and ask the same question.

Finally, point to the rectangle divided diagonally. Ask students whether they can tell you the area of one of the triangles formed by the diagonal. Again, it must be $42\,\text{m}^2$ because each triangle is exactly half of the rectangle.

Now ask students to work on completing the activities on pages 129–130 of the Student Book.

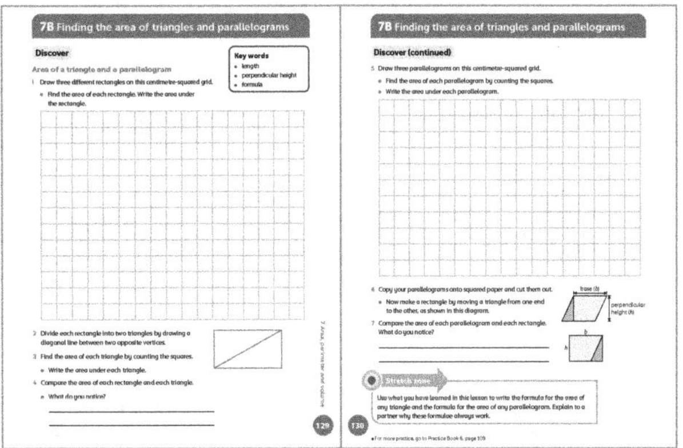

## Differentiation

**Supporting:** Help students to calculate the areas of rectangles, triangles and parallelograms.

**Consolidating:** Ask students to explain the findings of the activities and how to find the areas of triangles and parallelograms.

**Extending:** Ask students to use the formulae to find areas of triangles and parallelograms.

**Stretch zone:** *Use what you have learned in this lesson to write the formula for the area of any triangle and the formula for the area of any parallelogram. Explain to a partner why these formulae always work.*

Students should explain the formula to others and use examples to show how it works.

area of a triangle $= \frac{1}{2}$ (base $\times$ height)

area of a parallelogram $=$ base $\times$ height

 ## Reflection time

Invite students to share what they found out about the areas of rectangles, triangles and parallelograms. Establish that the areas of triangles and parallelograms can be found using their dimensions in similar ways to rectangles. *Are you surprised by this? Why?*

**Practice Book:** Students complete Practice Book page 109. They can do this directly after the Main activity, as homework, or as the focus of a separate mathematics session to help students consolidate their learning and build fluency.

Students should only complete the Practice Book activity at this point if they have worked through the Stretch zone activity. Otherwise, they can complete it once they have finished 7B Explore. The Practice Book activity depends on students being familiar with the formulae for the areas of triangles and parallelograms. Students sketch triangles and parallelograms with a given area. They use squared paper to help them.

| Differentiated outcomes | |
|---|---|
| **All students** | should find the areas of rectangles, triangles and parallelograms with support. |
| **Most students** | will find the areas of rectangles, triangles and parallelograms. |
| **Some students** | may find the areas of rectangles, triangles and parallelograms using formulae. |

## Answers

### Student Book pages 129–130

Students draw their own shapes and calculate the areas. They should notice that the areas of the triangles are half the areas of the corresponding rectangles and that the areas of the parallelograms are the same as the areas of the corresponding rectangles.

### Practice Book page 109

Students draw their own shapes. Check that they have calculated the areas correctly.

Stretch zone: This will come from a shape that students have made earlier, so check in each case that they have calculated the area of the parallelogram correctly.

---

# 7B Finding the area of triangles and parallelograms

## Explore    Student Book page 131 • Practice Book page 110

### Specific learning focus

- Use formulae to calculate the areas of triangles and parallelograms.

### Global skills

- **Creative skills:** exploring
- **Self-development skills:** reflecting on learning

### Key vocabulary

- perpendicular height, formula

### Resources

- squared paper

### Language support

Support students by verbalising the formulae and relating them to the shapes and their dimensions. For example, when saying 'half the base times the height', ask:

- *What is the base?*
- *What is the height?*

 ## Introductory activity

Look together at page 131 of the Student Book. Display on the IWB, if possible. Refer students to the Think back formulae. Discuss the two formulae for the areas of triangles and parallelograms and relate them to what was revealed in 7B Discover. Draw some triangles on the board and ask different students to come forward and mark on them the base and height. Use a range of triangles: right-angled, scalene, isosceles and equilateral. Point out that which side you choose as the base will determine where the height can be marked.

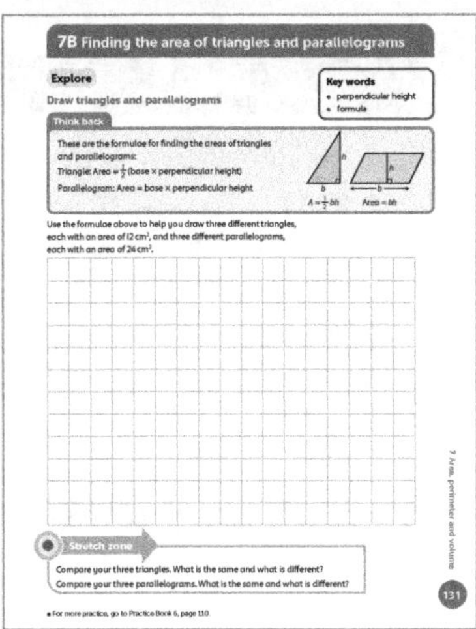

# Main activity

Write the formula for the area of a triangle on the board:

$$\text{area} = \frac{1}{2} \times \text{base} \times \text{height}$$

Say that you wish to draw a triangle with an area of 18 cm². *How can we decide what measurements to make the base and height? If half the base × height = 18, then what must the base × height be?* Ask students for possible dimensions and record them on the board, for example $1 \times 36$, $2 \times 18$, $3 \times 12$, $4 \times 9$ or $6 \times 6$. Ask students to comment on what they notice about the dimensions.

Repeat, using the formulae for the area of a parallelogram, and ask students for possible dimensions of the base and height for a parallelogram with an area of 48 cm².

Now ask students to complete the activities on page 131 of the Student Book.

## Differentiation

**Supporting:** Help students to draw triangles and parallelograms of a given area by using the formulae.

**Consolidating:** Ask students to draw triangles and parallelograms of a given area by using the formulae.

**Extending:** Ask students to explain how to use the formula to draw triangles and parallelograms of a given area.

**Stretch zone:** *Compare your three triangles. What is the same and what is different? Compare your three parallelograms. What is the same and what is different?*

Students should notice the dimensions as being related to factors of the area in each case.

 **Reflection time**

Invite students to share how they drew triangles and parallelograms of the given area using the formulae to help them decide on the dimensions of the bases and heights. Compare results and see how many different shapes were drawn that have the same area.

**Practice Book:** Students complete Practice Book page 110. They can do this directly after the Main activity, as homework, or as the focus of a separate mathematics session to help students consolidate their learning and build fluency.

Students are given triangles and parallelograms with dimensions marked. They need to use the formula to work out what the area of each shape is.

| Differentiated outcomes | |
| --- | --- |
| **All students** | should draw triangles and parallelograms of a given area with support. |
| **Most students** | will draw triangles and parallelograms of a given area using the formulae. |
| **Some students** | may explain how to draw triangles and parallelograms of a given area using the formulae. |

# Answers

## Student Book page 131

Students draw their own triangles and parallelograms. Check that they have the correct dimensions for the areas given.

## Practice Book page 110

**1** 6.25 cm²

**2** 12.5 cm²

**3** 8.75 cm²

**4** 12 cm²

**5** 24 cm²

**6** 22.5 cm²

**7** 11 cm²

**8** 24 cm²

Stretch zone: The shapes in questions 4 and 8 have the same bases and heights, but the triangle is half the area of the parallelogram.

# 7C Calculating areas of irregular shapes

## Discover
**Student Book page 132 • Practice Book page 111**

### Specific learning focus
- Calculate the area of shapes in a tangram puzzle.

### Global skills
- **Creative skills:** exploring

### Key vocabulary
- tangram, dimension, area, formula

### Resources
- Resource sheet 7.1: tangram cut into pieces – one set of pieces per student

### Language support
Remind students of these points:
- Perimeter is a length. We measure perimeter in centimetres.
- Area is a 'count' of the number of square centimetres covered by the shape.
- We measure area in square centimetres.

 ## Introductory activity

Look together at page 132 of the Student Book. Display on the IWB, if possible. Ensure that all students can see the **tangram**. Ask students to work in pairs to calculate the area of each piece of the tangram. Take feedback and write the area of each part on the board. Note that square D on the tangram is marked as 4 cm × 4 cm. This will allow students to work out the area of all the other pieces.

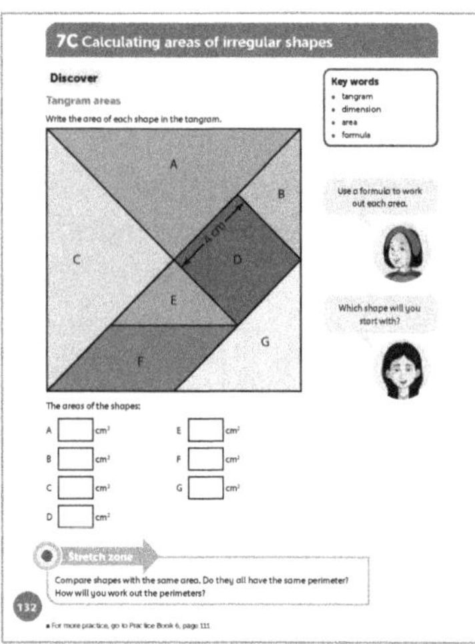

 ## Main activity

Ask each student to use Resource sheet 7.1. Ask students to use each of the seven pieces once to make two different shapes that have equal area. *Using the result from the Introductory activity, what will the area of each of the two shapes be?* (64 cm²) Ask students to draw their shapes. Repeat, asking students to find other pairs of shapes they can make that have the same area using each of the seven pieces once. Encourage students to create their shapes and then swap with a partner to check their answers.

### Differentiation
**Supporting:** Support students to find the areas of the different parts of the tangram.

**Consolidating:** Ask students to explain how they found the areas of the tangram pieces.

**Extending:** Ask students to find the fraction of the total area of the tangram that each shape makes.

**Stretch zone:** *Compare shapes with the same area. Do they all have the same perimeter? How will you work out the perimeters?*

Students explore finding the perimeters given that square D has sides of 4 cm. The perimeter of the right-angled triangles and parallelogram can only be calculated using Pythagoras' Theorem, which is beyond the expectation of Year 6. This is not expected but students could measure the sides of those they cannot work out.

 ## Reflection time

Work through the pieces of the tangram and ask different students to explain how they worked out the area of each piece. They should be encouraged to use comparative statements relating a piece to other pieces, perhaps comparing the edge lengths or the fact that one piece is a multiple or a fraction of another piece.

*Which pieces were easiest to find the area of? Which pieces were harder? Why?*

**Practice Book:** Students complete Practice Book page 111. They can do this directly after the Main activity, as homework, or as the focus of a separate mathematics session to help students consolidate their learning and build fluency.

Students draw composite shapes using rectangles, triangles and parallelograms. Then they find the area of each of their composite shapes by using what they know about the areas of the individual shapes that make up their shapes.

<table>
<tr><td colspan="2">Differentiated outcomes</td></tr>
<tr><td>All students</td><td>should find the areas of the tangram shapes with support.</td></tr>
<tr><td>Most students</td><td>will find the areas of the tangram shapes.</td></tr>
<tr><td>Some students</td><td>may explore fractions using the tangram pieces.</td></tr>
</table>

## Answers

### Student Book page 132

A 32 cm²     B 8 cm²     C 32 cm²     D 16 cm²
E 8 cm²     F 16 cm²     G 16 cm²

### Practice Book page 111

Answers will vary because students draw their own shapes on the squared paper. Check that the areas are correct.

Stretch zone: Answers will vary because students draw their own shapes on the squared paper. Check that the areas are all equal.

# 7C Calculating areas of irregular shapes

## Explore
Student Book page 133 · Practice Book page 112

### Specific learning focus

- Explore the areas of shapes made from tangram pieces.

### Global skills

- **Creative skills:** exploring

### Key vocabulary

- area, formula, composite shape

### Resources

- Resource sheet 7.1: tangram cut into pieces – one set of pieces per student

### Language support

Reinforce for students the language relating to areas of non-regular shapes. For example, help students to express the area formulae for squares, rectangles, triangles and parallelograms in words and explain what they mean for each shape.

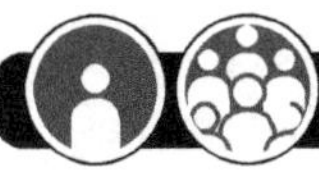

## Introductory activity

Ask each student to use two tangram pieces to make a new shape. Their shapes should be made from pieces with matching sides that they can line up. Ask students to make as many different shapes as they can from their chosen pieces.

Can they say what the areas of their new shapes are?

## Main activity

Ask students to work in pairs to use three tangram pieces and make as many new shapes as they can from them, in the same way as in the Introductory activity.

Ask students to look at the shapes they have made. *Do any of them have the same total area? What happens if you take three of the tangram pieces and make three different shapes using the same three shapes? What is the area of each **composite shape**? Why does this happen?*

Ask students to complete the activities on page 133 of the Student Book. As they are working, ask students questions to support their understanding, for example: *Are there any quick ways to find the area of your shapes? What formulae are you using?*

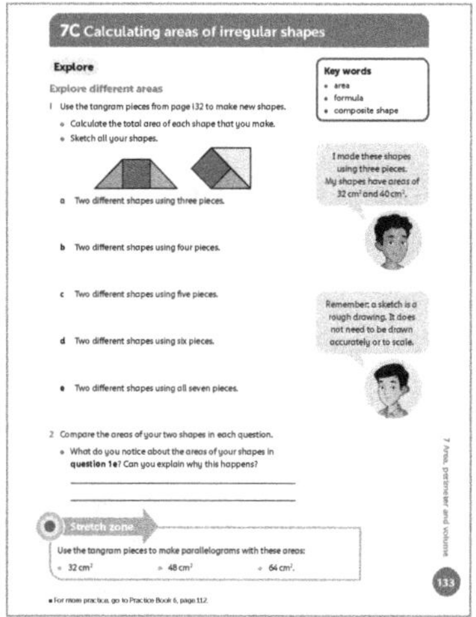

## Differentiation

**Supporting:** Support students to find the areas of the shapes they make.

**Consolidating:** Ask students to explain how they find areas.

**Extending:** Ask students to find the fraction of the total area of the square that each shape makes.

**Stretch zone:** *Use the tangram pieces to make parallelograms with these areas:*

- 32 cm²
- 48 cm²
- 64 cm².

Students should be encouraged to look back at the areas of the individual pieces to help them make parallelograms of the given areas.

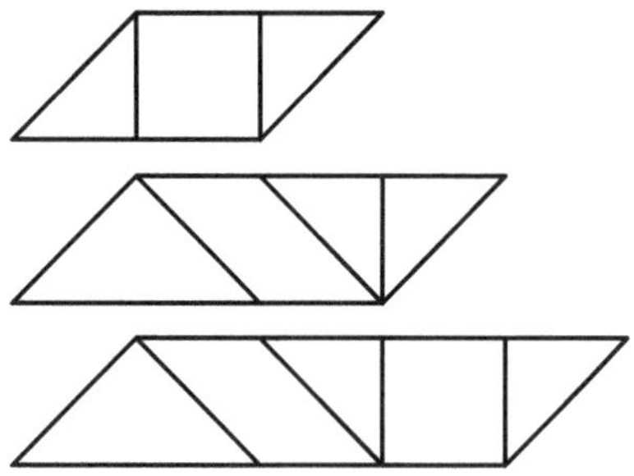

### Reflection time

Ask different students to share the shapes they made for each question and explain how they worked out the area of each one. They should be encouraged to use comparative statements relating a shape to other shapes, comparing one shape as a multiple or a fraction of another shape.

Ask them which shapes were easiest to find the area of. Which shapes were harder? Why?

**Practice Book:** Students complete Practice Book page 112. They can do this directly after the Main activity, as homework, or as the focus of a separate mathematics session to help students consolidate their learning and build fluency.

Students follow instructions to draw composite shapes made up of given shapes, with a specific combined area. *If the total area is 30 cm², what will the area of your square be? What will the area of the triangle then have to be?*

| Differentiated outcomes | |
|---|---|
| **All students** | should make a range of shapes using some of the tangram pieces and find the areas with support. |
| **Most students** | will find the areas of the shapes they make from some of the tangram pieces. |
| **Some students** | may explore fractions using the tangram pieces and the shapes they make from some of the pieces. |

### Answers

#### Student Book page 133

Students make their own arrangements of the tangram pieces. Check that the areas of students' shapes are correct.

Students should recognise that when the seven pieces are used to make different shapes, the area does not change.

#### Practice Book page 112

Students draw their own composite shapes. Check that the shapes are labelled and have the correct areas.

Stretch zone: Students draw their own composite shape. Check that each student's shape is labelled and has the correct area.

---

# 7D Calculating volume

**Discover** Student Book page 134 · Practice Book page 113

### Specific learning focus

- Compare volumes of different cuboids.

### Global skills

- **Creative skills:** exploring

### Key vocabulary

- volume, length, width, height

### Resources

- assorted packets that are cuboids
- sheets of A4 and A3 paper
- items to fill cuboids with, for example cubes, sweets, large marbles

### Language support

Model the use of the vocabulary associated with cuboids and their volumes. For example:

- *What is the **width**/length/**height** of the cuboid?*
- *What is its **volume** in cubic centimetres?*
- *Does this cuboid have a greater volume that that one? How do you know?*

###  Introductory activity

Put students into small groups and provide each group with a few different cuboid packets, such as cereal boxes. Ask groups to try to order their cuboids from smallest volume to largest. Can they explain how they decided on the order? What reasoning did they use to say that one cuboid had more volume than another?

###  Main activity

Students should now use cubes, or something similar, to test their order and see how many of the chosen item each cuboid will hold. Discuss which sort of item is easier to use to compare the volumes – large items or smaller items. *Would something like rice be useful for comparing volumes? Why?*

Look together at page 134 of the Student Book. Display on the IWB, if possible. Refer students to the Think back formula. Can students use this formula and measure their cuboids to calculate the volume of each one in cubic centimetres?

Give out sheets of A4 and A3 paper and ask students to complete the activities on page 134.

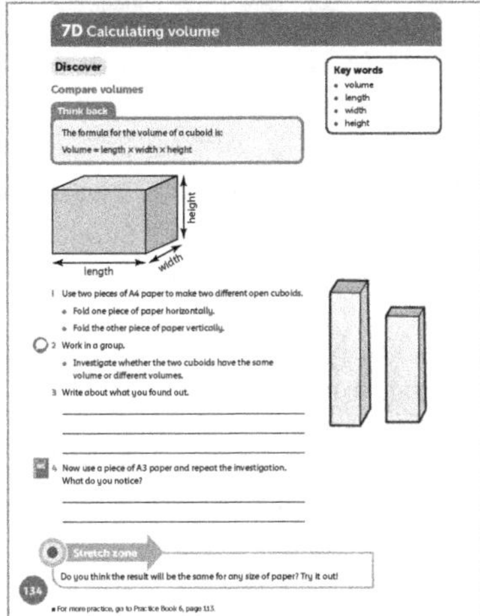

## Differentiation

Join in with group discussions and encourage all students to make a contribution. You should also encourage students to explain their ideas carefully so that all the other members of the group understand.

**Supporting:** Help students to order the cuboids by volume from smallest to largest by filling the containers with the same item and counting how many of the item each container holds.

**Consolidating:** Ask students to estimate the volume of each cuboid before measuring.

**Extending:** Challenge students to explain to the groups how to use the formula to find the volume. Can they say why the formula works?

**Stretch zone:** *Do you think the result will be the same for any size of paper? Try it out!*

Encourage students to use sheets of paper that are not in the international 'A' sizes.



## Reflection time

Take feedback from each group. Encourage them to talk through their results. Did they estimate correctly which cuboid would have greater volume? Ask them to explain how they checked – was it by filling each one, or measuring and calculating using the formula? *What did you find out about the volume of the cuboids you made by folding the paper? How did you measure the volume of each cuboid?*

**Practice Book:** Students complete Practice Book page 113. They can do this directly after the Main activity, as homework, or as the focus of a separate mathematics session to help students consolidate their learning and build fluency.

Students use the formula for finding volume to work out the volume of various cuboids, given their dimensions. They look for patterns between the dimensions and the volume of the cuboids.

| Differentiated outcomes | |
|---|---|
| **All students** | should estimate and order cuboids by volume with support. |
| **Most students** | will estimate and order cuboids by volume. |
| **Some students** | may calculate volumes of cuboids using the formula and explain why the formula works. |

## Answers

### Student Book page 134

**1–3** Students should find that folding an A size piece of paper horizontally will give a greater volume cuboid than folding it vertically. For an A4 sheet, the vertical fold gives a cuboid of volume $819\,cm^3$ and the horizontal fold gives a volume of $1158\,cm^3$.

**4** With an A3 piece of paper, the horizontal fold also has a greater volume than the vertical fold:

vertical = $23\,155\ cm^3$

horizontal = $32\,744\ cm^3$

### Practice Book page 113

| | Length (cm) | Width (cm) | Height (cm) | Volume (cm³) |
|---|---|---|---|---|
| **1** | 6 cm | 2 cm | 2 cm | 24 cm³ |
| **2** | 6 cm | 3 cm | 3 cm | 54 cm³ |
| **3** | 6 cm | 4 cm | 4 cm | 96 cm³ |
| **4** | 6 cm | 5 cm | 5 cm | 150 cm³ |
| **5** | 6 cm | 6 cm | 6 cm | 216 cm³ |
| **6** | 6 cm | 7 cm | 7 cm | 294 cm³ |

**7** The volumes increase by multiples of $6\,cm^3$.

**8** Each volume is 6 times a square number.

Stretch zone: False. Doubling the dimensions will give 8 times the volume, because $2 \times 2 \times 2 = 8$.

# 7D Calculating volume



### Specific learning focus

- Calculate volumes of cuboids.

### Global skills

- **Creative skills:** investigating

### Key vocabulary

- area, volume

### Resources

- a cuboid cardboard box, such as a cereal packet

### Language support

Use students who can act as good role models in Reflection time so that all students hear the appropriate terms and phrases from the key vocabulary. For example:

- The length is …
- The width is …
- The height is …
- The volume is …

 **Introductory activity**

Show students a cuboid packet. Ask them how they can find the volume of the packet. Recall with them the formula for volume from page 134 of the Student Book as volume = length × width × height.

Now say that you know that the volume of the packet is 160 cm³, and you know that the length and width of the packet are 8 cm and 5 cm. Without using a ruler to measure the height, ask students to discuss in pairs how they might calculate the height from the other information they have. Take feedback from pairs about their methods and establish, through discussion that the height is 4 cm, because 8 × 5 × 4 = 160. Ask students to say how they calculated it, perhaps by finding the missing factor in 8 × 5 × __ = 160 and then finding 160 ÷ (8 × 5) or 160 ÷ 40 = 4.

Write the volumes of three different cuboids on the board, along with two of the three dimensions, as follows:

Cuboid A:   volume = 180 cm³, height = 6 cm, length = 10 cm. (width = 3 cm)

Cuboid B:   volume = 240 m³, width = 2.5 m, length = 8 m. (height = 12 m)

Cuboid C:   height = 30 cm, volume = 36 000 cm³, width = 40 cm. (length = 30 cm)

Ask students to calculate the third dimension of each cuboid, either mentally or by a written method of their choice. Share answers and methods.

 **Main activity**

Tell students that a pet shop sells three different fish tanks. One is slightly longer than the other two, one is slighter wider and one is slightly taller. Ask them to use the dimensions to calculate the volumes of all three fish tanks to put them in order from smallest volume to largest.

Fish tank 1:   length = 60 cm, width = 36 cm, height = 34 cm. (volume = 73 440 cm³)

Fish tank 2:   length = 64 cm, width = 36 cm, height = 34 cm. (volume = 78 336 cm³)

Fish tank 3:   length = 60 cm, width = 40 cm, height = 32 cm. (volume = 76 800 cm³)

Look together at page 135 of the Student Book. Display on the IWB, if possible. Ask students to look at the illustration of the swimming pool. Explain that it is not drawn to scale. Explain how the swimming pool surround can be thought of as 4 separate cuboids, or as one large cuboid with the piece taken out of the middle. Students work in pairs to answer the questions on page 135.

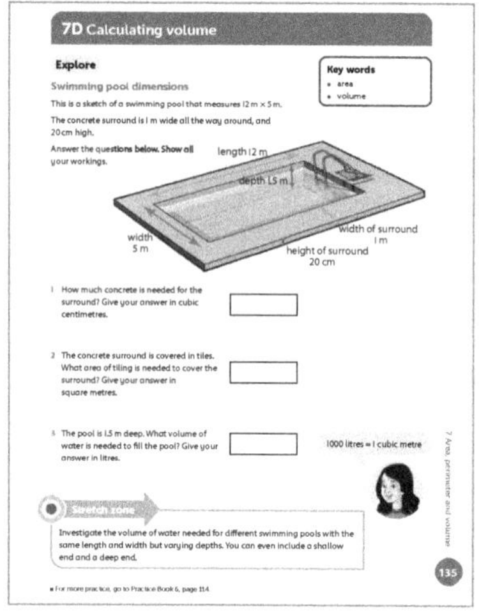

## Differentiation

**Supporting:** Help students to calculate the volumes and areas of the swimming pool.

**Consolidating:** Ask students to explain how they find areas and volumes. Make links to factors.

**Extending:** Ask students to create their own word problems about volumes of cuboids.

**Stretch zone:** *Investigate the volume of water needed for different swimming pools with the same length and width but varying depths. You can even include a shallow end and a deep end.*

Students might research the depth of the local swimming pool and calculate the volume of water it holds.

 **Reflection time**

Ask students to compare their answers. *Which method did you use to calculate the volume of concrete for the swimming pool surround and the tiled area?* Take feedback from different pairs.

**Practice Book:** Students complete Practice Book page 114. They can do this directly after the Main activity, as homework, or as the focus of a separate mathematics session to help students consolidate their learning and build fluency.

Students draw nets for cuboids and then work out what the volume of the folded nets would be. You may choose to do this activity following the Connect lesson.

| Differentiated outcomes | |
|---|---|
| **All students** | should calculate volumes of cuboids with support. |
| **Most students** | will calculate volumes of cuboids. |
| **Some students** | may calculate volumes of cuboids and explain their reasoning. |



**Student Book page 135**

**1** $7\,600\,000\,\text{cm}^3$

**2** $38\,\text{m}^2$

**3** $90\,000\,\text{l}$

**Practice Book page 114**

Answers will vary because students draw nets of boxes. Check that the surface area and the perimeter are correct.

Stretch zone: Various answers including halve the length, halve the width, halve the height.

# 7 Area, perimeter and volume

## Connect  Student Book page 136

### Big idea

I can find areas and volumes of a wide range of shapes by using what I know about simple shapes and by using formulae.

### Global skills

- **Creative skills:** problem solving
- **Self-development skills:** reflecting on learning

### Key vocabulary

- length, width, height, area, surface area, square centimetre ($\text{cm}^2$)

### Resources

- mini whiteboards and markers
- squared paper
- rulers
- cubes

### Language support

Encourage students to use the vocabulary of volume when working on the problem. Ask questions, for example:

- *What is the volume of each stock cube?*
- *What shape cuboid will look attractive to customers?*
- *How many ways can I design a box for 40 stock cubes?*

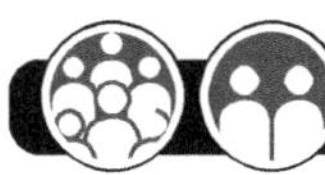 **Introductory activity**

Ask students to imagine that they need to pack a liquid that has a volume of $36\,\text{cm}^3$. *How many different cuboids can hold this volume of liquid?* Give each pair five minutes to find as many solutions as they can and write them on their whiteboards. Take feedback and select one of the answers. Ask a student to sketch the cuboid on the board. As a class, decide what the net for this cuboid is. Calculate the **surface area** and perimeter.

 **Main activity**

The Introductory activity allows you to identify students who may need initial support with the activity on page 136 of the Student Book. You may wish to group these students together and focus on them at the beginning of this part of the lesson. Some students may find that opening up boxes helps them to understand how to draw the net of a cuboid. Make sure that students have access to cubes so that they can use practical materials to support them.

Ask students to work in groups of four. Encourage them to try to interpret and begin the activity without any teacher input.

If you feel as though the class need additional support, ask students to calculate the volume of one stock cube and then look for factors of the total volume of stock cubes.

Groups sketch nets to represent the stock cube boxes. You could encourage students to make the nets too, so that they can try them using cubes. Encourage students to see this as a real-life activity and to give good reasons for the choices they make. During the activity, collect information from different students about their box designs and record this on the board. You can use this in Reflection time.

**Unit 7 Area, perimeter and volume**  179

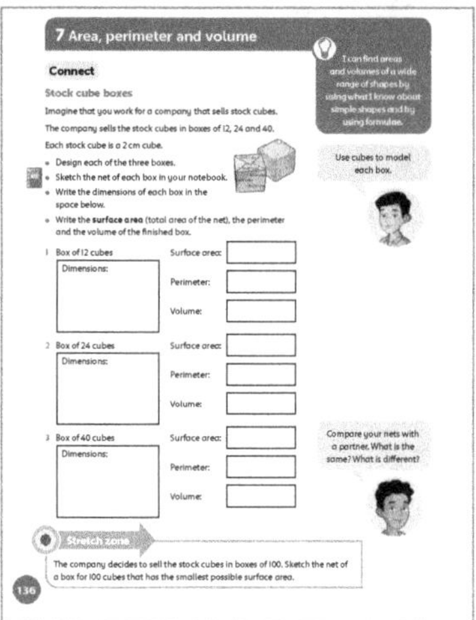

## Differentiation

Students should work in mixed-attainment groups to support one another in working through the activity on page 136 of the Student Book.

**Stretch zone:** *The company decides to sell the stock cubes in boxes of 100. Sketch the net of a box for 100 cubes that has the smallest possible surface area.*

Ask students why they think having the smallest surface area is desirable.

# 7 Area, perimeter and volume

## Review Student Book page 137 • Practice Book page 115

## Student Book

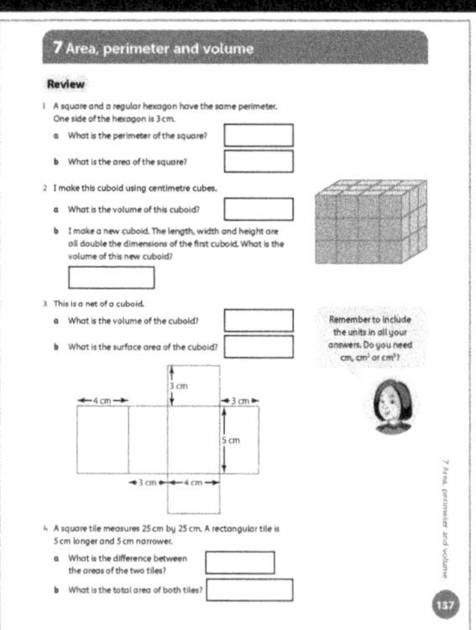

With young students, assessment activities are most effective when carried out as an everyday classroom activity. Students should be able to explore calculation of perimeters, areas and volumes in different contexts. As they work on the Student Book activity on page 137, they should be able to use their knowledge of measures and units and apply this to cuboids. Point out how measures link to other units in mathematics, such as adding and subtracting, and multiplying and dividing.

## Reflection time

Ask groups to present their solutions. Ask the class to decide which design for a box they prefer and to give reasons for this choice. Groups can also share their method for calculating the volumes and dimensions of their cuboids.

| Differentiated outcomes | |
| --- | --- |
| **All students** | should find surface areas and volumes with support. |
| **Most students** | will find surface areas and volumes. |
| **Some students** | may find surface areas and volumes and explain their reasoning. |

## Answers

### Student Book page 136

Students design their own boxes for holding different numbers of stock cubes. Check that their calculations are correct.

## Answers

### Student Book page 137

1   **a** 18 cm      **b** 20.25 cm²

2   **a** 36 cm³      **b** 288 cm³

3   **a** 60 cm³      **b** 94 cm²

4   **a** 25 cm²      **b** 1225 cm²

## Practice Book

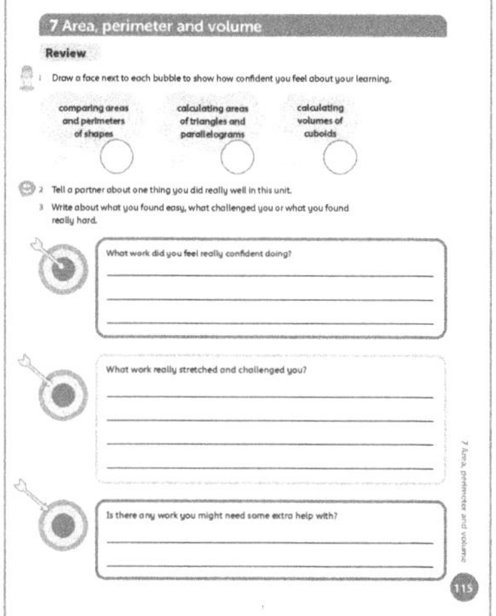

It is appropriate to complete this Practice Book Review as a whole-class discussion. You may choose to keep a record of the class discussion or a copy of the Review page for your own records. The Review provides an opportunity for students to reflect on their learning from the unit, to discuss any areas of mathematics that they feel went particularly well, and any areas that they feel less confident about. Ensure that all students have a copy of the Student Book as a reminder of the areas of mathematics that they have worked on in this unit.

Allow students plenty of time for discussion before asking them to complete the Practice Book page individually, and then, if appropriate, to share their responses with the rest of the class. If students complete this self-assessment at home, encourage them to discuss this with adults. Make a note of areas that students still feel unsure about. As measures are found in many everyday contexts, revisit these ideas regularly. Build estimating and measuring into other aspects of everyday practice, for example when filling containers or covering a surface.



There are additional end-of-unit assessments available on the *Oxford Owl for School* website.

# 8 Time



## Big idea

The Big idea for this unit is that time is measured using a variety of different units and we often have to convert between these units to solve problems. Students need to become familiar with how many seconds make up a minute, how many minutes make up an hour, how many hours make up a day and so on. They also need to become confident at reading time in 12-hour and 24-hour formats.

One new idea for students in this unit is that of different time zones. Usually, students who travel have an understanding of this concept but some may find it very difficult to understand that it is not the same time in every place in the world. It is helpful to remind students that, because the Earth turns on its axis, it is night in half the world when it is day in the other half. Midday in one part of the world is midnight in an opposite part of the world.

## Look out for

- **Students who think the time is same all around the world.** Use a globe to explain how it is daylight in a place when it is on the side of the Earth facing the sun, and night-time when it faces away from the Sun, meaning that it is day and night at different times for different places.

- **Students who cannot convert between different time units.** Support their knowledge of the vocabulary and facts, for example that 1 day = 24 hours, and 1 hour = 60 minutes.

- **Students who struggle in using the 24-hour clock.** Students are used to a decimal place-value system based on 10, so they often forget that an analogue clock is based on 12 hours, with 24 hours in one day. The best way to overcome this is to provide plenty of practice.

## Possible misconceptions

- **Students read 20:00 as 10 p.m. instead of 8 p.m. as they struggle to convert between 24-hour and 12-hour clock times (and vice versa).** Remind students how to work out 12-hour clock times from 24-hour clock times and vice versa. Explain that, from 13:00 onwards, we subtract 12 from the 24-hour clock times. Converting 23:00 hours to 12-hour clock time can be helpful when discussing this, because it cannot be expressed as a 12-hour clock time plus 10. It is 23:00 − 12 = 11 p.m.

- **Students may think that each time in a timetable is a different bus or train.** Help them to explore timetables to see that each column of times represents the stops for one bus or train.

## Key vocabulary

- digital clock, analogue clock, consecutive, 12-hour clock, 24-hour clock, millisecond, second
- hour, day, week, month, year, decade, century, millennium
- a.m., p.m.
- midday, midnight
- timetable, arrive, depart, time zones, time difference

## Coverage in lessons

| Learning objective | E | 8A | 8B | 8C | C | R |
|---|---|---|---|---|---|---|
| Solve problems involving the calculation and conversion of units of measure, using decimal notation up to three decimal places where appropriate. | ✓ | ✓ | ✓ | ✓ | ✓ | ✓ |
| Use, read, write and convert between standard units, converting measurements of length, mass, volume and time from a smaller unit of measure to a larger unit, and vice versa, using decimal notation up to three decimal places. | ✓ | ✓ | ✓ | ✓ | ✓ | ✓ |

# 8 Time

## Engage  Student Book page 138

### Big question

- How can I convert between units of time to solve problems and calculate time intervals?

### Global skills

- **Creative skills:** exploring
- **Real-world skills:** research
- **Interpersonal skills:** communication
- **Self-development skills:** reflecting on learning

### Key vocabulary

- millisecond, second, hour, day, week, month, year, decade, century, millennium, a.m., p.m., midday, midnight, timetable, arrive, depart, 24-hour clock, 12-hour clock, analogue clock, digital clock

### Resources

- large sheets of paper
- marker pens

### Language support

Work with each group in turn as they create their posters. Ask individual students to read out parts from the poster. This allows you to check their pronunciation and to model correct pronunciation. Refer back to the key vocabulary list to encourage students to use the correct vocabulary.

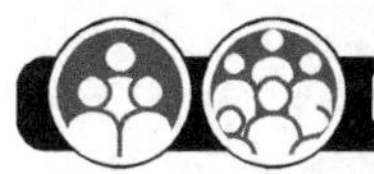 Introductory activity

Organise the class into mixed-attainment groups of four students. Ask them to write down as many examples as they can of ways they use time. Encourage them to include the key vocabulary. Ask them to try to classify this list using a mind map. After ten minutes, rotate the sheets of paper to the next group so that each group begins to add to each poster. After 30 minutes, select one of the posters. Ask the group to present this.

 Main activity

Look together at page 138 of the Student Book. Display on the IWB, if possible. Ask each group to select one of the discussion questions from the speech bubbles. Groups can choose to discuss the same question. Listen to students during their discussions. Make a note of the mathematical vocabulary they use. Write this on the board to refer to in Reflection time.

If students struggle to know where to start with their chosen question, help them by breaking down the question into smaller steps and asking, for example, *What do you need to find out first? Then what do you need to find out? What calculations do you need to do?*

### Differentiation

Students will all use and hear new vocabulary and hear pronunciation modelled by other students. As this is an Introductory lesson, listen carefully to students' discussions so that you can plan for your future teaching.

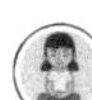 Reflection time

Look at the mathematical vocabulary you listed on the board. Ensure that it includes most of the key vocabulary shown opposite on this page. Ask groups to make posters with definitions of the words. They can refer to the posters throughout the unit.

# 8A Converting between units of time

## Discover  Student Book page 139 • Practice Book page 116

### Specific learning focus

- Convert times on digital and on analogue clocks.

### Global skills

- **Creative skills:** problem solving

### Key vocabulary

- 24-hour clock, 12-hour clock, digital clock, analogue clock, consecutive

### Resources

- analogue clock large enough for whole-class use
- mini whiteboards and markers
- analogue watches or clocks for individual use if possible

### Language support

Asking students to speak out their thinking in the Introductory activity and in Reflection time is important. Encourage students to develop extended responses, saying, for example, *Tell me a bit more about that. How do you know that is the correct answer?* Encourage students to use the 'analogue' form, for example:

- ten past one
- quarter to five.

Compare this with the digital forms, for example:

- one ten
- four forty-five.

 **Introductory activity**

*How many of you have analogue watches or **analogue clocks** at home?* Discuss the differences between analogue times and digital times, including the use of a.m. and p.m. Use the large analogue clock at the front of the class to show students a range of times. For every time you show, ask students to write on their whiteboards the same time using the digital **24-hour clock** form. Use the following times, moving the hands on by the specified time intervals:

- 09.24 a.m.
- 45 minutes later (10:09)
- 2 hours 15 minutes later (12:24)
- 1 hour 30 minutes later (13:54)
- 5 hours 20 minutes later (19:14)

After each time, ask, *How did you calculate the new time?* Use any errors as a teaching point. Work with students to identify where they made the mistake, modelling the correct workings using a number line to jump on the hours and minutes and splitting the minutes when crossing an hour boundary. This can also help other students to avoid making the same mistakes.

 **Main activity**

Look together at page 139 of the Student Book. Display on the IWB, if possible. Direct students to look at the activity title. *Does anyone know what '**consecutive**' means?* Agree on a definition for the term so that students are clear about the task of finding consecutive digits in digital times. Use the examples on the Student Book page to demonstrate that consecutive digits can go forward or back, for example 345 or 654.

Students work through the activities in pairs to support one another and to check answers. It may be helpful for students to have access to analogue clocks on their tables. They can use these to turn the hands to make the times that they write initially as digital times. You may need to remind students to look carefully at the position of the minute hand.

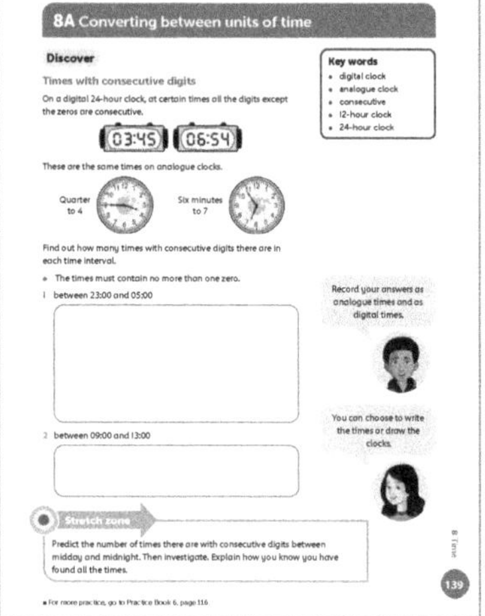

### Differentiation

**Supporting:** Ask students to say the times as analogue times as they write them down.

**Consolidating:** Ask students to find how long it is between some of the times they write down.

**Extending:** Ask students to find how long it is between all the times they write down and to look for any patterns.

**Stretch zone:** *Predict the number of times there are with consecutive digits between midday and midnight. Then investigate. Explain how you know you have found all the times.*

There are only two times with consecutive digits (using no more than one zero) between midday and midnight: 12:34 and 23:45.

 **Reflection time**

Ask students to discuss the following question with a partner: What are the most important things to remember when writing digital times as analogue times? After two minutes, take feedback from pairs. Students might suggest that they must remember to make every time have four digits and to check whether the time is in 12-hour format or 24-hour format, for example. List all the feedback somewhere where it can be referred to for the rest of the unit.

**Practice Book:** Students complete Practice Book page 116. They can do this directly after the Main activity, as homework, or as the focus of a separate mathematics session to help students consolidate their learning and build fluency.

Students are given a start time and a duration. They write the start time on a 24-hour digital clock, then work out the end time, given the duration, writing the end time on an analogue clock. Remind students to use a number line and jottings to work out the end times, drawing jumps for the hours and minutes.

| Differentiated outcomes | |
|---|---|
| **All students** | should convert times between digital and analogue clocks with support. |
| **Most students** | will convert times between digital and analogue clocks. |
| **Some students** | may convert times between digital and analogue clocks using mental methods. |

## Answers

### Student Book page 139

**1** 23:45, 01:23, 02:34, 03:21, 03:45, 04:32, 04:56

**2** 12:34

### Practice Book page 116

**1** 11:18; analogue clock face shows end time of forty-six minutes past two

**2** 16:35; analogue clock face shows end time of ten past ten

**3** 22:17; analogue clock face shows end time of quarter to three

**4** 23:55; analogue clock face shows end time of thirty-four minutes past seven

Stretch zone: For example, 09:15 and 12:40, 11:42 and 13:07, 17:02 and 20:27

# 8A Converting between units of time

**Explore** Student Book pages 140–141; Practice Book page 117

### Specific learning focus

- Solve problems using units of time.

### Global skills

- **Creative skills:** problem solving

### Key vocabulary

- millisecond, decade, century, millennium

### Resources

- flashcards with one of the following 'units of time' vocabulary on each of them: millennium, century, decade, year, month, week, day, hour, minute, second, millisecond
- pot of lollipop sticks with each student's name on a separate stick
- calculators
- calendar

### Language support

As you work with students, encourage them to explain their working. Ask, for example:

- *How are you doing the calculation?*
- *Why are you multiplying by 24 (or 60)?*

Encourage students to talk to one another about their answers so that they can check that their answers make sense. Students are all probably a similar number of days old, for example. They may ask one another:

- 'How many days old are you?'
- 'How many milliseconds are there in a second?'

### Introductory activity

Pick a lollipop stick out of the pot. Ask the named student to come to the front and pick a flashcard. Repeat with other students. When each student picks a flashcard, ask them to stand in a row, in ascending order (with **'millisecond'** on the left of the row and **'millennium'** on the right of the row). Each time a student picks a card, ask them to state one fact they know about that unit of time, for example: 'There are 100 years in a **century**.' or 'There are 60 seconds in 1 minute.'

Ask students to work in pairs to discuss and complete the table on page 140 of the Student Book. When students have completed the table, go through the answers so that all students have the correct information before they move on to the questions. Check any errors with students to find out what mistakes they made.

You may choose to group students by prior attainment for the questions on pages 140–141 so that you can support them in carrying out the calculations. Students can use calculators when answering the questions. Refer students to the speech bubbles. Remind them to account for an extra day in leap years.

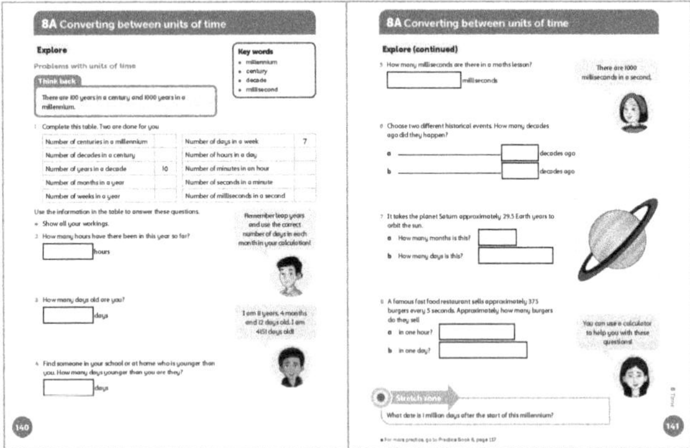

## Differentiation

**Supporting:** Model the calculations and explain how to carry out each calculation.

**Consolidating:** Support students in carrying out the calculations.

**Extending:** Ask students to share their strategies for each question.

**Stretch zone:** *What date is 1 million days after the start of this millennium?*

Students may calculate the year or they may try to be more precise and find the date. There are, on average 365.24 days in a year. So in 1 000 000 days, there are 2737.9 years (to one decimal place). This means that, when we add this number of years to the 2000 years we have already had before the start of this millennium, the year would be 4737, although close to 4738.

 ## Reflection time

Put students into mixed-attainment pairs. Pick two flashcards at random. *How many of the smaller unit are there in the larger unit?* The more-confident student should explain to the less-confident student how to work this out. Take feedback from one of the pairs, asking, *How did you carry out the calculation?*

Repeat as many times as you think is necessary.

**Practice Book:** Students complete Practice Book page 117. They can do this directly after the Main activity, as homework, or as the focus of a separate mathematics session to help students consolidate their learning and build fluency.

Students solve problems that involve converting between units of time. Allow them to use a calculator where necessary as some of the calculations are complex.

| Differentiated outcomes | |
| --- | --- |
| **All students** | should solve time problems in different units with support. |
| **Most students** | will solve time problems in different units. |
| **Some students** | may solve time problems in different units and explain their strategies. |

## Answers

### Student Book pages 140–141

**1**

| | |
| --- | --- |
| Number of centuries in a millennium | 10 |
| Number of **decades** in a century | 10 |
| Number of years in a decade | 10 |
| Number of months in a year | 12 |
| Number of weeks in a year | 52 |
| Number of days in a week | 7 |
| Number of hours in a day | 24 |
| Number of minutes in an hour | 60 |
| Number of seconds in a minute | 60 |
| Number of milliseconds in a second | 1000 |

**2–6** Answers will vary. Check that students' answers are reasonable.

**7 a** 354 months   **b** 10 768 approximately

**8 a** 270 000   **b** 6 480 000

### Practice Book page 117

**1** Answers will vary. Check that students' answers are reasonable.

**2** 2080 weeks

**3** 119 decades

**4** 2737 (approximately)

Stretch zone: Answers will vary. Check that students' answers are reasonable. For example, if a student is at school for 6.5 hours each day, they will be there for 23 400 000 milliseconds.

# 8B Using the 24-hour clock and timetables

## Discover  Student Book page 142 • Practice Book page 118

### Specific learning focus

- Read and use timetables using the 24-hour clock system.

### Global skills

- **Creative skills:** problem solving
- **Real-world skills:** interpreting information

### Key vocabulary

- 24-hour clock, timetable

### Resources

- mini whiteboards and markers
- lollipop sticks with students' names (reuse from 8A Explore)

### Language support

Model the use of the 24-hour clock saying, for example: 'oh five seventeen' for 05:17.

 Introductory activity

Ask each pair to write down on their whiteboards two times that relate to an event from yesterday. Ask them to use the 24-hour clock format and to include one time from the morning and one time from the afternoon or evening. The event could be, for example, the time I woke up: 07:00, the time I left home to go to school: 08:15, the time I finished my homework: 16:25, the time I went to football practice: 18:15.

Use the lollipop sticks to select six students. Ask them to arrange themselves in a line at the front of the class in ascending order, from the earliest event to the latest event. Ask students to work in pairs to calculate the time gaps between the events. Take feedback from pairs asking them to describe their strategies.

 Main activity

Discuss the idea of a **timetable**. Ask students to say where they might see a timetable and what it is for. Provide some examples: school timetables, bus or train timetables or a timetable of a day's activities. Look together at page 142 of the Student Book. Display on the IWB, if possible. Direct students to the worked example. Ask what they think this timetable shows. *Why do you think someone might use a timetable for a day trip?* Students work in pairs to create timetables for the activity on page 142 of the Student Book.

Encourage students to use times that are not simply half and quarter hours on the school day and the holiday timetables.

### Differentiation

**Supporting:** Help students to create timetables to the nearest quarter-hour.

**Consolidating:** Ask students to create timetables to the nearest five minutes.

**Extending:** Challenge students to create timetables to the nearest minute.

**Stretch zone:** *Write a two-part word problem based on one of your timetables. Give your problem to a partner to solve.*

As an additional activity, ask students to keep diaries of the day's or week's events. They can create a timetable to show the activities and use these to write some word problems.

### Reflection time

Ask one pair with an interesting holiday timetable to come to the front. Ask them to write their timetable on the board. Each pair should create a question based on this timetable and swap it with another pair.

**Practice Book:** Students complete Practice Book page 118. They can do this directly after the Main activity, as homework, or as the focus of a separate mathematics session to help students consolidate their learning and build fluency.

Students create their own school timetable based on their ideal week at school. They write the days that they attend school along with the times for each lesson. They should use the 24-hour clock for each time.

| Differentiated outcomes | |
|---|---|
| **All students** | should create timetables to the nearest quarter hour. |
| **Most students** | will create timetables to the nearest five minutes. |
| **Some students** | will create timetables to the nearest minute. |

**Student Book page 142**

Students create their own timetables.

**Practice Book page 118**

Students write their own school timetables. Check that the facts they wrote about the timetables are appropriate.

Stretch zone: Check that students have included sensible answers and referred to times.

# 8B Using the 24-hour clock and timetables

## Explore  Student Book page 143 • Practice Book page 119

### Specific learning focus

- Read and use timetables using the 24-hour clock.

### Global skills

- **Creative skills:** problem solving

### Key vocabulary

- 24-hour clock, timetable

### Resources

- large sheets of paper

### Language support

Display real-life bus timetables to see the sorts of places that students could include on their timetable. Make a list, in their first language, of all the places they could include and then write the English translation.

### Introductory activity

Look together at page 143 of the Student Book. Display on the IWB, if possible. Work with the whole class to model the activity. First, list six possible places that need a bus stop in the local area (one of which is the 'bus station'). Draw a simple sketch map starting (and ending) at a bus station and joining these places together. Take suggestions from students. *How long does it take a bus to travel between … and …?* Write these times on the sketch map. Then, starting at the bus station with a starting time of 08:45, write down the times that the bus arrives at, and departs from, each stop.

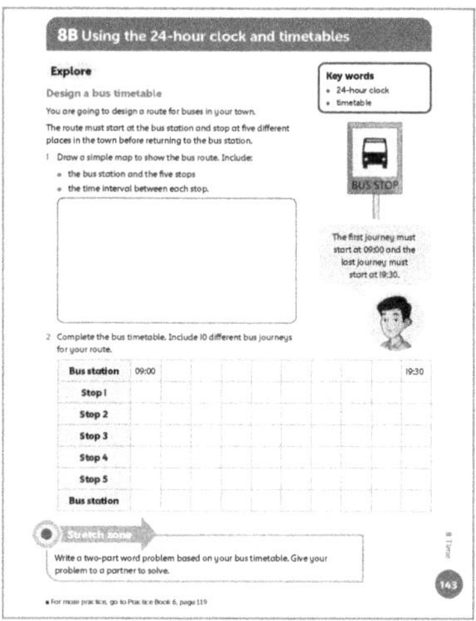


### Main activity

Arrange for students to work in mixed-attainment groups. Using a large sheet of paper, each group draws a sketch map of their bus route and writes the bus timetable for their route. Encourage students to give specific feedback on the routes. Students can look at other groups' routes and offer feedback before each group completes a final version based on the comments of other students. They can check, for example, that timings are accurate based on the sketch maps or that the bus stops at the most important places on the map. When they are sure that they have an accurate map and timetable, ask them to describe their routes and timetables to you. They can then each make an individual copy on page 143 of the Student Book.

### Differentiation

**Supporting:** Help students to construct the timetable. Allow them to work in their first language before translating for them.

**Consolidating:** Ask students to explain how they know what the time intervals should be.

**Extending:** Ask students to create word problems based on the timetable.

**Stretch zone:** *Write a two-part word problem based on your bus timetable. Give your problem to a partner to solve.*

As a follow-up activity, ask students to find bus timetables for the local area. They can then write problems, based on these timetables, for their partner to solve.

### Reflection time

Choose one group to present their bus route and timetable to the rest of the class. Encourage other students to ask questions and to give feedback on the effectiveness of the route.

**Practice Book:** Students complete Practice Book page 119. They can do this directly after the Main activity, as homework, or as the focus of a separate mathematics session to help students consolidate their learning and build fluency.

Students design a flight timetable, based on given information. Encourage them to plan out their timetable in their notebook or on a piece of paper before writing the final version in the Practice Book. *What information from the list will you start with?*

| Differentiated outcomes | |
|---|---|
| **All students** | should understand how a timetable is constructed. |
| **Most students** | will use the timetable to calculate time intervals. |
| **Some students** | may create word problems based on the timetable. |

## Answers

### Student Book page 143

Students design their own bus timetables. Check that these seem reasonable.

### Practice Book page 119

Students design their own flight timetables. Check that these seem reasonable.

Stretch zone: Accept any reasonable answers, for example those that meet all the rules, or that have flights spread out fairly evenly throughout the day.

---

## 8C Time zone problems

### Discover Student Book page 144 · Practice Book page 120

#### Specific learning focus

- Appreciate how the time is different in different time zones around the world.

#### Global skills

- **Creative skills:** problem solving
- **Real-world skills:** interpreting information

#### Key vocabulary

- time zone, time difference

#### Resources

- large map of the world showing time zones
- mini whiteboards and markers
- lollipop sticks with students' names (reuse from 8A Explore)

#### Language support

Model the use of the language of comparison, for example:

- *How much earlier is that?*
- *How much later is it?*
- *How many hours before … ?*
- *How many hours after … ?*

 **Introductory activity**

Ask students, in pairs, to list on their whiteboards all the different places in the world that they have visited or where their family members live. Select students by picking lollipop sticks. Ask each selected student to choose a place from their list. Find the place on the large world map and work out the current time in that place using the international **time zones**. Keep a list of all the times and places on the board.

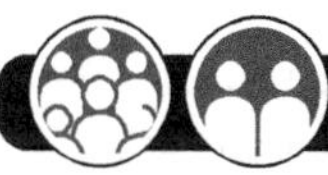 **Main activity**

Look together at page 144 of the Student Book. Display on the IWB, if possible. Refer students to the time zone map and then to the two speech bubbles. These give some interesting facts based on **time differences** around the world. Work through each example, explaining, or asking a student to explain, how they know these facts are correct. Use the large map with time zones to refer to.

Ask students to work in the same pairs as for the Introductory activity to complete the activities on page 144 of the Student Book. Encourage them to write down a range of facts. Share any interesting facts with the whole class as a model of good practice.

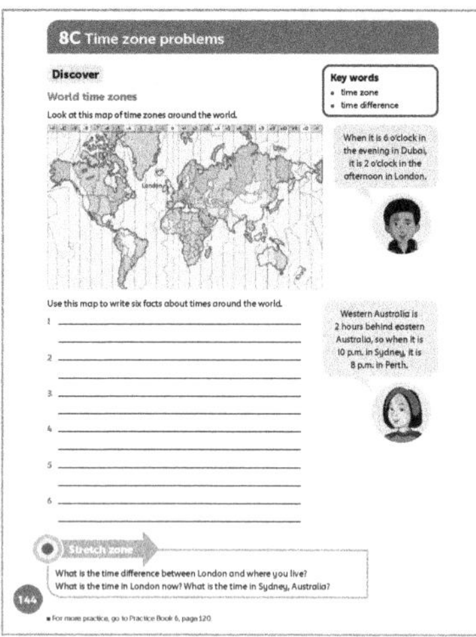

## Differentiation

**Supporting:** Help students who have no experience of time zones.

**Consolidating:** Ask students to share their strategies for finding time differences.

**Extending:** Ask students to create two-step word problems based on their facts.

**Stretch zone:** *What is the time difference between London and where you live? What is the time in London now? What is the time in Sydney, Australia?*

As a follow-up activity, ask students to look at airline timetables to explore time zones. In particular, ask them to look at journeys when you appear to arrive before you leave!

 **Reflection time**

Use the lollipop sticks again to select students. Ask each student to share one of their facts. Ask other students to check that this fact is correct, using the large display map. Then ask each pair to create a word problem based on their facts. They should exchange this problem with another pair.

**Practice Book:** Students complete Practice Book page 120. They can do this directly after the Main activity, as homework, or as the focus of a separate mathematics session to help students consolidate their learning and build fluency.

Students solve word problems relating to time zones, having been given times in different cities around the world. Explain that all the times on the clocks show the time, using 24-hour format, in that city at the same time as the other cities. For example, when it is 10:25 in Mexico City, it is 17:25 in Cairo.

| Differentiated outcomes | |
|---|---|
| **All students** | should appreciate how the time is different in different time zones around the world. |
| **Most students** | will find time differences in different time zones. |
| **Some students** | may create word problems based on different time zones. |

## Answers

### Student Book page 144

Accept any sensible answers.

### Practice Book page 120

**1**  10:00

**2**  18:45

**3**  13:45

**4**  21:25

**5**  21:00

**6**  20:30

**7**  01:00

Stretch zone: Answers will vary depending on students' location.

# 8C Time zone problems

## Explore
Student Book page 145 • Practice Book page 121

### Specific learning focus
- Appreciate how the time is different in different time zones around the world.

### Global skills
- **Creative skills:** problem solving
- **Real-world skills:** interpreting information

### Key vocabulary
- time zone, time difference

### Resources
- large map of the world showing time zones
- lollipop sticks with students' names (reuse from 8A Explore)

### Language support
When students talk in pairs during Reflection time, listen to their conversations. You can use students who are confident in English to act as models for others in the class.

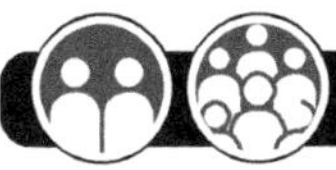 **Introductory activity**

Use the list of times that you created in 8C Discover, or repeat the activity to create a new set of times and places. Ask pairs:

- *Imagine that you want to telephone (select a city from your list, for example Beijing) now.*
- *What might be happening there now?*
- *What might be happening there 8 hours later than now?*

Repeat for three different cities from your list, all in different time zones.

 **Main activity**

Students should work in the same pairs to complete the time zone activities on page 145 of the Student Book. Where possible, these should include a student with experience of time zones. Students may need to use the map of the world to locate places within time zones. Encourage students to show all their workings and then to share answers with one another to check that they are carrying out calculations accurately.

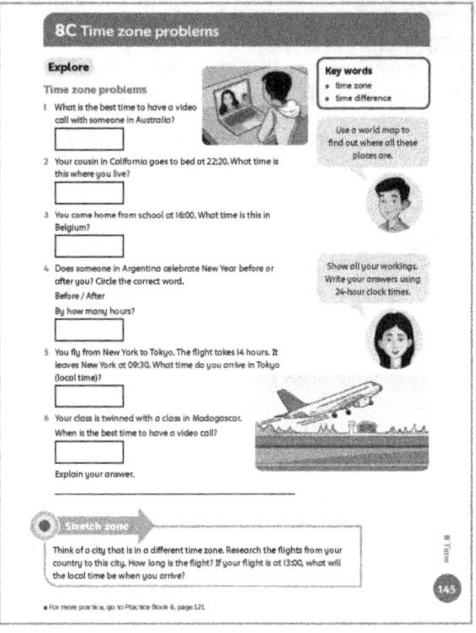

## Differentiation

**Supporting:** Help students who have no experience of time zones.

**Consolidating:** Ask students to share their strategies for finding time differences.

**Extending:** Challenge students to create two-step word problems.

**Stretch zone:** *Think of a city that is in a different time zone. Research the flights from your country to this city. How long is the flight? If your flight is at 13:00, what will the local time be when you arrive?*

Remind students to factor in both the duration of the flight and the time difference in the arrival country.

 **Reflection time**

Ask students to discuss the following in pairs:

- *What have you learned in 8C Discover and 8C Explore?*
- *What are you surprised about?*
- *What did you already know?*

Take feedback from each pair in turn.

As an additional activity, if your school has a link to another school or a town in a different country, ask students to keep diaries of their activities during the day and link them to those of students in other countries. For example, ask, *What are students in Sudan doing when you are going to school?*

**Practice Book:** Students complete Practice Book page 121. They can do this directly after the Main activity, as homework, or as the focus of a separate mathematics session to help students consolidate their learning and build fluency.

Students complete a flight timetable based on flights from Dubai. For each journey in the timetable they are given the flight departure time and the duration. They must write the arrival time, in local time, taking into account the time difference and the flight duration.

<table>
<tr><td colspan="2">Differentiated outcomes</td></tr>
<tr><td>All students</td><td>should appreciate how the time is different in different time zones around the world.</td></tr>
<tr><td>Most students</td><td>will find time differences in different time zones.</td></tr>
<tr><td>Some students</td><td>may create their own word problems based on different time zones.</td></tr>
</table>

## Answers

### Student Book page 145

**1–4** Accept any sensible answers, based on the local time in your country.

**5** 12.30 the next day

**6** Accept any sensible answer.

### Practice Book page 121

| | |
|---|---|
| Dhaka | 17:40 |
| Tokyo | 00:55 (the next day) |
| Nairobi | 18:00 |
| Phnom Penh | 01: 35 (the next day) |
| Kuala Lumpur | 06:40 (the next day) |
| Toronto | 04:10 (the next day) |
| Anchorage | 03:55 (the next day) |
| London | 02:05 (the next day) |

Stretch zone: 03:40

---

# 8 Time

## Connect  Student Book page 146

### Big idea

I can convert between units of time using multiplication or division. I can use time conversion facts and timetables to solve real-life time problems.

### Global skills

- **Creative skills:** problem solving
- **Real-world skills:** interpreting information
- **Interpersonal skills:** teamwork
- **Self-development skills:** reflecting on learning

### Key vocabulary

- hour, day, week, a.m., p.m. timetable, arrive, depart, 24-hour clock, 12-hour clock

### Resources

- large sheets of paper

### Language support

As the groups prepare their presentations, remind them of key vocabulary they can use. You could display some key language structures on the board for them to use, For example:

- We will leave school at …
- We will spend … hours on the island.
- We will leave the island at …

 ## Introductory activity

Arrange for the class to work in mixed-attainment groups of four students for the activity on page 146 of the Student Book. Keep the introduction short. It is important that students try to understand the problem for themselves, including thinking about how they might represent it and the solution. After students have had five minutes interpreting the problem, you may wish to clarify the meaning of any of the vocabulary that students do not understand.

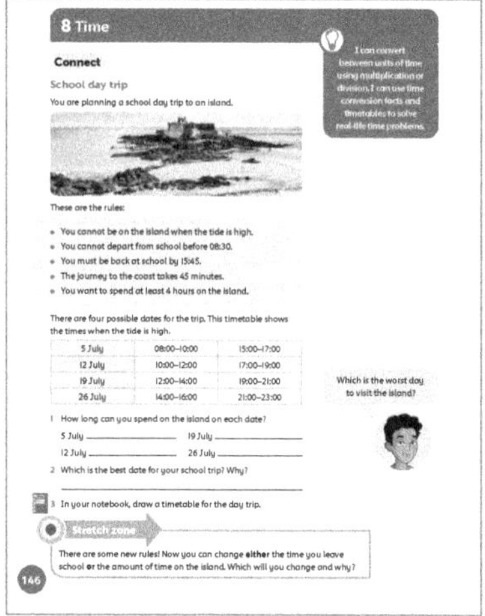

 ## Main activity

For groups who are struggling to start their work on the problem, suggest that they first draw out a timetable for the day, in hour intervals, so they can 'map out' the visit. Students may need help in understanding low and high tide if this is a concept that they have not met before. You could shade in the times, on their timetable for the day, for when there is a high tide. You could use a different colour to show the travelling time. As you work with the class, select two groups with different solutions to present in Reflection time.

# Differentiation

Students should work in mixed-attainment groups to support one another in working through the investigation on page 146 of the Student Book.

**Stretch zone:** *There are some new rules! Now you can change either the time you leave school or the amount of time on the island. Which will you change and why?*

 **Reflection time**

Ask the two selected groups to present their recommendations. Other students can then decide which timetable they think is best.

| Differentiated outcomes | |
| --- | --- |
| **All students** | should participate in a group discussion. |
| **Most students** | will participate in the discussion and calculate possible trip dates. |
| **Some students** | may calculate the best possible trip and give reasons why. |

# 8 Time

## Review  Student Book page 147 • Practice Book page 122

### Global skills

- **Self-development skills:** reflecting on learning

### Student Book

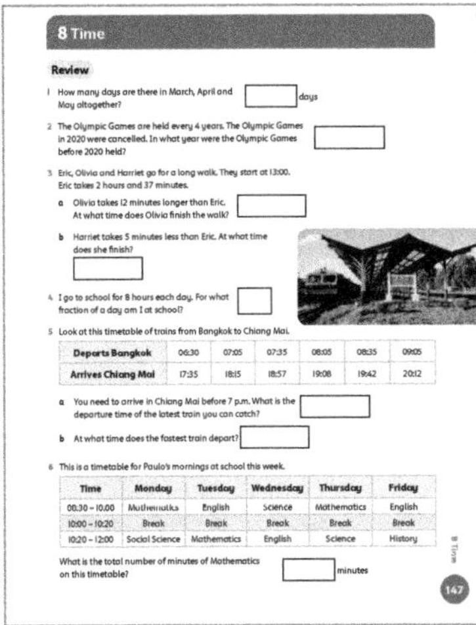

With young students, assessment activities are most effective when carried out as an everyday classroom activity. Students should have timetables, calendars and both analogue and digital clocks available to support them. Watch as students calculate the number of months, days, hours and minutes for the questions, and observe their strategies for working them out.



### Student Book page 146

1  5 July – 5 hours

 12 July – 3 hours

 19 July – 2 hours 45 mins

 26 July – 4 hours 45 mins

2  5 July: this day allows 5 hours to explore the island.

3  Timetables will vary slightly, departing school any time between 08:30 and 09:15, to get to the destination between 09:15 and 10:00. For example:

 09:15 – depart from school

 10:00 – arrive on island

 15:00 – depart from island

 15:45 – arrive at school

## Answers

### Student Book page 147

1  92 days

2  2016

3  a  15.49    b  15:32

4  $\frac{1}{3}$

5  a  07:35    b  08:05

6  4 hours 40 minutes

## Practice Book

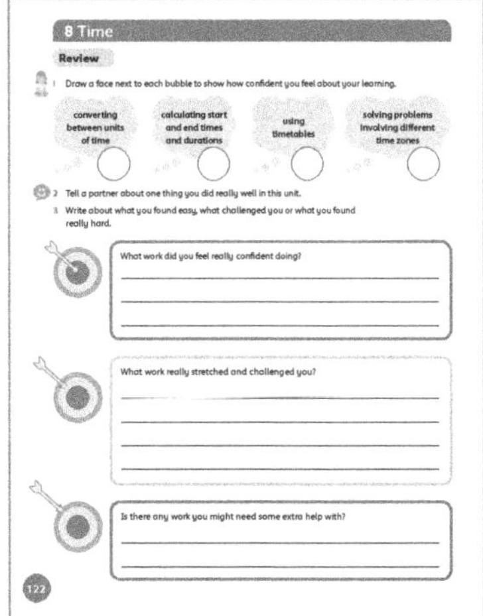

It is appropriate to complete this Practice Book Review as a whole-class discussion. You may choose to keep a record of the class discussion or a copy of the Review page for your own records. The Review provides an opportunity for students to reflect on their learning from the unit, to discuss any areas of mathematics that they feel went particularly well, and any areas that they feel less confident about. Ensure that all students have a copy of the Student Book as a reminder of the areas of mathematics that they have worked on in this unit.

Allow students plenty of time for discussion before asking them to complete the Practice Book page individually, and then, if appropriate, to share their responses with the rest of the class. If students complete this self-assessment at home, encourage them to discuss this with adults. Make a note of areas that students still feel unsure about, for example counting on days on a calendar, or working out the time difference between the start and end of events. Give students plenty of opportunities to calculate durations of journeys and time differences between cities.



There are additional end-of-unit assessments available on the *Oxford Owl for School* website.

# 9 Geometry – properties of shapes

## Overview

### Big idea

The Big idea in this unit is using the properties of 2D and 3D shapes to classify and name shapes. Exploring what shapes have in common and what is different about them helps students to understand these properties. You can do this through visualisation and through explanation in discussion. It is vital that students spend time talking about what they notice about shapes. All the activities in this unit require students to describe what they see. Students will encounter the names of parts of a circle in this unit for the first time, so discussion and examples will be important.

It is also important that students see angles both as a dynamic measure, formed through movement, and as a static property of shapes. Use classroom examples of acute, right, obtuse and reflex angles and develop students' understanding of a definition for each. Geometric knowledge about angles on a straight line (180°) and around a point (360°) can be developed through examples.

### Look out for

- **Students who may not understand the definitions of shape properties for regular or irregular shapes.** For example, a hexagon is any 6-sided shape, and this includes regular hexagons that have 6 equal sides and 6 equal angles. Regular hexagons have 6 lines of symmetry and rotational symmetry order 6, but irregular hexagons do not have these symmetries.
- **Students who confuse the vocabulary of 2D and 3D shapes.** For example, students may call a cube a '3D square' or a sphere a '3D circle'. Have the correct vocabulary available and displayed and always reinforce the correct names when dealing with 2D or 3D shapes.

### Possible misconceptions

- **Students think angles are bigger if the lines that make them are longer.** Use dynamic examples of turning to emphasise the concept of angle and ask students 'how far it has turned'.
- **Students think that the more angles there are on a straight line, or around a point, the greater the sum of the angles will be.** Use paper angles to show that two angles that fit together to form an angle of 180° can be cut into multiple angles and will still all fit together to form a straight line.

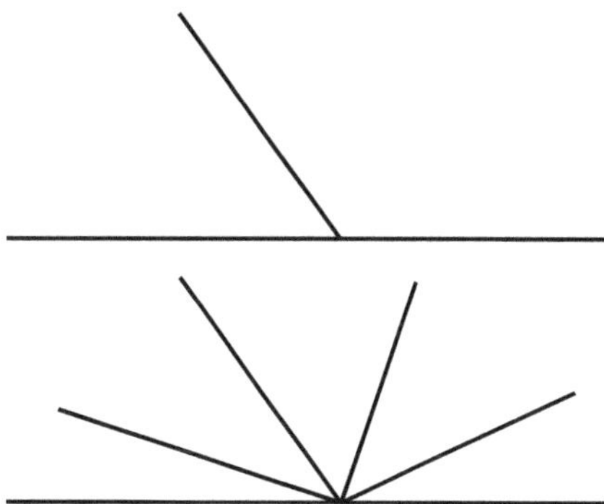

### Key vocabulary

- 2-dimensional (2D), polygon, triangle, quadrilateral, circle, triangle, equilateral triangle, isosceles triangle, scalene triangle, pentagon, hexagon, heptagon, octagon, kite, parallelogram, rhombus, rectangle, square, circle, centre, diameter, radius, circumference,
- 3-dimensional (3D), polyhedron, cube, cuboid, pyramid, sphere, cone, cylinder, prism, tetrahedron, octahedron, dodecahedron
- net, dimensions, volume, isometric paper, isometric drawing, elevation, front elevation, side elevation, plan view
- properties, edge, face, vertex, vertices, How many edges/vertices/faces does that shape have?
- base, concave, convex, open, closed, cross-section
- regular, irregular, parallel, perpendicular, rotational symmetry, order of rotational symmetry, tessellating
- acute angle, obtuse angle, reflex angle, right angle, missing angle, internal (or interior) angle
- protractor
- predict

| Learning objective | E | 9A | 9B | 9C | 9D | 9E | 9F | 9G | C | R |
|---|---|---|---|---|---|---|---|---|---|---|
| Draw 2D shapes using given dimensions and angles. | | ✓ | ✓ | | | ✓ | ✓ | | ✓ | ✓ |
| Recognise, describe and build simple 3D shapes, including making nets. | ✓ | | | ✓ | ✓ | | | | ✓ | ✓ |
| Compare and classify geometric shapes based on their properties and sizes and find unknown angles in any triangles, quadrilaterals and regular polygons. | ✓ | ✓ | ✓ | ✓ | ✓ | ✓ | ✓ | | | ✓ |
| Illustrate and name parts of circles, including radius, diameter and circumference and know that the diameter is twice the radius. | | | | | | | | ✓ | | ✓ |
| Recognise angles where they meet at a point, are on a straight line, or are vertically opposite, and find missing angles. | | | ✓ | | | | ✓ | | ✓ | ✓ |

# 9 Geometry – properties of shapes

## Engage  Student Book page 148

### Big question

- How can I describe, draw and make 2D and 3D shapes?

### Global skills

- **Creative skills:** exploring
- **Self-development skills:** reflecting on learning

### Key vocabulary

- edge, vertex, vertices, face, straight, curved, base, regular, irregular, parallel, perpendicular, concave, convex, open, closed, cross-section, 2D, 3D, 2-dimensional, 3-dimensional, polygon, circle, triangle, equilateral triangle, isosceles triangle, scalene triangle, pentagon, hexagon, heptagon, octagon, quadrilateral, kite, parallelogram, rhombus, rectangle, square, cube, cuboid, pyramid, sphere, cone, cylinder, prism, tetrahedron, polyhedron, octahedron, dodecahedron, acute angle, obtuse angle, reflex angle, right angle, interior angle

### Resources

- mini whiteboards and markers
- index cards
- selection of magazines and newspapers that students or their families read regularly
- large sheets of paper, glue or sticky tape

### Language support

Use the posters students make in the Introductory activity for the rest of the unit. It is important that students present the posters to the rest of the class, to enable them to practise saying the words. Remind students how to pronounce the more unfamiliar words. Students should complete the write-in glossary at the back of their Student Book so they always have access to the definitions.

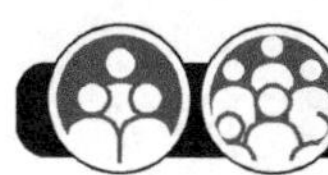 Introductory activity

Look together at page 148 of the Student Book. Display on the IWB, if possible. Arrange the class into mixed-attainment groups of between four and six students. They should look at the image on page 148 of the Student Book and list on their whiteboards as many shapes as they can see. After five minutes, ask groups to give feedback. List all the shapes on the board. Ask each group to write the different shapes on an index card, one shape per card. Then ask the groups to classify the shapes in at least three different ways (for example: 2-dimensional/3-dimensional, regular/irregular, equal sides, equal angles, straight sides, curved). Ask them to record each classification.

 Main activity

Give each group a selection of newspapers and magazines and ask the group to cut out as many images of shapes as they can find in them. Then ask students to stick these shapes onto a large piece of paper to make a poster. They should label the shapes using as much of the language of shape as they can remember.

### Differentiation

All students should contribute to the group discussion and use the vocabulary they can remember. All students will also hear vocabulary that is new to them. Encourage students to ask for definitions of any words other students use that they don't understand.

**Supporting:** Help students to locate the names of the shapes they discover (if they don't know them) by referring to books or the internet.

**Consolidating:** Ask students to describe properties of the shapes they can identify.

**Extending:** Challenge students to sort the shapes and explain their reasoning.

 Reflection time

Ask each group to share their poster with the class. Encourage other students to suggest vocabulary that can be added to the poster. Display the posters for the rest of the unit. When new vocabulary is introduced, students can add it to the posters.

# 9A Classifying 2D shapes

## Discover
Student Book pages 149–150 · Practice Book page 123

### Specific learning focus
- Classify different polygons and understand whether a 2D shape is a polygon or not.

### Global skills
- **Creative skills:** exploring

### Key vocabulary
- edge, vertex, vertices, polygon, properties

### Resources
- large tissue-paper equilateral triangle
- large tissue-paper square
- card, scissors
- poster of shape names from the Engage lesson
- index cards or mini whiteboards and markers

### Language support
The Reflection time game asks students to use the technical vocabulary of shape. Model a wide range of vocabulary and its pronunciation. Encourage students to use the posters from the Engage activity as support and to add any new vocabulary.

 **Introductory activity**

Ask two students to come to the front of the class. Ask one student to hold the large tissue-paper equilateral triangle and one to hold the large tissue-paper square. Then ask them to gradually overlap the two shapes. Ask them to stop on several occasions. Each time they stop, ask the other students, in pairs, to write down as many properties as they can for the shape they can see in the overlap. Give them one minute to do this. Ask pairs to feed back after each overlap.

 **Main activity**

Look together at page 149 of the Student Book. Display on the IWB, if possible. Ask students to use the scissors and card to make the shapes they need to carry out the activity on pages 149–150 of the Student Book. Encourage pairs to use the posters from the Engage lesson to help them classify any shapes they do not know, or they need reminding of, for example: hexagon, heptagon, octagon. The Think back information is also a reminder of some of the shapes and their names.

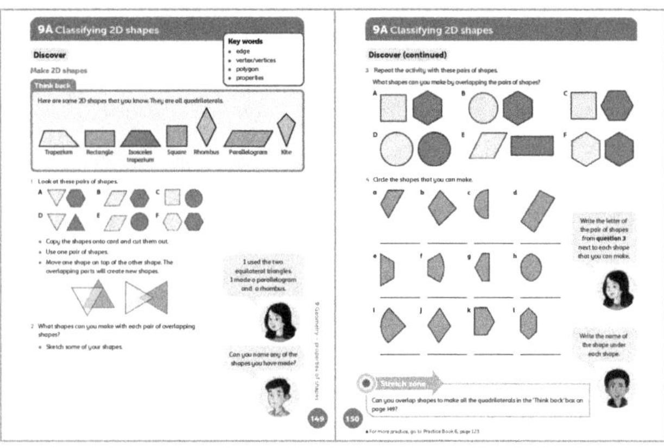

## Differentiation

**Supporting:** Use posters to support students in naming shapes. Ask for **properties** such as number of sides and curved and straight **edges**.

**Consolidating:** Ask for properties such as number of sides and curved and straight edges and symmetry properties.

**Extending:** Encourage students to use dictionaries to find names of shapes they are not familiar with. They should describe a wide range of properties of these shapes.

**Stretch zone:** *Can you overlap shapes to make all the quadrilaterals in the 'Think back' box on page 149?*

Students write instructions to allow someone to make an exact copy of a shape they are thinking about.

 **Reflection time**

Ask each student to draw, on an index card or on their whiteboard, one of the shapes they made using the overlap in question 1. Ask them to move around the room (or the class could go outside to play this game). When you say, 'Stop!', ask students to partner with the person next to them. In pairs, ask them to find one thing that their shapes have in common and one difference. Continue this procedure. Encourage students to use a wide range of vocabulary.

**Practice Book:** Students complete Practice Book page 123. They can do this directly after the Main activity, as homework, or as the focus of a separate mathematics session to help students consolidate their learning and build fluency.

Students create **polygons** by folding and then cutting a single piece of paper. They then describe some of the polygons they made by sketching each one, naming it and listing two of its properties. *Do any of your polygons have the same properties? Do they look the same? Can you say one different thing about each shape?*

| Differentiated outcomes | |
|---|---|
| **All students** | should recognise and name common shapes and describe their properties. |
| **Most students** | will recognise and name most shapes and describe their properties, including symmetry. |
| **Some students** | may recognise and name all the shapes and describe their properties, including symmetry and angle properties. |

## Answers

### Student Book pages 149–150

Answers will vary. Accept any accurate answers. Ask students to explain how they know the name of the overlapping parts of the shapes.

### Practice Book page 123

Answers will vary because students make their own shapes. Check that the names and the properties the students list match their sketches of their polygons.

Stretch zone: Students will explore the possibilities. Check that they have found ways of making shapes with one cut.

# 9A Classifying 2D shapes

## Explore  Student Book page 151 • Practice Book page 124

### Specific learning focus

- Classify different polygons and understand whether a 2D shape is a polygon or not.

### Global skills

- **Creative skills:** exploring, problem solving

### Key vocabulary

- regular, irregular, parallel, internal (or interior) angle

### Resources

- rope (or similar) cut into 12 m lengths
- large sheets of paper

### Language support

Encourage students to explain their thinking and to say the vocabulary aloud. Use questions, for example:

- *Are these shapes exactly the same?*
- *What is this shape?*
- *Is this shape a polygon? Why? Why not?*

## Introductory activity

Take the class outdoors or into a large space such as a hall. Ask the class to work in groups of 12 students if possible. Give each group a piece of rope. Explain that they are going to make various shapes by holding the rope in different ways and using their bodies and the rope to form the shape. Ask them to make:

- a square
- a rectangle
- a different rectangle
- a triangle
- an equilateral triangle
- a right-angled triangle
- an isosceles triangle
- a scalene triangle
- a hexagon.

After each shape, ask one student from each group to sketch, on a large piece of paper, the shape the group made. At the end of the activity, ask groups to share their results. *Are your shapes exactly the same? If not, what is the same and what is different?*

 ## Main activity

Remind students that a polygon is a closed 2-dimensional (2D) shape with straight sides. Any shape with a curved side, or any shape with a gap in the perimeter (an open shape) is not a polygon. If necessary, remind students about **parallel** sides and perpendicular sides, and that perpendicular sides meet at a right angle.

Ask students to work, in pairs, on the activities on page 151 of the Student Book. Encourage pairs to discuss their answers and to draw and name various triangles and polygons separately first before drawing them in the correct section of the classification tables.

Remind students that they should be specific when they name the shapes. For example, they may use an 'equilateral triangle' for a **regular** triangle or a 'square' for a regular quadrilateral. A regular pentagon should be written simply as 'regular pentagon' to emphasise that a 'regular pentagon' is a subset of the set 'pentagon'.

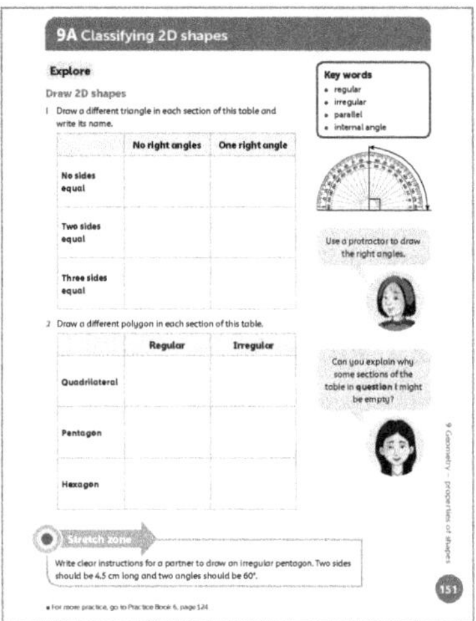

## Differentiation

**Supporting:** Use posters to support students in naming shapes. Ask for properties such as number of sides and number of vertices.

**Consolidating:** Ask for properties such as number of sides or vertices and symmetry properties.

**Extending:** Encourage students to use dictionaries to find names of shapes they are not familiar with. They should describe a wide range of properties of these shapes.

**Stretch zone:** *Write clear instructions for a partner to draw an **irregular** pentagon. Two sides should be 4.5 cm long and two angles should be 60°.*

## 🧑 Reflection time

Ask students to take photographs of shapes around the school. They can use the tables in questions 1 or 2 to classify them.

Draw a large table on the board like this:

|  | Polygon | Not a polygon |
|---|---|---|
| Contains only right angles |  |  |
| Contains at least one obtuse angle |  |  |
| Contains at least one reflex angle |  |  |

Ask students to come up to the board one at a time and to draw a shape in one of the boxes on the table. Remind students of the meanings of the terms 'acute', 'obtuse' and 'reflex'.

**Practice Book:** Students complete Practice Book page 124. They can do this directly after the Main activity, as homework, or as the focus of a separate mathematics session to help students consolidate their learning and build fluency.

Students write three properties for each shape from a given list of 2D shapes. Encourage them to use the poster they created in the Engage lesson to remind them of some of the vocabulary they might use to describe the properties of shapes.

| Differentiated outcomes | |
|---|---|
| **All students** | should classify common shapes and describe their properties with support. |
| **Most students** | will classify a wide range of shapes and describe their properties, including symmetry. |
| **Some students** | may recognise and name all the shapes and describe their properties, including symmetry and angle properties. |

## Answers

**Student Book page 151**

1

|  | No right angles | One right angle |
|---|---|---|
| **No sides equal** | scalene triangle | right-angled triangle |
| **Two sides equal** | isosceles triangle | right-angled isosceles triangle |
| **Three sides equal** | equilateral triangle | There is no equilateral triangle that also contains a right angle, as all the angles must be 60° |

2

|  | Regular | Irregular (examples only) |
| --- | --- | --- |
| Quadrilateral | square | parallelogram |
| Pentagon | regular pentagon | irregular pentagon |
| Hexagon | regular hexagon | irregular hexagon |

## Practice Book page 124

Answers will vary because students choose properties for each shape. Check that the names and the properties that students identify are correct. For example, a scalene triangle has three different-length sides, three different-sized angles and no reflective symmetry.

Stretch zone: Regular and irregular polygons have the same number of sides. A regular pentagon and an irregular pentagon both have five sides. They are different because in a regular polygon the sides are all the same length and the angles are all the same size.

# 9B Properties of 2D shapes

## Discover 1 — Student Book page 152 • Practice Book page 125

### Specific learning focus

- Classify 2D shapes by their properties.

### Global skills

- **Creative skills:** investigating

### Key vocabulary

- polygon, regular, irregular, parallel

### Resources

- mini whiteboards and markers
- rulers, set squares
- sets of 2D shapes

### Language support

Ask students questions to support their learning, for example:

- *What can you tell me about this polygon? Is there anything else?*
- *Can you point to an example of this shape in the classroom?*
- *Can you explain why this shape is irregular?*

 Introductory activity

Draw on the board the following 2D shapes:

rectangle       equilateral triangle       rhombus

Ask students, in pairs, to discuss the properties of the sides of each shape in turn. Ask them to look for parallel sides and equal sides. Discuss together the properties they found, and ask questions such as:

- *How many pairs of parallel sides does the rhombus have?*
- *Which shape has the most equal sides?*
- *In what ways are the shapes the same?*
- *In what ways are the shapes different?*

 Main activity

Give pairs of students a set of mixed 2D shapes. Ask them to sort the shapes by how many pairs of equal sides they have. They can use rulers to check. Then ask them to sort the shapes by the number of parallel sides they have. As they work, listen to their discussions and make sure that they are using the correct vocabulary.

Ask questions to prompt the discussion, such as:

- *Can a polygon have three equal sides?* (yes, equilateral triangle or irregular hexagon for example)
- *Can a polygon have more than two pairs of parallel sides?* (yes, regular hexagon, octagon and so on)

Look together at page 152 of the Student Book. Display on the IWB, if possible. Ask students to complete the table to identify shapes with different combinations of equal sides and parallel sides.

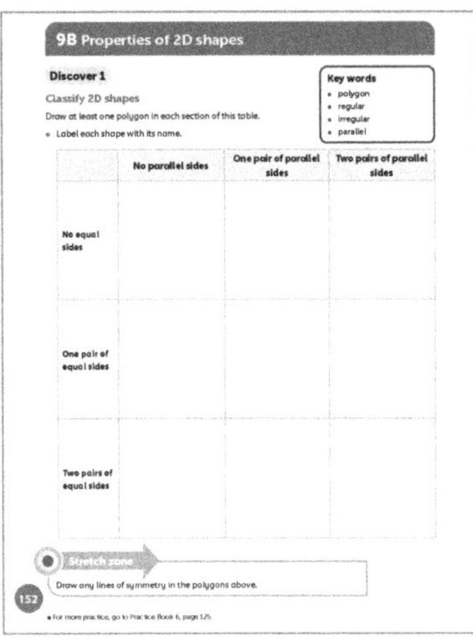



## Differentiation

**Supporting:** Help students to identify and count pairs of equal or parallel sides.

**Consolidating:** Ask students to explain how they identified the properties of each shape.

**Extending:** Challenge students to find at least two shapes for each box in the table.

**Stretch zone:** *Draw any lines of symmetry in the polygons above.*

Students should be encouraged to find lines of symmetry at any angle and not just vertical or horizontal.

 ## Reflection time

Using their completed tables of properties, ask students to share at least one example from each box in the table. *Were there any boxes that had several different possible shapes? How many did you find? Did any box have no examples? Why?*

**Practice Book:** Students complete Practice Book page 125. They can do this directly after the Main activity, as homework, or as the focus of a separate mathematics session to help students consolidate their learning and build fluency.

Students sketch 2D shapes from given properties, including side length and perpendicular height. They then name each shape they have drawn.

| Differentiated outcomes | |
| --- | --- |
| **All students** | should identify and sort 2D shapes with support. |
| **Most students** | will identify and sort 2D shapes. |
| **Some students** | may find a range of examples of 2D shapes for each set of properties. |

## Answers

### Student Book page 152

Examples are as follows, but this is not an exhaustive list.

| | No parallel sides | One pair of parallel sides | Two pairs of parallel sides |
| --- | --- | --- | --- |
| **No equal sides** | scalene triangle | non-isosceles trapezium | irregular hexagon |
| **One pair of equal sides** | irregular quadrilateral | isosceles trapezium | irregular hexagon |
| **Two pairs of equal sides** | kite | irregular hexagon | rectangle |

### Practice Book page 125

1 parallelogram, for example

2 rhombus, for example (must be different from question 1)

3 trapezium, for example

4 irregular quadrilateral, for example (must be different from question 3)

Stretch zone: Students choose their own polygons. Check that they have listed the correct properties.

## Discover 2 — Student Book page 153

### Specific learning focus

- Find rotational symmetry in 2D shapes.

### Global skills

- **Creative skills:** exploring

### Key vocabulary

- rotational symmetry, order of rotational symmetry

### Resources

- tracing paper
- mini whiteboards and markers

### Language support

Support students by asking appropriate questions and emphasising the vocabulary relating to rotational symmetry, for example:

- *How many ways does this shape fit its outline?*
- *Does this shape have rotational symmetry?*
- *What is the order of rotational symmetry for this shape?*

 **Introductory activity**

Draw a rectangle on the board and ask students to recall its properties. They will most likely say that it has four straight sides, two pairs of opposite equal sides and two pairs of parallel sides. Some students may say that it has four right angles, and some may include that it has line symmetry, both vertically and horizontally.

Ask students what shape they would see if the rectangle was rotated, around its centre point, 180° or two right angles. They may see that the rectangle would look the same. *Are there any other rotations that would leave it looking identical to how it started?* Agree that there are only two orientations that will look identical: the start/end position and a rotation of 180°. Explain that this means the rectangle has **rotational symmetry**, and because it appears to fit its own outline in two positions, it has rotational symmetry is of 'order 2'.

**Main activity**

Look together at page 153 of the Student Book. Display on the IWB, if possible. Refer students to the worked example. As a class, follow the instructions and allow students to verify for themselves that the equilateral triangle has rotational symmetry of order 3. *Can you explain why?* They should refer to the triangle fitting into its own outline in three different orientations.

Students now use tracing paper to complete the activities on page 153, finding the **orders of rotational symmetry** of different 2D shapes.

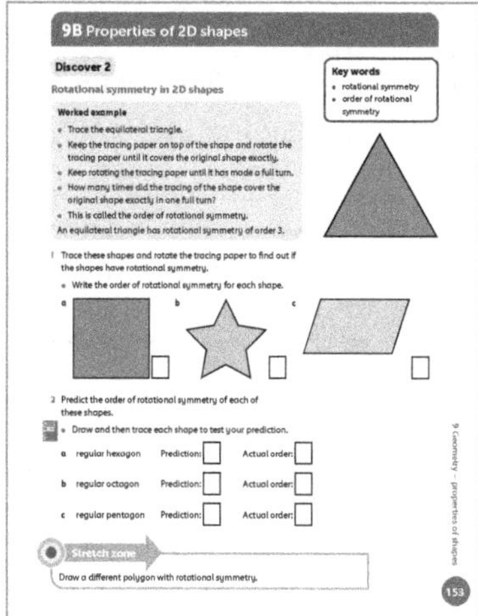

## Differentiation

**Supporting:** Model for students how to use the tracing paper and rotate it around the centre of a shape to find how many positions the shape fits its outline.

**Consolidating:** Ask students to explain how they predicted and found the order of rotational symmetry of 2D shapes.

**Extending:** Challenge students to predict and check the orders of rotational symmetry for a range of regular and irregular 2D shapes.

**Stretch zone:** *Draw a different polygon with rotational symmetry.*

Students can explore regular and irregular shapes and make deductions about whether they have rotational symmetry.

 **Reflection time**

Ask students to write on their whiteboards any shapes they found with rotational symmetry of order 3. Ask them to share what shapes have this order of rotational symmetry.

Now repeat for any shapes with order 4 rotational symmetry. Can students say what is the same and what is different about the shapes with order 3 and order 4 rotational symmetry?

Check for shapes with rotational symmetry of orders 2, 5 and 6. What conclusions can students make about the number of sides on a regular shape and its order of rotational symmetry?

**Practice Book:** There is no corresponding Practice Book page for this lesson.

<table>
<tr><th colspan="2">Differentiated outcomes</th></tr>
<tr><td>All students</td><td>should find the order of rotational symmetry for 2D shapes with support.</td></tr>
<tr><td>Most students</td><td>will predict and find the order of rotational symmetry for some 2D shapes.</td></tr>
<tr><td>Some students</td><td>may predict and find the order of rotational symmetry for a wide range of 2D shapes.</td></tr>
</table>

## Answers

**Student Book page 153**

**1 a** 4     **b** 5     **c** 2

**2 a** 6     **b** 8     **c** 5

# 9B Properties of 2D shapes

**Explore 1**   Student Book pages 154–155 • Practice Book page 126

### Specific learning focus

- Draw 2D shapes with given side lengths and angles.

### Global skills

- **Creative skills:** investigating

### Key vocabulary

- polygon, regular, irregular, parallel

### Resources

- rulers
- protractors

### Language support

As students draw each shape, make sure that they understand what shape it is from the language related to the properties given. For example, if it is a scalene triangle, they should know that it will have three different sides and three different angles. You may need to revise some of the vocabulary in order for them to be able to draw each shape.

###  Introductory activity

Give students protractors and rulers and explain that they will be drawing various polygons, following precise measurements. The first shape will be an equilateral triangle with sides of 8 cm.

Ask students to draw a baseline of 8 cm as accurately as possible. *What do you know about the angles in an equilateral triangle?* Agree that each angle is 60° and then talk through with students how to draw an angle of 60° at each end of the 8 cm baseline. Remind students

of the two scales on the protractor – the inner scale and the outer scale – and make sure that they read from the correct zero marking once they have placed the centre point of the protractor on the end of the base.

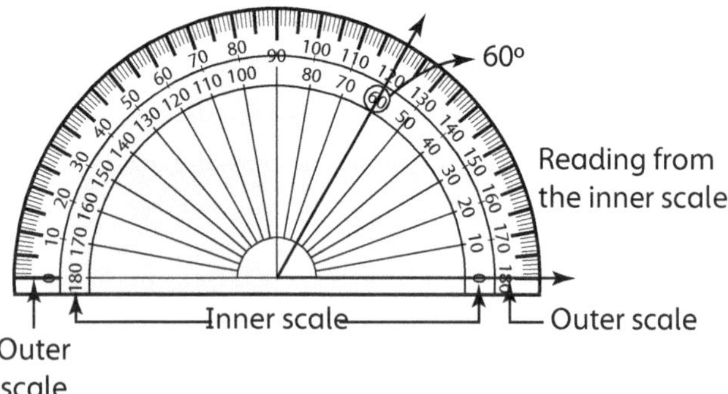


Model the process on the board and show students how to mark the direction of 60° at each end of the baseline, then extend the lines until they cross to form the triangle.

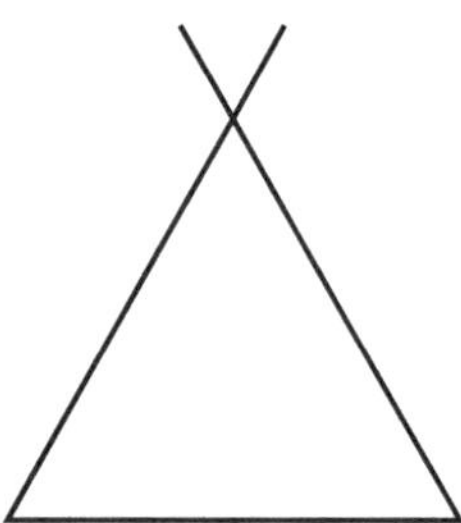

Discuss the fact that the other sides do not need to be measured as they will be the correct length at the crossing point if the angles are correct, although students can check this with a ruler.

###  Main activity

Look together at pages 154–155 of the Student Book. Display on the IWB, if possible. Explain that students should draw each shape to the specification given, measuring the sides and angles accurately to draw each shape. Review students' knowledge of vocabulary, for example acute and obtuse angle, regular and irregular polygons and the names of the different types of triangle.

When they have completed the activity, they should write down properties of their shapes to share in Reflection time.

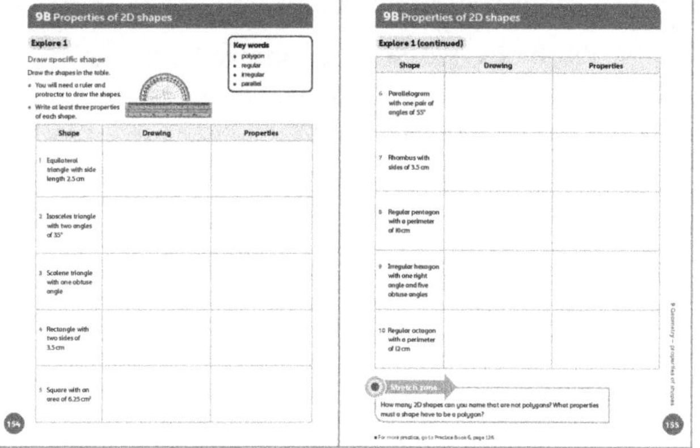

## Differentiation

**Supporting:** Model how to use the protractor to get the correct angles at the ends of line segments for the polygon sides.

**Consolidating:** Ask students to explain how they are using the protractor and how they know which scale to use.

**Extending:** Challenge students to think of as many properties as they can about each shape.

**Stretch zone:** *How many 2D shapes can you name that are not polygons? What properties must a shape have to be a polygon?*

Students should recognise the properties of a polygon as 'any 2D shape with three or more straight sides'. Ask them to name some 2D shapes that are not polygons such as a circle and a semi-circle. They could also draw some 'non-polygons', for example:

### Reflection time

Ask several students to come forward and explain how they drew the different shapes. Ask them to describe precisely how they started each shape and what choices they made about where to mark the angles. For question 5, how did they know what size lengths and angles to draw? For questions 8 and 10, how did they know what size angles to draw for the pentagon and octagon? You can show them how to divide a polygon into triangles by joining one vertex to each of the others, then totalling the angles for each triangle and dividing the total number of degrees in all the triangles by the number of vertices of the polygon. For example, a regular pentagon can be divided into three triangles, so $3 \times 180° = 540°$, then $540° \div 5 = 108°$ for each angle.

**Practice Book:** Students complete Practice Book page 126. They can do this directly after the Main activity, as homework, or as the focus of a separate mathematics session to help students consolidate their learning and build fluency.

Students draw their own set of ten 2D shapes. They then sort their shapes into a Carroll diagram, first choosing the sorting criteria. Remind students of the format for the headings in a Carroll diagram: 'X' and 'not X'.

| Differentiated outcomes | |
|---|---|
| **All students** | should construct 2D shapes from given properties with support. |
| **Most students** | will construct 2D shapes from given properties. |
| **Some students** | may construct 2D shapes from given properties and recognise other properties of the shapes. |

## Answers

### Student Book pages 154–155

1  three equal sides, each of 2.5 cm, three equal angles of 60°

2  two equal sides and two equal angles of 35°

3  three unequal sides, three unequal angles

4  two pairs of opposite parallel and equal sides, with one pair of sides being 3.5 cm, and 4 right angles

5  four equal sides, each of 2.5 cm, 4 right angles

6  two pairs of opposite parallel sides, two pairs of equal sides, two pairs of opposite equal angles, with one pair being 55° and the other being 125°

7  four equal sides, each of 3.5 cm, two pairs of opposite, parallel sides, two pairs of opposite equal angles

8  five equal sides, each of 2 cm, five equal angles of 108°

9  six sides not all equal, six angles not all equal (five of which are greater than 90° and one is 90°)

10  eight equal sides, each of 1.5 cm, eight equal angles of 135°

### Practice Book page 126

Students choose their own ten shapes and the sorting criteria for their Carroll diagram. Check that they have completed the Carroll diagram correctly for their shapes.

Stretch zone: Check that the shapes have been correctly sorted in a Venn diagram.

# 9B Properties of 2D shapes

## Explore 2   Student Book page 156 • Practice Book page 127

### Specific learning focus

- Draw shapes having a given order of rotational symmetry.

### Global skills

- **Creative skills:** exploring

### Key vocabulary

- rotational symmetry, order of rotational symmetry

### Resources

- collection of 2D shapes
- tracing paper

### Language support

Support students by asking appropriate questions, as they draw their shapes, to support their vocabulary. Ask, for example:

- *How do you know that this pattern has order 4 rotational symmetry?*
- *Can you change this shape to give it rotational symmetry?*

###  Introductory activity

Ask students to give you the name of a shape that has rotational symmetry of order 4 and explain how they know. *Can you give me the name of a shape that does not have rotational symmetry of order 4 and explain why?*

Using a squared grid on the board, start to build up a pattern using coloured squares. As you add each new square, ask students where to place more squares so the pattern keeps its rotational symmetry.

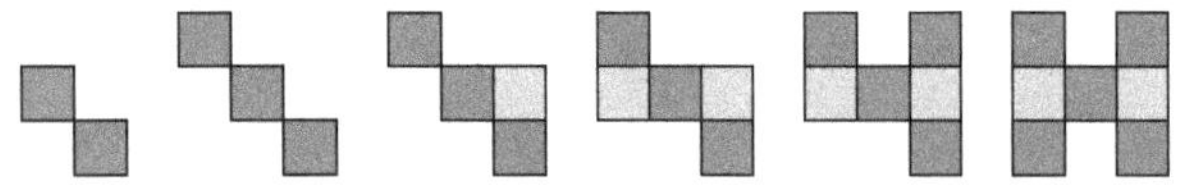

###  Main activity

Discuss the connection between rotational symmetry and reflection symmetry. Ask students whether they can see a link between a shape with rotational symmetry of order 4 and a shape with 4 lines of symmetry. *How can this help us make a shape or pattern with rotational symmetry of order 4?*

Look together at page 156 of the Student Book. Display on the IWB, if possible. Students should try to draw shapes or patterns with rotational symmetry of each different order. Refer students to the first speech bubble and encourage them to draw more than one of each shape or pattern.

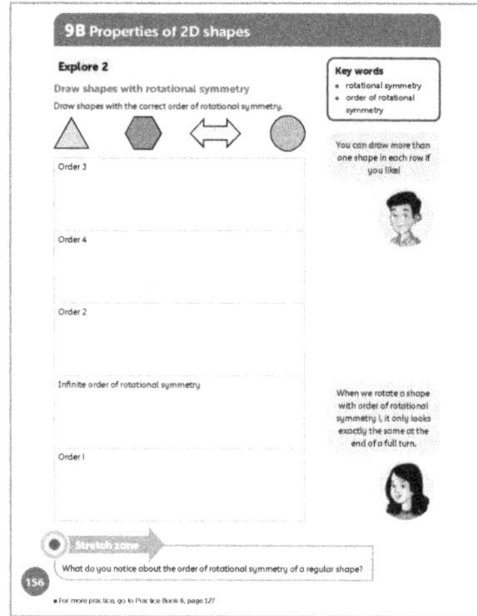

## Differentiation

**Supporting:** Help students to decide how to extend a shape or pattern so that it keeps its rotational symmetry.

**Consolidating:** Ask students to explain how they knew their shapes or patterns would have the correct order of rotational symmetry.

**Extending:** Challenge students to design a pattern with rotational symmetry of order 5.

**Stretch zone:** *What do you notice about the order of rotational symmetry of a regular shape?*

Students should be encouraged to make links between the order of rotational symmetry and the number of sides in the regular shape.

### Reflection time

Ask different students to share their shapes or patterns. Ask the class to comment on each shape or pattern and say why they think it has the correct order of symmetry. After looking at several students' shapes and patterns, can the class think of a way to describe the connection between the number of lines of symmetry and the order of rotational symmetry?

**Practice Book:** Students complete Practice Book page 127. They can do this directly after the Main activity, as homework, or as the focus of a separate mathematics session to help students consolidate their learning and build fluency.

Students draw shapes or patterns on a grid, with a given order of rotational symmetry. They are encouraged to start with one shape and then gradually add others to complete their patterns with the correct order of symmetry.

<table>
<thead>
<tr><th colspan="2">Differentiated outcomes</th></tr>
</thead>
<tbody>
<tr><td>All students</td><td>should draw shapes and patterns with rotational symmetry with support.</td></tr>
<tr><td>Most students</td><td>will draw shapes and patterns with rotational symmetry.</td></tr>
<tr><td>Some students</td><td>may draw shapes or patterns with rotational symmetry.</td></tr>
</tbody>
</table>

# 9C Properties of 3D shapes

**Discover**  Student Book pages 157–158 • Practice Book page 128

## Specific learning focus

- Visualise and describe the properties of 3D shapes.

## Global skills

- **Creative skills:** investigating

## Key vocabulary

- edge, vertex, vertices, face, cross-section

## Resources

- modelling clay
- modelling straws

## Language support

Encourage students to use the appropriate technical vocabulary, for example: **vertices**, **edges** and **faces**. Remind students that the plural of **vertex** is vertices.

It is important that students realise that there are lots of different prisms, pyramids and cylinders, and that these are classified by the shape of their base. Remind students that the base of a prism is its **cross-section**.

Ask questions such as:

- *How many vertices/edges/faces does that shape have?*
- *What shape is the base?*

## Answers

### Student Book page 156

Examples include:

Order 3: equilateral triangle

Order 4: square

Order 2: rectangle, rhombus, parallelogram

Infinite order: circle

Order 1: irregular quadrilateral

### Practice Book page 127

Check that students have drawn appropriate shapes or patterns of the correct order of rotational symmetry.

Stretch zone: Students can print, display and label the examples they find.

## Introductory activity

Look together at page 157 of the Student Book. Display on the IWB, if possible. Refer students to the worked example, which provides an image of a model of a cube. Ask one student to come to the front of the class. The other students are going to give this student instructions to make a cube, using modelling clay and straws. Each student can give only one instruction; then it is the next student's turn.

Write the key instructions on the board as a model for students to follow when they complete the activity on page 157, for example: Attach two straws to one ball of clay. Put the straws at right angles to one another…

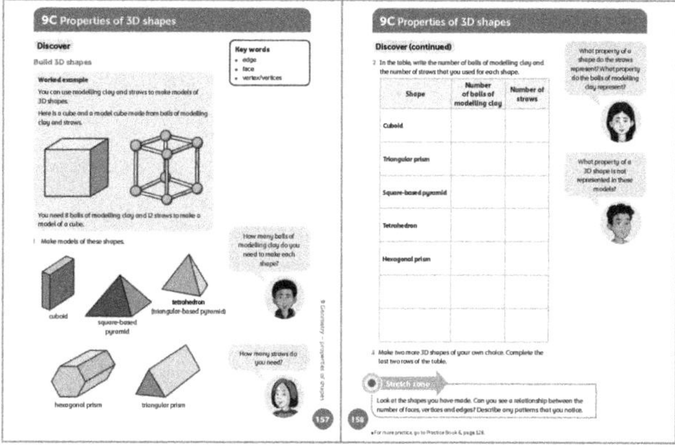

## Main activity

Ask students to work, in pairs, on the activities on pages 157–158 of the Student Book, to encourage discussion.

*What do you notice about the relationship between the number of straws and balls of clay and the numbers of vertices and edges of the shape?* Students should name the shapes they are making and describe other properties of the shapes.

There are two additional rows in the table on page 158 for students to make their own 3D shapes and complete the table for the number of straws and balls of modelling clay.

## Differentiation

**Supporting:** Ask students to show you how they are counting vertices, edges and faces.

**Consolidating:** Ask students to name the new shapes they are making and to tell you their properties.

**Extending:** Encourage students make complex shapes of their own and find their names.

**Stretch zone:** *Look at the shapes you have made. Can you see a relationship between the number of faces, vertices and edges? Describe any patterns that you notice.*

As an additional activity, ask students to bring in examples of 3D shapes from home to share with the class. Alternatively, they can take photographs of 3D shapes in the locality to create posters.

 ## Reflection time

Ask one of the pairs to describe, but not name, one of the 3D shapes that they created in the bottom two rows of the table on page 158. They need to describe it so that you can copy the shape exactly. As they describe the shape, ask questions and model phrases using the vocabulary of shape to demonstrate the correct pronunciation. For example, ask, *How many faces does this shape have? This shape has five faces. What shape are the faces? The two faces at each end are equilateral triangles. They are the same size.*

**Practice Book:** Students complete Practice Book page 128. They can do this directly after the Main activity, as homework, or as the focus of a separate mathematics session to help students consolidate their learning and build fluency.

Students find examples of 3D shapes in the environment. They draw the shapes (or they could take photographs and stick these into their books) and then write the number of faces, vertices and edges that each shape has.

| Differentiated outcomes | |
|---|---|
| **All students** | should make the shapes and count faces, angles and vertices. |
| **Most students** | will make shapes of their own and find their names. |
| **Some students** | may make complex shapes of their own and find their names. |

## Answers

### Student Book page 158

2

| Shape | Number of balls of clay | Number of straws |
|---|---|---|
| Cuboid | 8 | 12 |
| Triangular prism | 6 | 9 |
| Square-based pyramid | 5 | 8 |
| Tetrahedron | 4 | 6 |
| Hexagonal prism | 12 | 18 |

### Practice Book page 128

Answers will vary because students choose their own shapes to analyse. Check that they have recorded the correct numbers of faces, vertices and edges.

Stretch zone: Students will find out the name of a shape they did not know. Check that they have named their shape correctly.

# 9C Properties of 3D shapes

## Explore  Student Book page 159 • Practice Book page 129

### Specific learning focus
- Visualise and describe the properties of 3D shapes.

### Global skills
- **Creative skills:** exploring

### Key vocabulary
- edge, vertex, vertices, face, polyhedron

### Resources
- a range of different 3D shapes
- mini whiteboards and markers

### Language support

Encourage students to use the appropriate technical vocabulary, for example vertices, edges and faces. Remind students that vertices is the plural of vertex.

For Reflection time, students need to ask questions that require only a yes or no answer, for example:
- Does the shape have six faces?
- Does the shape have any curved faces?

### Introductory activity

Place a 3D shape in a container or bag on top of your desk. Describe one property of the shape. For example, for a cube you can say, *It has six faces.* Ask pairs to write on their whiteboards any possible shapes with this property. Now say another property and students cross out all the shapes on their list that don't now 'fit' both properties. Repeat with other properties until students correctly guess the shape. Repeat two or three times using different 3D shapes. Then encourage students to choose a shape and describe the properties themselves with other students guessing the shape.

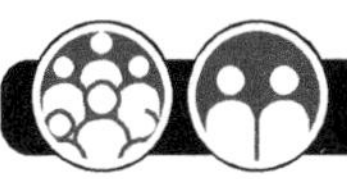

### Main activity

Show students a hexagonal prism. Ask a student to count the number of faces it has (8), then ask another student to count its edges (18) and finally a third student to count the vertices (12). Review the definitions of faces, edges and vertices, then ask students to work in pairs on the activity on pages 159 of the Student Book to encourage discussion.

Encourage students to find a 3D shape and see where it fits in the grid rather than trying to fill in the grid from top to bottom and left to right. You could bring a range of books into the class for students to explore for examples of the 3D shapes, including more unusual or complex ones such as tetrahedron, octahedron, dodecahedron.

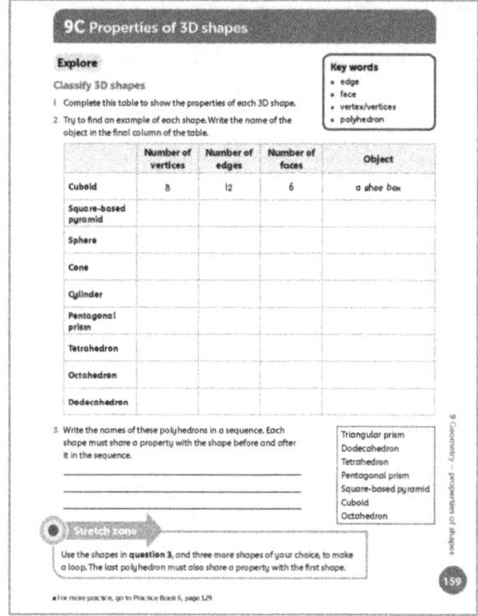

## 9C Properties of 3D shapes

**Explore**

Classify 3D shapes

1 Complete this table to show the properties of each 3D shape.

2 Try to find an example of each shape. Write the name of the object in the final column of the table.

**Key words**
- edge
- face
- vertex/vertices
- polyhedron

| | Number of vertices | Number of edges | Number of faces | Object |
|---|---|---|---|---|
| Cuboid | 8 | 12 | 6 | a shoe box |
| Square-based pyramid | | | | |
| Sphere | | | | |
| Cone | | | | |
| Cylinder | | | | |
| Pentagonal prism | | | | |
| Tetrahedron | | | | |
| Octahedron | | | | |
| Dodecahedron | | | | |

3 Write the names of these polyhedrons in a sequence. Each shape must share a property with the shape before and after it in the sequence.

Triangular prism
Dodecahedron
Tetrahedron
Pentagonal prism
Square-based pyramid
Cuboid
Octahedron

Use the shapes in **question 3**, and three more shapes of your choice, to make a loop. The last polyhedron must also share a property with the first shape.

159

For more practice, go to Practice Book 6, page 129.

## Differentiation

**Supporting:** Focus on accurately counting the faces, vertices and edges of basic 3D shapes.

**Consolidating:** Encourage students to find examples of all the 3D shapes in the table and count faces, edges and vertices.

**Extending:** Encourage students to find examples of any 3D shapes that do not appear in the table and to describe their properties.

**Stretch zone:** *Use the shapes in question 3, and three more shapes of your choice, to make a loop. The last **polyhedron** must also share a property with the first shape.*

**cube** → (both have 6 faces) → **cuboid** → (both have 12 edges) → **octahedron** (both have at least one triangular face) → **square-based pyramid** → (both have at least one triangular face) → **tetrahedron** → (both have at least one triangular face) → **icosahedron** → (both have at least one triangular face) → **triangular prism** → (both have at least one rectangular face) → **hexagonal prism** → (both have at least one rectangular face) → **pentagonal prism** → (both have at least one pentagonal face) → **dodecahedron** → (both have all the same shape faces) → **cube**

### Reflection time

Ask a student to come to the front of the class and to choose one of the 3D shapes they can see in the room without saying which it is. Ask the other students to ask questions (for example: 'Does it have six faces?'), one at a time, to help them guess the shape. The student can only answer 'yes' or 'no' to each question.

**Practice Book:** Students complete Practice Book page 129. They can do this directly after the Main activity, as homework, or as the focus of a separate mathematics session to help students consolidate their learning and build fluency.

Students consolidate their learning on 3D shape properties by completing a table listing three properties for each of a given shape. They then draw a sequence of shapes, linking the shapes by a shared property.

| Differentiated outcomes | | 
| --- | --- |
| **All students** | should find examples of 3D shapes and count faces, edges and vertices. |
| **Most students** | will find examples of all the 3D shapes in the table and count faces, edges and vertices. |
| **Some students** | may find examples of all the 3D shapes in the table as well as others and count faces, edges and vertices. |

## Answers

### Student Book page 159

1

| | Number of vertices | Number of edges | Number of faces |
| --- | --- | --- | --- |
| Cuboid | 8 | 12 | 6 |
| Square-based pyramid | 5 | 8 | 5 |
| Sphere | 0 | 0 | 1 |
| Cone | 1 | 1 | 2 |
| Cylinder | 0 | 2 | 3 |
| Pentagonal prism | 10 | 15 | 7 |
| Tetrahedron | 4 | 6 | 4 |
| Octahedron | 6 | 12 | 8 |
| Dodecahedron | 20 | 30 | 12 |

2 Answers will vary. Students will find a variety of objects.

3 Check that students have listed a sequence of 3D shapes linked by properties.

### Practice Book page 129

Answers will vary because students choose their own properties for each shape. Check that they have recorded the correct numbers of faces, vertices and edges, and that any other properties are correct for the shape. For example, for a hemisphere, they might say that it has one flat face, one curved surface and one curved edge.

Stretch zone: False. A hemisphere is a 3D shape because it occupies 3D space and has a volume.

# 9D Making 2D representations of 3D shapes

**Discover** Student Book pages 160–162 • Practice Book page 130

### Specific learning focus

- Recognise and make 2D representations of 3D shapes including nets.

### Global skills

- **Creative skills:** exploring

### Key vocabulary

- net, cube, dimensions, volume

### Resources

- squared paper
- large number of boxes including different-shaped cuboids, different-shaped cylinders, prisms and pyramids
- card, scissors, sticky tape

### Language support

Throughout this activity, ask students to explain their answers. Model the use of the key mathematical vocabulary, for example: length, width, height, faces, edges, as well as the names of shapes.

 Introductory activity

Ask students to find as many different ways as possible of arranging five squares, with at least one edge of each square touching the edge of another square. For example:

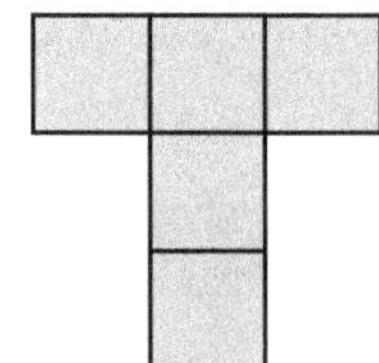 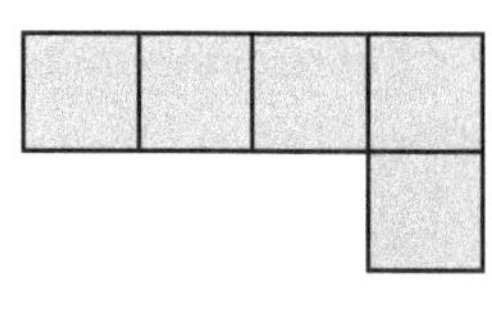

Students draw examples on squared paper. After 10 minutes, collect as many examples as you can. Ask students to draw examples on the board. Then ask them to decide: *Which can you fold up into an open box? Which can you not fold up into an open box?* Ask students to explain their answers. Remind them that the shapes that can be folded to form an open cube are called **nets** of an open cube.

Give pairs of students three different boxes to examine. *Can you think what shape net was used to make each box? Can you sketch it?* Ask students to cut open their boxes carefully along the edges and to open up the net of each box. *Does it match your sketch? Could you have cut it a different way to match your net?*

Look together at page 160 of the Student Book. Display on the IWB, if possible. Ask students to work in pairs on the activities on pages 160–162. Encourage discussion between paired students as they draw their nets for **cubes** and make scale drawings of one of the nets. Ask them to refer to the Think back examples on page 160 for some starter ideas for their cube nets. Before they start question 5 on page 162 of the Student Book, remind students of their work on ratio and scale drawings and models from Unit 4. *What scale factor will you use so that your scale model is half the size of the original box? So, what will the length of your sides be? Can you work out how the **volume** of your scale model will differ from the volume of the original box? What calculation do you need to do?*

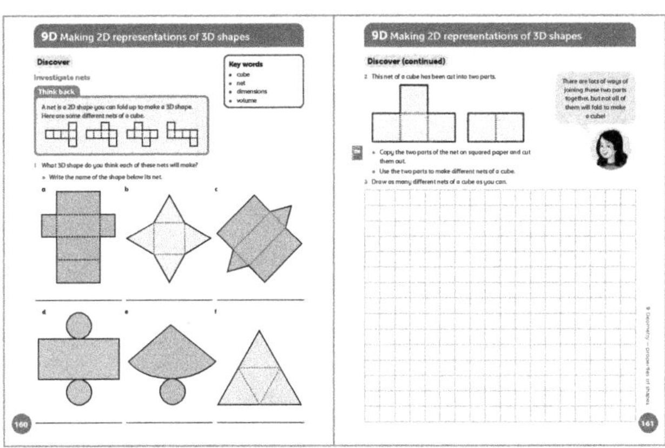

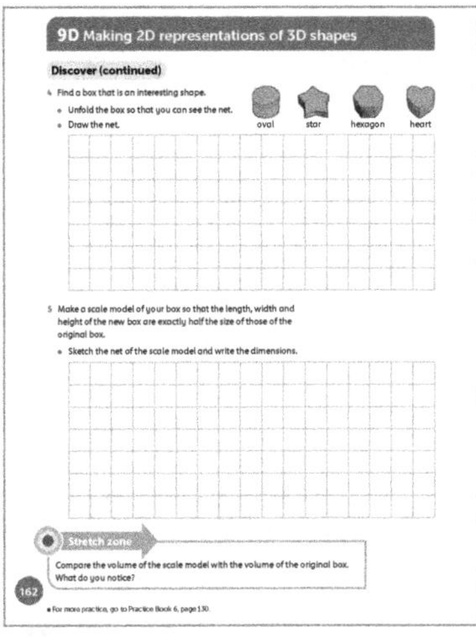

## Differentiation

**Supporting:** Encourage students to find all possible arrangements of the squares to make an open cube, and to describe the properties of their net.

**Consolidating:** Ask students to describe the process of making a scale model.

**Extending:** Ask students to explore a general rule to predict which will form nets of cubes and general rules linking lengths, areas and volumes.

**Stretch zone:** *Compare the volume of the scale model with the volume of the original box. What do you notice?*

Students can look for connections between the **dimension**s of both the box and the model. They will notice that if the dimensions of a box are halved, the total volume will be 8 times smaller than the original box, or $2^3$ times smaller.

### 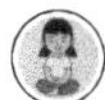 Reflection time

Take feedback from students. *What did you notice about the difference in volume between the scale model and the original box?* Write the following sentences on the board and ask students to complete them by predicting what would happen if they doubled the dimensions of the original box, instead of halving them.

If I double the dimensions of a box, the new length is ___ times the original length.

If I double the dimensions of a box, the new area of a face is ___ times the original area of a face.

If I double the dimensions of a box, the new volume is ___ times the original volume.

**Practice Book:** Students complete Practice Book page 130. They can do this directly after the Main activity, as homework, or as the focus of a separate mathematics session to help students consolidate their learning and build fluency.

Students complete scale drawings of the nets of three different boxes. They then label the side lengths of the nets and use these drawings to work out the volume of each box.

| Differentiated outcomes | |
|---|---|
| **All students** | should find a range of ways to arrange the squares and draw a net of their open box. |
| **Most students** | will find all the ways to arrange the squares and will visualise which are cubes of nets and make accurate scale models. |
| **Some students** | may find a general rule to predict which arrangements will form nets of cubes and find general rules linking lengths, areas and volumes. |

### Student Book pages 160–162

1 **a** cuboid

  **b** square-based pyramid

  **c** triangular prism

  **d** cylinder

  **e** cone

  **f** tetrahedron

### 2–5

Answers will vary because students draw their own nets and make their own models. Accept any correct answers. While students make their models, ask them what they notice about the size of their models compared to the size of the original shapes.

### Practice Book page 130

Answers will vary because students use different boxes to make their own 3D shapes. Accept any correct answers.

Stretch zone: Students calculate the volumes for their own Check that their answers are correct.

## 9D Making 2D representations of 3D shapes

### Explore
Student Book pages 163–164 · Practice Book page 131

### Specific learning focus

- Recognise and make 2D representations of 3D shapes including nets.

### Global skills

- **Creative skills:** exploring

### Key vocabulary

- cube, dimensions, isometric paper, volume, elevation

### Resources

- linking cubes in different colours
- large box
- isometric paper
- squared paper

### Language support

The Introductory activity demands a high level of language skills. You may need to model the use of relevant vocabulary to help students with their descriptions of their models, for example to describe the relative positions of the cubes:

- The black cube is ____________ the white cube.

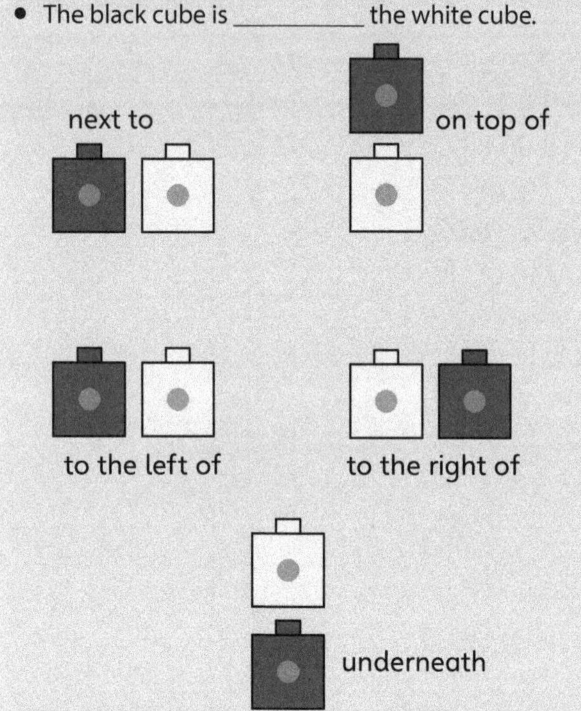


### Introductory activity

Ask two students to come to the front. Arrange it so that one student can make a model without any other students seeing. Ask this student to make a model out of five linking cubes. Put it in the large box so that the other students cannot see it. Then ask the student to give instructions so that their partner exactly replicates the model. This should include matching colours. Repeat this activity three times with different students.

Use the last model from the Introductory activity to draw a front **elevation**, then a side elevation and finally a plan view on the board. Write these terms on the board. For support, students can make copies of the models so that they can see the three views clearly for themselves.

Look together at page 163 of the Student Book. Display on the IWB, if possible. Refer students to the example drawings of the cube model in the worked example. Students can use this as a reference as they draw their own models. Ask students to work in pairs on the activity on pages 163–164 of the Student Book. Encourage discussion among pairs. Students may need additional support in carrying out the isometric drawing. Use peer support for this.

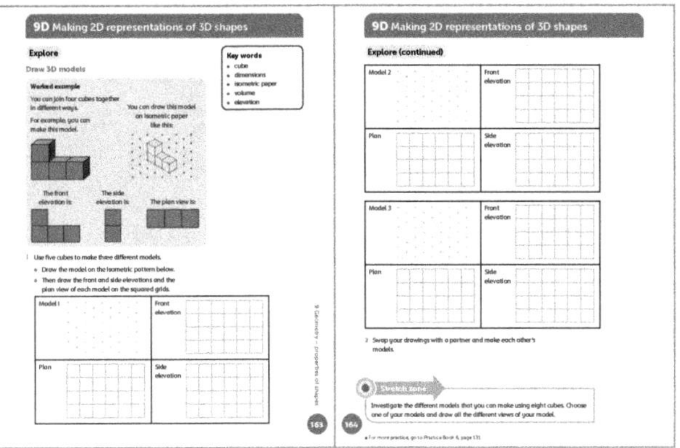

## Differentiation

**Supporting:** Help students to use **isometric paper** to draw the views of the models.

**Consolidating:** Ask students to explain how to use isometric paper to other students.

**Extending:** Ask students to use eight cubes and to draw more complex plans using isometric paper.

**Stretch zone:** *Investigate the different models that you can make using eight cubes. Choose one of your models and draw all the different views of your model.*

As a follow-up activity, students can use isometric paper to sketch a wide range of 3D shapes.

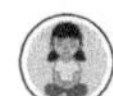 **Reflection time**

Ask one pair to come to the front and draw the plan view for one of their shapes. Ask other students: *Is this enough information to replicate the model?* Then ask the pair to draw the front elevation. Other students make any changes to their models that they need to. Finally, ask the pair to draw the side elevation. Ask other students, *Is your model an exact copy?* Repeat this activity if necessary.

**Practice Book:** Students complete Practice Book page 131. They can do this directly after the Main activity, as homework, or as the focus of a separate mathematics session to help students consolidate their learning and build fluency.

Students use two cardboard boxes to create a 3D model. They then draw, on isometric paper, the plan view, side elevation and front elevation of their model. *Can you use your drawings to work out the volume of your 3D model?*

| Differentiated outcomes | |
|---|---|
| **All students** | should recognise the different views and use isometric paper with support. |
| **Most students** | will recognise the different views and support other students to use isometric paper. |
| **Some students** | may draw more complex plans with more cubes. |

## Answers

### Student Book pages 163–164

Answers will vary because students use the cubes to make their own shapes. Check that students' drawings match their models. Observe students while they work and note who is able to record their shapes on isometric paper and who needs more practice.

### Practice Book page 131

Answers will vary because students draw their own plans and elevations. Check that students have drawn the different views correctly.

Stretch zone: Check that students have written suitable instructions for calculating the volume of their models.

# 9E Angles in shapes

## Discover
Student Book page 165 • Practice Book page 132

### Specific learning focus

- Estimate, recognise and draw acute and obtuse angles and use a protractor to measure to the nearest degree.

### Global skills

- **Creative skills:** exploring

### Key vocabulary

- triangle, quadrilateral, polygon

### Resources

- 9-peg circular pinboards and elastic bands (if possible) or Resource sheet 9.1: circles with nine dots equally spaced around the circumference
- large protractor

### Language support

Challenge students to make shapes with particular properties.

Ask, for example:

- *Can you draw a shape with an obtuse angle?*
- *Can you draw a shape with a reflex angle?*
- *Can you draw a shape with a right angle?*

Encourage students to classify the triangles using examples of equilateral, isosceles and scalene triangles.

Support students with the pronunciation of the names of shapes.

 Introductory activity

Draw five 9-peg circular pinboards on the board, or display Resource sheet 9.1 so all students can see it. Ask five students to come to the front of the class and draw a triangle. Each triangle should be different from the previous one. Ask pairs of students to discuss the properties and the names of the triangles. Take feedback and label, on the board, the different types of triangles and the different angles. Students can refer to these throughout the Main activity.

If any of the key vocabulary relating to triangles is not listed, ask a student to draw a triangle that will give an example of that vocabulary.

## Main activity

Give out the pinboards and elastic bands. If this is not possible, then students can use Resource sheet 9.1 to experiment with shapes before they record their final shapes in their Student Book. Ask each student in the group to make a different shape on their pinboard. Ask them to record the shapes in the Student Book on page 165. As you work with groups, encourage them to look carefully at the shapes to check that they are all different shapes and not the same shape in a different orientation. They can check this by measuring angles.

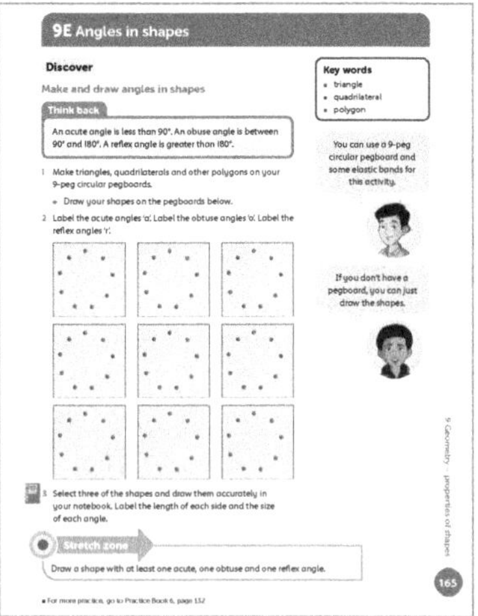

### Differentiation

**Supporting:** Model how to draw examples of shapes that include acute, obtuse and reflex angles.

**Consolidating:** Encourage students to name examples of shapes that include acute, obtuse and reflex angles.

**Extending:** Encourage students to measure and label the angles in their shapes.

**Stretch zone:** *Draw a shape with at least one acute, one obtuse and one reflex angle.*

Students use the pinboards to explore different shapes. For example, *How many **quadrilaterals** can they make with each of the different angles? How many pentagons?*

### Reflection time

Ask a student from each group to come to the front and sketch one of their shapes on the board. Use a large protractor to check that the interior angles in the triangles add up to 180° and the angles in the quadrilaterals add to 360°. This helps to remind students how to use a protractor, which they need to do in 9E Explore. Try to find examples of each of the different types of triangle and quadrilateral, equilateral, isosceles and scalene triangles, and irregular quadrilaterals, which all include different types of angles.

**Practice Book:** Students complete Practice Book page 132. They can do this directly after the Main activity, as homework, or as the focus of a separate mathematics session to help students consolidate their learning and build fluency.

Students use a different arrangement of pins on a pinboard to draw 12 different triangles, including right-angled triangles, isosceles triangles and scalene triangles. They label the angles they have made.

| Differentiated outcomes | |
| --- | --- |
| **All students** | should make and name a range of triangles and quadrilaterals. |
| **Most students** | will make all the different triangles, quadrilaterals and other polygons and label angles correctly. |
| **Some students** | may understand that interior angles of all triangles add up to 180° and all quadrilaterals add up to 360°, by estimating and measuring angles. |

# 9E Angles in shapes

## Explore    Student Book page 166 • Practice Book page 133

### Specific learning focus

- Estimate, recognise and draw acute and obtuse angles and use a protractor to measure to the nearest degree.

### Global skills

- **Creative skills:** exploring

### Key vocabulary

- acute angle, obtuse angle, reflex angle, protractor, tessellating

### Resources

- protractors
- strips of card
- scissors
- paper fasteners

### Language support

As you work with students, ask them to classify the angles. Ask, for example:

- *What type of angle is that?*
- *Is it acute, obtuse, reflex or a right angle?*
- *How do you know?*



## Answers

### Student Book page 165

Answers will vary because students draw their own shapes on the pinboards. Check that students have labelled the acute angles, obtuse angles and reflex angles correctly.

### Practice Book page 132

Answers will vary because students make their own triangles on the pinboards. Accept any correct answers.

Stretch zone: Check that students have measured correctly for the shape they chose.

## Introductory activity

Ask each student to make an angle measurer by putting a paper fastener through the corner of two pieces of card. It will look like this:

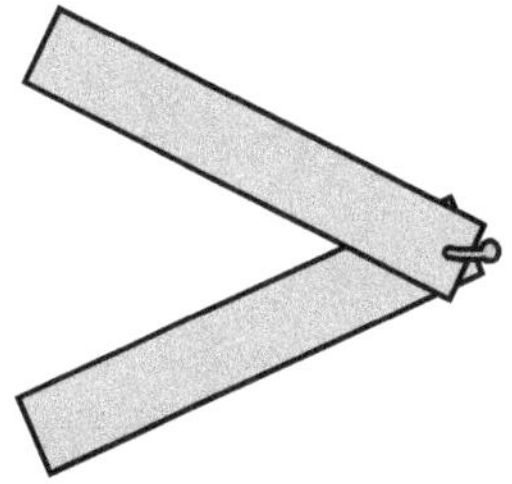

Ask students to show you an angle of roughly 60°, then 140° and finally 230°. They should check their angles with a partner first and agree who is closest (if they have different angles). Remind students of the meanings of the terms '**acute angle**', '**obtuse angle**' and '**reflex angle**'. Hand out **protractors** to pairs of students. Ask one of the pair to challenge the other to make a specific angle with their angle measurer. Then the first student checks the angle with their protractor. Remind students of the importance of using the cross as the zero point.

## Main activity

Ask students to work in the same pairs on the activity on pages 166 of the Student Book. Students could explore the environment around the classroom or outside to give them more choice of angles and triangles. They should take it in turns to complete a column in the table on page 166. Refer them to the speech bubble, which gives an example of a reflex angle that they might see in the classroom. Give students some more clues if you think they might need some help. For example, you could set them a trail of clues from the classroom door to the play equipment outside. You could also tell them to look for objects where the angle alters depending on the amount of turn, such as a door, or an open pair of scissors.

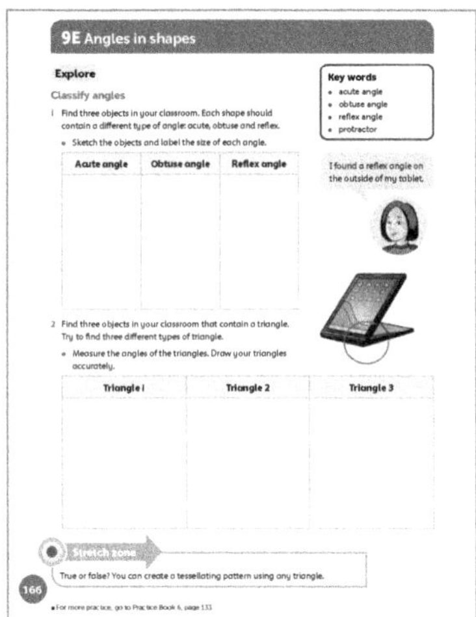

## Differentiation

**Supporting:** Help students to draw examples of acute, obtuse and reflex angles.

**Consolidating:** Ask students to identify and draw examples of acute, obtuse and reflex angles.

**Extending:** Challenge students to estimate, measure and label the different types of angles in shapes.

**Stretch zone:** *True or false? You can create a **tessellating** pattern using any triangle.*

Students should draw different triangles and then show how they tessellate.

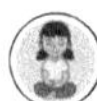 Reflection time

Ask a student from each group to come to the front and sketch one of their triangles from question 2 on the board. Check that the interior angles add up to 180°. Try to find examples of equilateral, isosceles, right-angled and scalene triangles that include all the different types of angles. Ask students to share the examples of objects they found that included reflex angles.

**Practice Book:** Students complete Practice Book page 133. They can do this directly after the Main activity, as homework, or as the focus of a separate mathematics session to help students consolidate their learning and build fluency.

Students find 2D shapes on objects at school or at home. They then measure the sides and angles of the objects and make scale drawings of the objects. Refer students to the hint on page 133 of the Practice Book. This reminds them that angles remain the same size in scale drawings, even when the lengths of the sides are changed.

| Differentiated outcomes | |
| --- | --- |
| **All students** | should find and sketch different angles and triangles and understand the terms 'acute', 'obtuse' and 'reflex'. |
| **Most students** | will measure angles accurately with support. |
| **Some students** | may estimate angles accurately and support other students in using protractors. |

## Answers

### Student Book page 166

Answers will vary because students find different angles and triangles in the classroom. Check that students' examples seem reasonable and that each angle type fits the measured size of the angle.

### Practice Book page 133

Answers will vary because students draw their own scale drawings of 2D shapes. Check that students have used a suitable scale and drawn the shapes correctly.

Stretch zone: Check that students have correctly drawn a scale version of their composite shape.

 **Unit 9** Geometry – properties of shapes

# 9F Missing angles

## Discover    Student Book page 167 • Practice Book page 134

### Specific learning focus

- Find missing angles in triangles.

### Global skills

- **Creative skills:** problem solving

### Key vocabulary

- missing angle, protractor, predict

### Resources

- protractors

### Language support

Reinforce the language for calculating missing angles, for example:

- Two angles add up to 140°, so the missing angle is___?
- The sum of the angles in a triangle is 180°.

 **Introductory activity**

Draw three different triangles on the board. Ask two students to come forward and measure two of the angles in the triangle. Write these on the board.

Ask students to recall the sum of the angles in a triangle and agree that it is 180°. If students are unsure of this, demonstrate by cutting out a triangle and cutting off the corners and then putting them together to meet along a straight line, which is also 180°.

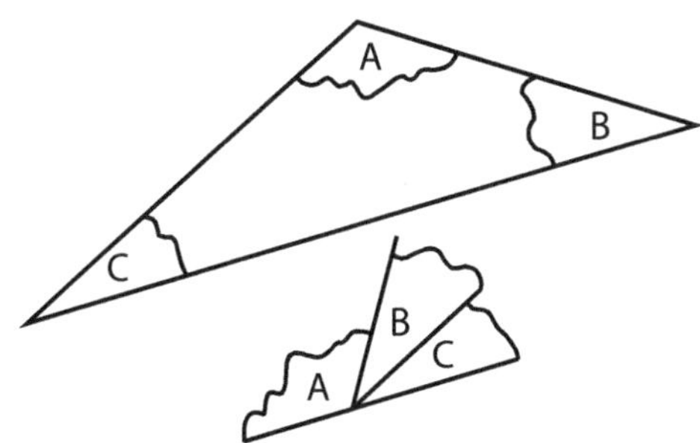

Ask students to subtract the total on the board from 180° to find the **missing angle**. Ask a student to come out and measure it to check using the protractor.

 **Main activity**

Look together at page 167 of the Student Book. Display on the IWB, if possible. Direct students to the Think back statement as a reminder of what has been verified in the Introductory activity. Explain that students are going to use their protractors to draw different triangles from two given angles, then calculate the size of the missing angle and check it with the protractor.

When students have completed the activities on page 167 ask them to check each other's answers in pairs.

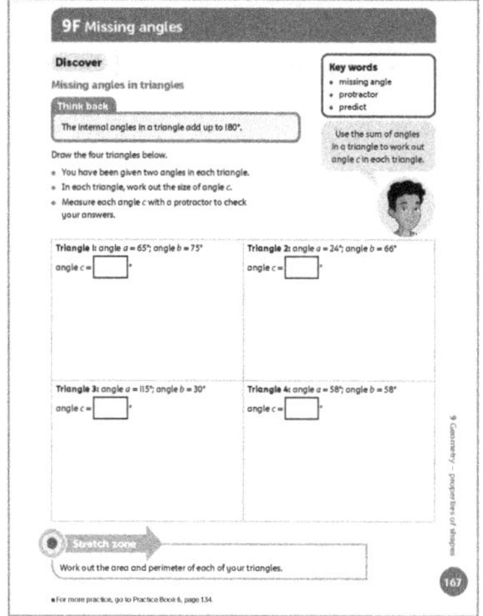

## Differentiation

**Supporting:** Help students who find it difficult to construct triangles using a protractor.

**Consolidating:** Ask students to explain how they constructed their triangles and calculated the missing angles.

**Extending:** Challenge students to use their knowledge of the properties of isosceles and right-angled triangles to find missing angles.

**Stretch zone:** *Work out the area and perimeter of each of your triangles.*

Students should be referred back to using the formulae for area and perimeter of triangles in Unit 7, 7B.

 **Reflection time**

Ask pairs of students to share their answers and explain how they constructed the triangles and calculated the missing angles. *Did you draw the triangle first, or did you work out the missing angle first? Which triangle was easiest to draw? Why?*

**Practice Book:** Students complete Practice Book page 134. They can do this directly after the Main activity, as homework, or as the focus of a separate mathematics session to help students consolidate their learning and build fluency.

Students complete a series of angles-in-triangles facts. They then use these facts to work out the size of some missing angles in different types of triangles.

| Differentiated outcomes | |
| --- | --- |
| **All students** | should construct triangles and calculate missing angles with support. |
| **Most students** | will construct triangles and calculate missing angles. |
| **Some students** | may calculate missing angles in a range of triangles by using what they know about the triangles' properties. |

### Student Book page 167

Check that students have constructed the triangles accurately.

1  $c = 40°$

2  $c = 90°$

3  $c = 35°$

4  $c = 64°$

### Practice Book page 134

1  60° because 180 is divided by 3 as the angles are equal

2  80° because the base angles add to 100° and the total is 180°

3  right-angled triangle

4  60°

5  65°

6  70°

7  150°

Stretch zone: A quadrilateral can always be divided into two triangles.

---

## 9F Missing angles

### Explore  Student Book page 168 • Practice Book page 135

#### Specific learning focus

- Explore angles in pentagons.

#### Global skills

- **Creative skills:** exploring

#### Key vocabulary

- missing angle, internal angle, predict

#### Resources

- protractors

#### Language support

Ask students questions to support them as they work through the activity in the Student Book. For example:

- *How many acute angles does your pentagon have?*
- *How large is the largest angle? How did you work that out?*
- *The sum of the angles in your pentagon is how much?*

 **Introductory activity**

Use the same triangle artwork from the previous lesson to show that angles on a straight line always add to 180°.

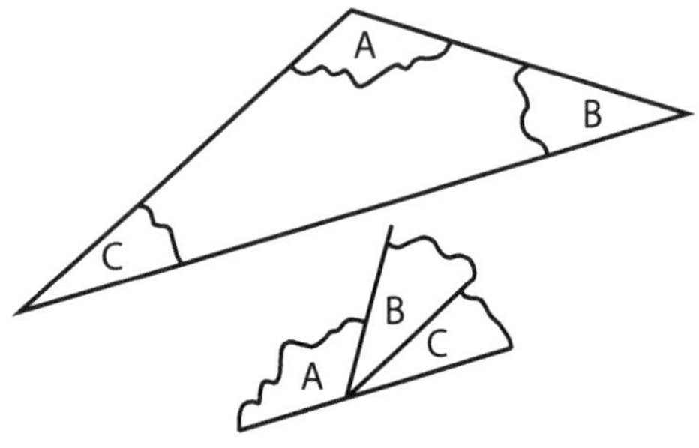


Ask students to predict what the sum of the angles around a point is.

Show this diagram of angles around a point.

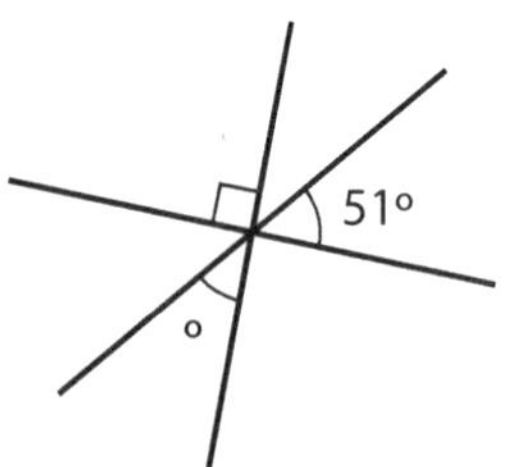


*If we know that all these angles will total 360°, how can we work out the missing angles?*

List the following rules.

- Angles on a straight line sum to 180°.
- Angles around a point sum to 360°.
- Vertically opposite angles are equal.

Point out the right-angle symbol. *So, what is the size of this angle? What size is the angle opposite it?*

Point at the angle labelled 51°. *Which other angle must also be 51°?* (Point at the vertically opposite angle.)

Then ask how we could work out the other two angles. *What do we know about them?* (They must be equal because they are opposite.)

Students may suggest using the angles on a straight line to work out one of the missing angles: 180 − (90 + 51) = 39° or they may suggest looking at all the angles around the point and working out the sum of the two missing angles, before dividing that by 2 to find the size of one of the missing angles:

$$\frac{360 - (180 + 102)}{2} = 39°$$

 **Main activity**

Draw a quadrilateral on the board. *How many degrees do the angles add up to?* Students may recall that it is 360°. Draw a diagonal across the quadrilateral to show that it makes two triangles. *How many degrees are there in each triangle?* Agree that the answer is 180°.

Ask students to discuss why they think the internal angles of every triangle sum to 180° and why they sum to 360° in every quadrilateral.

Look together at page 168 of the Student Book. Display on the IWB, if possible. Draw a few different pentagons on the board. They can be regular or irregular. Follow the example pentagon on page 168 of the Student Book to divide all your pentagons into three triangles, using the same format. Ask students whether they can **predict** what the sum of the **internal angles** of each pentagon will be. Explain that in this lesson, students are going to investigate how many degrees are in the internal angles of a pentagon. Students can now complete the activities on page 168 of the Student Book.

As they work, ask students to measure, and then label, the size of each angle so they can total them at the end. After students have drawn their first pentagon, pause the lesson and discuss what they have discovered so far. Agree that the sum of the internal angles of the pentagon are 540° (because there are three triangles and 3 × 180° is 540°) and use this to make predictions about the sizes of the angles in the second pentagon they draw.

Before students measure the final angle of their pentagon, ask, *What must the size of the final angle be in the pentagon? How do you know?*

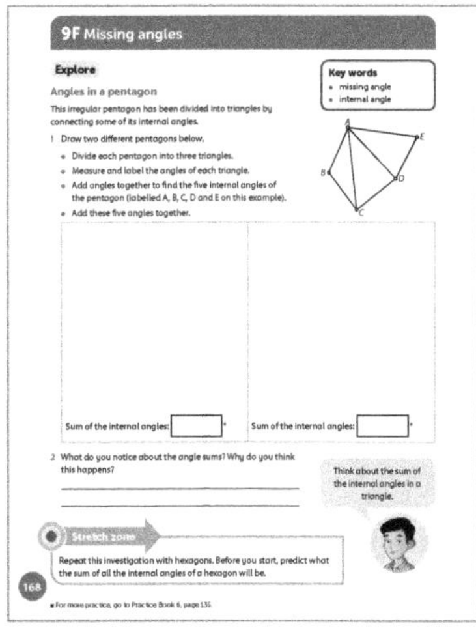

## Differentiation

**Supporting:** Help students to divide the pentagons into triangles first and then calculate missing angles by asking what the final angle in the pentagon will be.

**Consolidating:** Ask students to explain their predictions of angles.

**Extending:** Challenge students to predict what the internal angles of other polygons might total.

**Stretch zone:** *Repeat this investigation with hexagons. Before you start, predict what the sum of all the internal angles of a hexagon will be.*

Prompt students to think about how many triangles a hexagon can be divided into.

 **Reflection time**

Ask students to think back over their calculations of missing angles and internal angles. Ask them to share what they have learned about angles in a pentagon, for example using the fact that there are 180° in a triangle to find that the sum of the angles in any pentagon is 540°.

**Practice Book:** Students complete Practice Book page 135. They can do this directly after the Main activity, as homework, or as the focus of a separate mathematics session to help students consolidate their learning and build fluency.

Students find missing angles on a straight line, around a point and in vertically opposite angles. They consolidate all their learning from the last two lessons.

| Differentiated outcomes | |
| --- | --- |
| **All students** | should measure or calculate angles in a pentagon with support |
| **Most students** | will predict and measure or calculate angles in a pentagon. |
| **Some students** | may predict and measure or calculate angles in a pentagon and give reasoning. |

## Answers

### Student Book page 168

Check that students have drawn different pentagons and calculated their angle sums to be 540°.

### Practice Book page 135

**1 a** $= 215°$

**2 b** $= 133°$

**3 c** $= 33°$

**4 d** $= 128°$

**5 e** $= 125°$

**6 f** $= 180°$, **g** $= 131°$

**7 h** $= 58°$

**8 i** $= 124°$, **j** $= 56°$, **k** $= 124°$

Stretch zone: $56° + i = 180°$, and $j + k = 180°$, hence the angles at a point equal 360°.

# 9G Circles



## Specific learning focus

- Draw, label and name parts of circles.

## Global skills

- **Creative skills:** exploring

## Key vocabulary

- centre, diameter, radius, circumference

## Resources

- pairs of compasses
- string
- circular items – discs, tin lids, saucers
- large sheets of paper

## Language support

Use questions to support students' circle knowledge, for example:

- *How many times larger than the radius is the diameter?*
- *What is the perimeter of a circle called?*

### Introductory activity

Show students a diagram with parts of the circle labelled. You could display the one on Student Book page 169. Ask them, in pairs, to think about how they would explain what a radius, diameter and circumference are to someone who cannot see the diagram. Take feedback and agree definitions for each.

Ask students to describe the relationship between the **diameter** and **radius** as a ratio (2 to 1, 2 : 1; in other words, the diameter is twice the radius).

*If you draw a circle with a radius of 5 cm, what will the diameter be? If you measure a diameter of a circle to be 18 cm, what is the radius?*

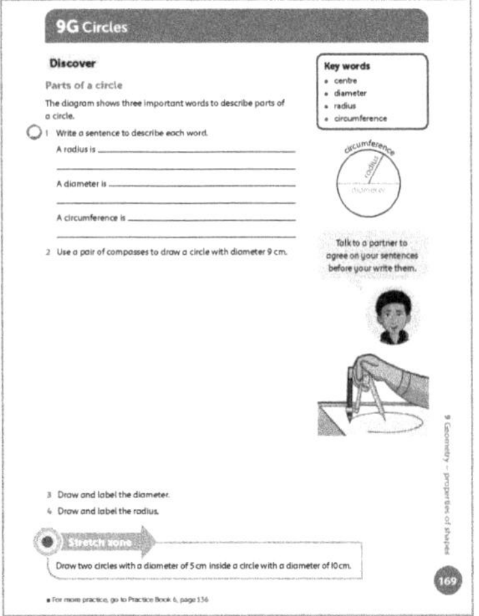

### Main activity

Give out circular items to each pair of students and ask them to estimate the diameter of each one. Measure the diameters accurately and use this measurement to calculate the radius of each circle.

*How could you measure the **circumference**?* Give pairs lengths of string to and suggest that they use it to curl round the circumference of their objects, mark where it meets the end, then measure this distance with a ruler.

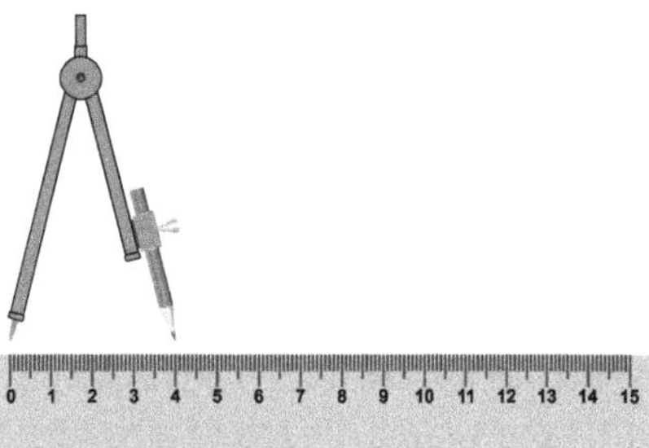

The distance between the point and the pencil point is the radius of the circle.

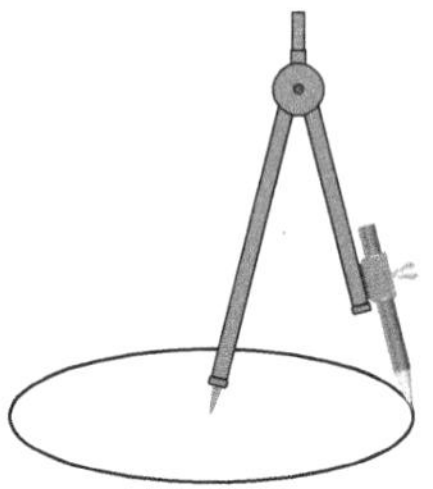

Hold the handle at the top and twist to turn the pencil around the point of the compasses.

Demonstrate how to use the compasses to draw circles if any students are not familiar with using them. Students should then complete the activities on page 169 of the Student Book.

### Differentiation

**Supporting:** Help students to measure the circumference of circular objects using string and a ruler.

**Consolidating:** Ask students to define the terms 'radius', 'diameter' and 'circumference'.

**Extending:** Challenge students to estimate the circumference of circular objects.

**Stretch zone:** *Draw two circles with a diameter of 5 cm inside a circle with a diameter of 10 cm.*

Students may need some guidance to line up the smaller circles on a diameter of the larger one.

### Reflection time

First, ask different students to share their definitions of radius, diameter and circumference. Then ask students to look at the diameters and circumferences of each circle they measured. Ask them to see whether they can see a link between the numbers. Prompt them by asking, *How many diameters are approximately equal to the circumference?* Students should notice that the circumference is a little more than 3 times the diameter in every case. Ask them to divide the circumference by the diameter for each circle and record the result. Explain that this result is the same for any circle.

**Practice Book:** Students complete Practice Book page 136. They can do this directly after the Main activity, as homework, or as the focus of a separate mathematics session to help students consolidate their learning and build fluency.

Students label the parts of a circle. They may need to refer to the internet or a mathematics dictionary to help them as there are terms that they have not come across before, such as 'arc', 'chord', 'sector', and 'segment'. This gives students greater depth of knowledge about the parts of a circle.

| Differentiated outcomes | |
| --- | --- |
| **All students** | should recognise and measure the radius, diameter and circumference of a circle with support. |
| **Most students** | will recognise and measure the radius, diameter and circumference of a circle. |
| **Some students** | may recognise and measure the radius, diameter and circumference of a circle and find the ratio of the circumference to the diameter. |



### Student Book page 169

**1** A radius is the distance from the centre to the circumference of a circle.

A diameter is the distance across a circle through the centre.

A circumference is the perimeter of a circle.

**2–4** Check that students have drawn and labelled the circle correctly.

### Practice Book page 136

| | | | |
| --- | --- | --- | --- |
| **1** | centre | **6** | arc |
| **2** | diameter | **7** | sector |
| **3** | circumference | **8** | segment |
| **4** | radius | **9** | tangent |
| **5** | chord | | |

The radius is half the length of the diameter.

The diameter is twice the length of the radius.

Stretch zone: Check that students have written correct definitions of the words they chose. They can use the internet to research the vocabulary.

---

# 9G Circles

## Explore   Student Book page 170 • Practice Book page 137

### Specific learning focus
- Draw and label circles.

### Global skills
- **Creative skills:** problem solving

### Key vocabulary
- diameter, radius, circle, segment, sector, arc, chord

### Resources
- ruler
- pairs of compasses
- large sheets of paper

### Language support

Reinforce the newly introduced terms for parts of a circle, emphasising the terms through labelling and questioning. Ask, for example:

- *What is the angle of that sector?*
- *How long is the arc for that segment?*

 ### Introductory activity

Display or draw a circle on the board or on a large piece of paper, using a pair of compasses.

Draw two radii (plural of radius) that form an angle of 90°.

Explain to students that part of a circle between two radii is called a **sector**, and that the part of the circumference along the perimeter of the sector is called an **arc**.

Ask one student to draw a line across their circle that does not pass through the centre. Tell them that this line is called a **chord**, and it divides the circle into **segments**. Segments also have part of their perimeter on the circumference, and it is also called an arc. Label each part of the circle as you introduce each new term. Ask students to tell you what angle the sector is that the two radii have created.

 **Main activity**

Remind students how to use a pair of compasses to draw a circle with given radius. For example, remind them how to align the point with 0 on a rule and the pencil tip with, say, 5 cm. Ask them to draw a circle of radius 5 cm and mark onto it the radius.

Look together at page 170 of the Student Book. Display on the IWB, if possible. Students should work in pairs to draw the circles. They check the accuracy of each other's drawings before working together on the Stretch zone activity of making a poster about circles.

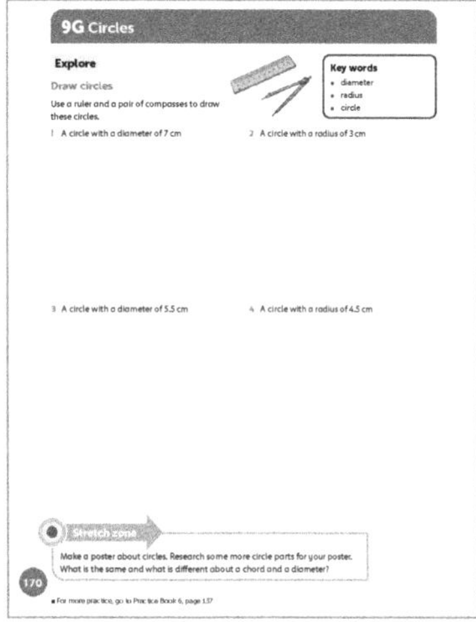

## Differentiation

**Supporting:** Help students to measure and draw circles accurately using a pair of compasses.

**Consolidating:** Ask students to describe to you how they can set up the compasses to draw the correct circles.

**Extending:** Challenge students to draw a set of nested circles with the same centres.

**Stretch zone:** *Make a poster about circles. Research some more circle parts for your poster. What is the same and what is different about a chord and a diameter?*

Students use examples of circles from their surroundings to enhance their posters.

 **Reflection time**

Ask pairs who completed the Stretch zone activity to present their posters to the class. They can explain what they have learned about circles from completing the poster, including any new vocabulary they found. Ask them to share the examples of circles they found in their surroundings.

**Practice Book:** Students complete Practice Book page 137. They can do this directly after the Main activity, as homework, or as the focus of a separate mathematics session to help students consolidate their learning and build fluency.

Students draw circles with a given radius or diameter. They then label all the parts of a circle they know. *What is the difference between a sector and a segment? What is the same and what is different about a radius and a diameter? Can you label a chord on either of your circles?*

| **Differentiated outcomes** | |
| --- | --- |
| **All students** | should draw circles to given measurements with support. |
| **Most students** | will draw circles to given measurements accurately. |
| **Some students** | may use accurate drawings of circles to create patterns. |

## Answers

### Student Book page 170

Check that students have used compasses accurately to draw each circle.

### Practice Book page 137

Check that students have drawn the circles and labelled the parts correctly.

**Stretch zone:** Students make their own pattern of circles. Accept any suitable pattern (the circles must all have radius 2.5 cm).

# 9 Geometry – properties of shapes



## Big idea

I can describe, draw and make a wide range of 2D and 3D shapes accurately, using equipment such as a protractor to measure angles, a ruler to measure length and a pair of compasses to draw circles.

## Global skills

- **Creative skills:** problem solving
- **Interpersonal skills:** teamwork
- **Self-development skills:** reflecting on learning

## Key vocabulary

- regular, irregular, 2-dimensional (2D), 3-dimensional (3D), acute angle, obtuse angle, straight line

## Resources

- squared paper
- protractors, rulers
- modelling materials
- card, scissors, glue, sticky tape, string, coloured pencils

## Language support

The presentation is an important way to ensure that students are articulating the key vocabulary. Work with students to prepare these presentations. Focus on pronunciation of the key vocabulary from this unit.

## Introductory activity

Either visit a local play park or use a video of children playing in a playground. Arrange the class into each mixed-attainment group of four to six students. Ask each group to make a list of important features in a successful playground. Take feedback on this to support the designs that groups will make. List the features identified by groups and display this list throughout the activity.

## Main activity

Ask students to work in their small groups to design a playground following the instructions on page 171 of the Student Book. Give them a variety of materials to work with as suggested in the Resources. It may be appropriate to extend this activity over a number of of lessons.

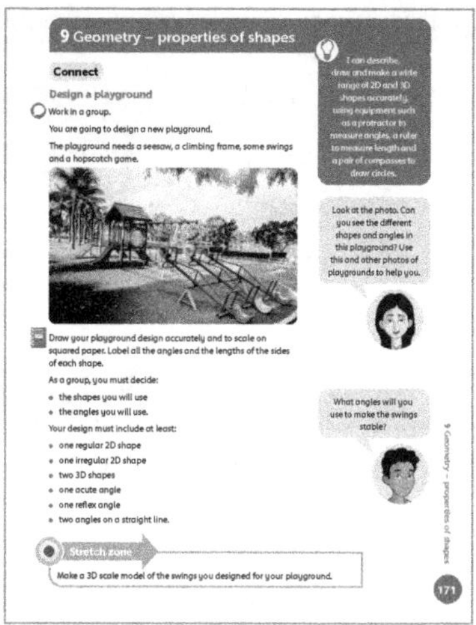

## Differentiation

Students work in mixed-attainment groups to support one another for the investigation on page 171 of the Student Book.

**Stretch zone:** *Make a 3D scale model of the swings you designed for your playground.*

Students could explore other architectural features. For example, they could find designs that are predominantly circular or triangular.

 ## Reflection time

At the end of the activity, ask each group to present a report on their design. As a class, agree a set of criteria you can use to assess the designs. Use the list you created in the Introductory activity as a basis for the criteria.

| Differentiated outcomes | |
| --- | --- |
| **All students** | should take part in the group planning and draw some shapes for the design with support. |
| **Most students** | will take part in the group planning and draw accurate shapes for the design. |
| **Some students** | may design the playground, drawing and labelling angles and lengths accurately. |

## Answers

### Student Book page 171

Answers will vary because students design their own playground and play apparatus. Check that students' answers seem reasonable – the shapes they use should be appropriate for each piece of play apparatus and the angles should be clearly marked and be realistic for each piece of play apparatus.

# 9 Geometry – properties of shapes



## Global skills

- **Self-development skills:** reflecting on learning

## Student Book

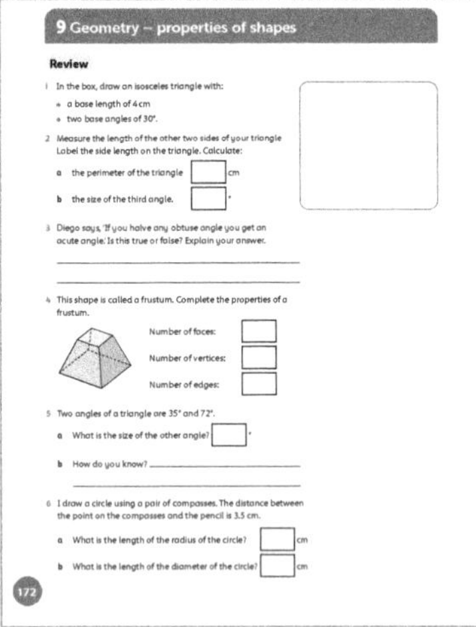

With young students, assessment activities are most effective when carried out as an everyday classroom activity. Students should have pencil, paper, rulers and protractors available to support them. Watch as students draw shapes with given side lengths and angles, estimate and measure side lengths and angles, identify types of angles and lines of symmetry, and calculate missing angles, and observe their strategies for working them out.

Students should complete the Review activity on page 172 of the Student Book as individuals as it is a summative assessment.

## Answers

### Student Book page 172

**1** Check that students have drawn the isosceles triangle accurately.

**2 a** 8.6 cm      **b** 120°

**3** True. Any angle between 90° and 180° when halved will be between 45° and 90°

**4** 6 faces, 8 vertices, 12 edges

**5 a** 73°      **b** 73° = 180 − (35° + 72°)

**6 a** 3.5 cm      **b** 7 cm

## Practice Book

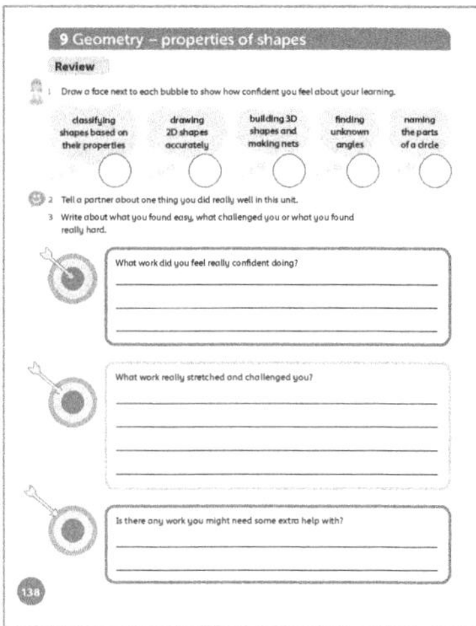

It is appropriate to complete this Practice Book Review as a whole-class discussion. You may choose to keep a record of the class discussion or a copy of the Review page for your own records. The Review provides an opportunity for students to reflect on their learning from the unit, to discuss any areas of mathematics that they feel went particularly well, and any areas that they feel less confident about. Ensure that all students have a copy of the Student Book as a reminder of the areas of mathematics that they have worked on in this unit.

Allow students plenty of time for discussion before asking them to complete the Practice Book page individually, and then, if appropriate, to share their responses with the rest of the class. If students complete this self-assessment at home, encourage them to discuss this with adults. Make a note of areas that students still feel unsure about, for example using a protractor correctly to measure angles, using knowledge of angle sums to calculate missing angles and identifying lines of symmetry in 2D shapes. Give students plenty of opportunities to estimate and calculate angles and draw shapes with varying properties.

## Additional material

There are additional end-of-unit assessments available on the *Oxford Owl for School* website.

# 10 Geometry – position and direction



## Big idea

The Big idea in this unit is how we use mathematics to describe the position or the movement of an object on a coordinate grid. Coordinates describe position. Transformations (reflections, rotations and translations) describe movement. After an object moves, the coordinates that describe its position are different. It is vital that students are able to learn actively. They need to experience the movement in order to notice how position changes. Instead of using mirrors for reflection, for example, you can make cut-outs of the shapes to be rotated, translated or reflected. This allows students to physically move the objects.

## Look out for

- **Students who inaccurately plot coordinates when reflecting a shape.** They may need help with making sure that every point and its reflection are the same distance from the reflection line.

- **Students who confuse reflections with translations.** They may simply translate a shape to the other side of a mirror line without reversing its orientation. Help them to understand that translated shapes keep the same orientation but reflected shapes are 'reversed'.

## Possible misconceptions

- **Students forget the order of coordinates.** Some teachers use the phrase 'You go along the hall and then up the stairs.' This reminds students that the first coordinate (the $x$-coordinate) describes position along the horizontal ($x$-axis), and the second coordinate (the $y$-coordinate) describes the position up and down the vertical axis ($y$-axis).

- **Students reflect shapes vertically or horizontally when the mirror line is at an angle.** This may be because of students' early experiences with reflections in vertical and horizontal lines only. They will need guidance on orienting the reflection correctly, making sure that every point is reflected the same distance from the mirror line and perpendicular to it.

## Key vocabulary

- coordinate grid, quadrant, origin, coordinates
- axis/axes, first, second, third, fourth quadrants
- symmetry, line of symmetry, axis of symmetry, symmetrical, pattern, vertex, vertices
- triangle, square, hexagon, rectangle
- parallel, perpendicular
- mirror line, reflect, reflection, rotate, rotation, transformation
- clockwise, anti-clockwise
- translate, translation, slide

## Coverage in lessons

| Learning objective | E | 10A | 10B | C | R |
|---|---|---|---|---|---|
| Describe positions on the full coordinate grid (all four quadrants). | ✓ | ✓ | ✓ | | ✓ |
| Draw and translate simple shapes on the coordinate plane and reflect them in the axes. | ✓ | ✓ | ✓ | ✓ | ✓ |

# 10 Geometry – position and direction



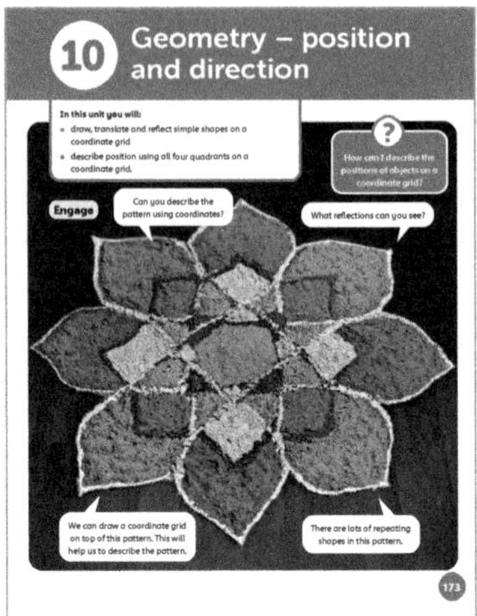

## Big question

- How can I describe the positions of objects on a coordinate grid?

## Global skills

- **Creative skills:** problem solving/exploring/ investigating
- **Interpersonal skills:** communication/teamwork/ leadership
- **Self-development skills:** reflecting on learning

## Key vocabulary

- reflect, reflection, pattern, coordinates, coordinate grid, origin, first, second, third, fourth quadrants, symmetrical, line of symmetry, hexagon, triangle, square, rectangle

## Resources

- magazines or books of illustrations that use symmetry
- digital camera
- large sheets of paper

## Language support

As students give you the instructions in the Introductory activity, make a note on a large piece of paper of all the vocabulary they use. Include all the words from the key vocabulary listed above. Use this list as the start of a display poster that they can refer to throughout the unit.

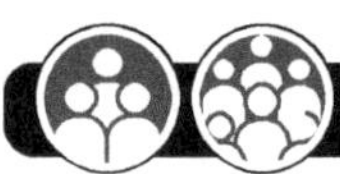 Introductory activity

Look together at page 173 of the Student Book. Display on the IWB, if possible. Arrange the class into mixed-attainment groups of four to six students. Ask each group to look at the image on page 173 of the Student Book. Tell them to try to describe the shape as clearly as possible. They should imagine that they are having a telephone conversation with someone who cannot see the image but who wants to make an exact copy. You may need to use a vocabulary box on the board to help students with the language (see Key vocabulary section).

Ask one group to read their instructions to the class. As they read their instructions, another student should try to recreate the image on the board.

 Main activity

Ask students, working in the small groups, to find an example of a Rangoli pattern, a piece of Islamic art or any appropriate design or pattern that uses symmetry. Alternatively, they can find photographs of patterns that use symmetry (either natural patterns or patterns within architectural designs around the school). Then tell groups to use these designs to create posters explaining the key vocabulary listed in the overview.

### Differentiation

**Supporting:** Model key vocabulary that you do not hear students using.

**Consolidating:** Ask students to describe shapes and patterns using the vocabulary listed on the poster.

**Extending:** Challenge students to draw a pattern from given instructions.

 Reflection time

Ask groups to share their posters with the class. Decide on one poster to display for the rest of the unit. Students can use this poster and their own ideas to complete the position and movement words in the glossary at the back of the Student Book.

# 10A Reading and plotting coordinates



## Specific learning focus

- Read and plot coordinates in all four quadrants.

## Global skills

- **Creative skills:** exploring

## Key vocabulary

- origin, coordinate grid, quadrant

## Resources

- tape measures
- squared paper

## Language support

Model the language of coordinates by asking questions such as:

- *Which students are in the third **quadrant**?*
- *Where is the **origin** in the classroom?*
- *How do you remember the order of coordinates?*

 **Introductory activity**

Ask students, in pairs, to list all the facts they can remember about coordinates and plotting coordinates on a **coordinate grid**. After five minutes, take feedback from the class. Make sure that students can recall key features such as:

- drawing and labelling the axes
- writing and reading coordinates as number pairs
- using the first number to move horizontally along the x-axis, then the second number to move vertically to the correct point.

 **Main activity**

Tell students that they are going to make a plan of the classroom. As a group, they need to decide what is the best scale factor to use for their drawing. Use this checklist for what students need to remember to do:

1  Choose a suitable scale factor for your drawing.

2  Draw on all the large objects from the classroom, for example desks and the board.

3  Add smaller details such as people, shelves and a sink.

Tell students to work in small groups on the activity on page 174 of the Student Book. This ensures that students discuss the work. Students can also check one another's decisions. This activity depends on students' previous experience so you may choose to group students by prior attainment and ask some groups to create a scale plan and others simply to plot points.

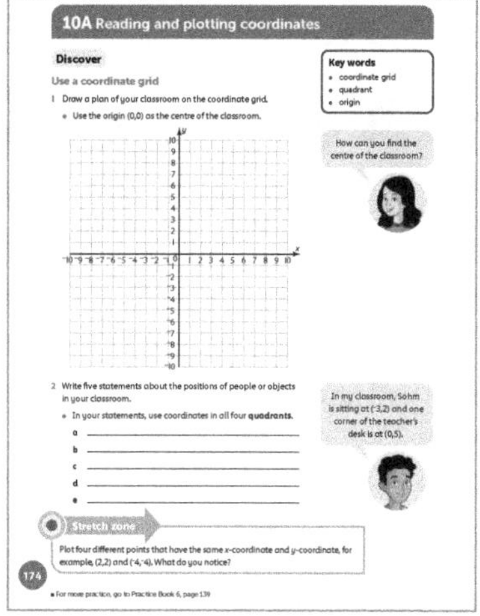

## Differentiation

**Supporting:** Help students to plot points on the grid that represent objects in the classroom.

**Consolidating:** Ask students to draw a plan that represents the main objects in the classroom.

**Extending:** Challenge students to draw a plan on the grid to represent objects in the classroom to scale.

**Stretch zone:** *Plot four different points that have the same x-coordinate and y-coordinate, for example (2,2) and (⁻4,⁻4). What do you notice?*

Students should recognise the linear arrangement of the points.

 **Reflection time**

Ask each group to read one of their five statements about the positions of people or objects in the classroom, for example as follows.

- Ahmed is sitting at point (⁻3, 4).
- There are eight students sitting in the first quadrant.

**Practice Book:** Students complete Practice Book page 139. They can do this directly after the Main activity, as homework, or as the focus of a separate mathematics session to help students consolidate their learning and build fluency.

Students draw plans of a room at home similar to the plan they drew in the main lesson. They then write the coordinates of four objects in that room.

| Differentiated outcomes | |
| --- | --- |
| **All students** | should plot points accurately in the grid with support. |
| **Most students** | will create a plan of the classroom, using points on the grid. |
| **Some students** | may create a scale plan of the classroom, using points on the grid. |

## Answers

### Student Book page 174

Answers will vary because students draw plans of the classroom on the coordinate grids. Check that their statements about the positions of objects and people in the plan are correct and that the coordinates are accurate.

Stretch zone: Check that the four points have been correctly plotted and that students have noted the linear arrangement to identify the coordinates for the centre of the square.

### Practice Book page 139

Answers will vary because students draw plans of a room in their home on the coordinate grids. Check that the coordinates that they give for items in the room are accurate.

Stretch zone: Check that an object has been correctly plotted at (⁻1.5, ⁻4.5)

---

# 10A Reading and plotting coordinates

**Explore** Student Book pages 175–176 • Practice Book page 140

## Specific learning focus

- Read and plot coordinates in all four quadrants.

## Global skills

- **Creative skills:** exploring

## Key vocabulary

- origin, coordinate grid, quadrant

## Resources

- two polygons drawn on a large Cartesian grid, which you have prepared before the activity (one regular and one irregular shape)
- Resource sheet 10.1: blank coordinate grids for four quadrants
- mini whiteboards and markers

## Language support

As students are working, ask questions to encourage the use of the language of properties of polygons, for example:

- *Are there any parallel/perpendicular lines?*
- *What sort of symmetry does that shape have?*
- *Are there any acute/obtuse/reflex angles?*

## Introductory activity

Use the two polygons that you prepared before the activity. Read the coordinates of each vertex in turn. Ask students to mark the vertices on their coordinate grid to recreate the polygon. Then ask them to talk to a partner and to list as many properties of the shape as they can, including its name, on their whiteboards. Repeat for your second shape. Use one regular and one irregular shape.

## Main activity

Explain that students are going to draw their own polygons on coordinates and then list the coordinates. They can plot any polygons as long as they have at least one vertex in each quadrant. Ask students to work in pairs on the activity on pages 175–176 of the Student Book. Each student prepares four different polygons and writes the coordinate of each vertex, which they then share with their partner. Their partner then plots the points of each vertex and draws the shapes on their own grids. They may need to refer to the posters from Unit 9 to remind them of the names of different polygons.

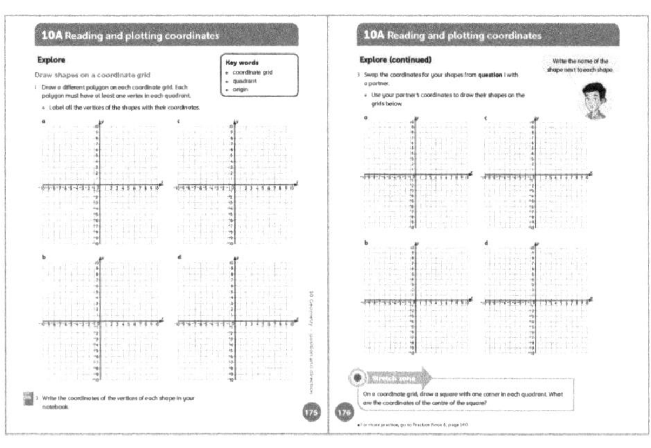

# Differentiation

**Supporting:** Help students to label the coordinates of points on their polygons.

**Consolidating:** Ask students to label the coordinates of points and the polygons.

**Extending:** Ask students to label the coordinates of points and describe the properties of the shapes.

**Stretch zone:** *On a coordinate grid, draw a square with one corner in each quadrant. What are the coordinates of the centre of the square?*

 ## Reflection time

Ask one student to give you instructions to allow you to recreate they shape they drew. Plot the coordinates on a grid on the board as the student gives you the coordinates. Then model the use of language by naming as many properties of the shape as you can.

Choose another student to give instructions for plotting their shape and this time ask the class to tell you the shape's properties.

**Practice Book:** Students complete Practice Book page 140. They can do this directly after the Main activity, as homework, or as the focus of a separate mathematics session to help students consolidate their learning and build fluency.

Students plot coordinates on a grid to create shapes. They predict whether a list of given coordinates will create a hexagon when joined.



| Differentiated outcomes | |
| --- | --- |
| **All students** | should use coordinates to describe the position of their shapes with support. |
| **Most students** | will use coordinates to describe the position of their shapes and use a range of shapes that they can name. |
| **Some students** | may use coordinates to describe the position of their shapes and use a wide range of shapes that they can name. |

## Answers

### Student Book pages 175–176

Answers will vary because students choose their own shapes to draw on the coordinate grids. Check that students have labelled the vertices of each shape with the correct coordinates.

### Practice Book page 140

Answers will vary because students choose where to draw a hexagon on the coordinate grids. While students are working, observe who can read and plot coordinates accurately and who requires more practice.

On the second grid, the points cannot be the six vertices of a hexagon because three of the vertices have the $x$-coordinate $^-5$, which means that these will all be on a straight line. The six points can be joined to form a pentagon.

Stretch zone: Check that the hexagon is similar to the previous ones and that the coordinates are listed correctly.

# 10B Translations and reflections

## Discover  Student Book page 177 • Practice Book page 141

## Discover  Student Book page 177 • Practice Book page 141

### Specific learning focus

- Recognise where a polygon will be after one reflection or one translation.

### Global skills

- **Creative skills:** investigating

### Key vocabulary

- coordinates, axis, translate, slide

### Resources

- large cut-out rectangle and equilateral triangle
- reusable adhesive
- glue, card, scissors
- prepared Cartesian grid (axes labelled from ⁻10 to 10) on the board
- prepared cards with the following instructions:
  - Reflect about the x-**axis**.
  - Reflect about the y-axis.
  - **Translate** 3 units in the positive x-direction.
  - Translate 4 units in the negative y-direction.

### Language support

As students are working, ask them to talk about the type of transformation they are using. Ask, for example:

- *Is that a reflection or a translation?*
- *Where is the mirror line for that reflection?*

### Introductory activity

Draw a grid on the board and attach the cut-out rectangle to it. Draw around the rectangle to mark its position. Ask a student to come to the front. Tell them to pick a card, carry out the instruction and mark the new position of the rectangle on the grid. Repeat this three more times with different students, each time moving the previous shape. Repeat with the triangle. This time let the first student select a starting position for the triangle. Move the triangle four times. This creates a pattern of rectangles and triangles.

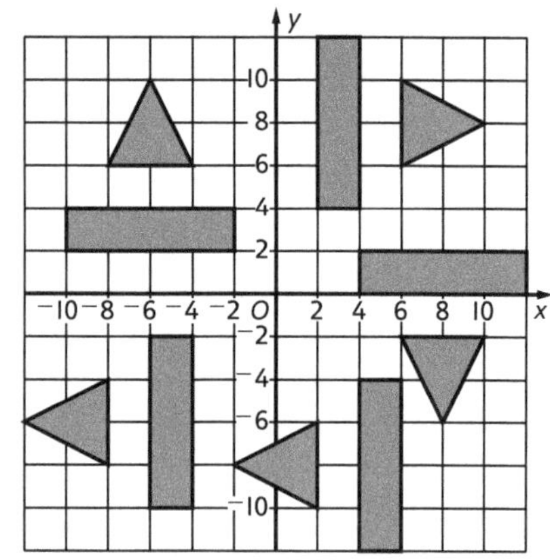

### Main activity

Ask students to create their own artworks using simple polygons as the base. First, ask them to cut out the shapes they are using as the basis of the artwork. Then tell them to use different transformations – translations and reflections – of these shapes to create their final artwork. Students work in pairs on the activities on page 177 of the Student Book to support discussion.

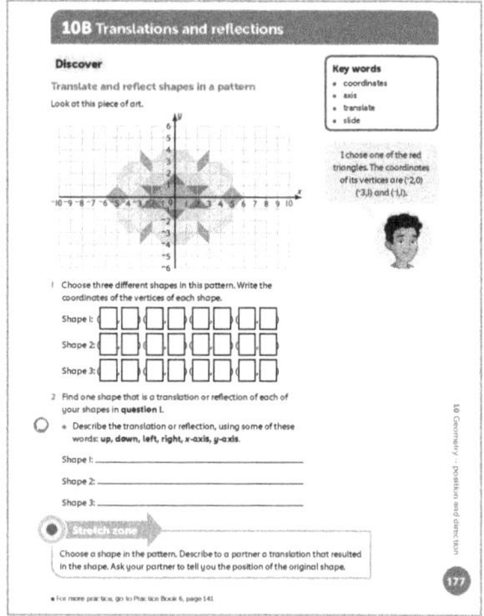

### Differentiation

**Supporting:** Point out the transformations so that you model the vocabulary.

**Consolidating:** Ask students to describe the transformations to you.

**Extending:** Encourage students to use multiple transformations.

**Stretch zone:** *Choose a shape in the pattern. Describe to a partner a translation that resulted in the shape. Ask your partner to tell you the position of the original shape.*

Students can then find similar patterns at home or in the local environment and write descriptions of any reflections or translations they can see.

 ## Reflection time

Select some of the final artworks to share. Model the use of language by describing what you can see. Say, for example:

- *I can see a reflection in the x-axis here.*
- *I can see that you translated this shape here from there.*

**Practice Book:** Students complete Practice Book page 141. They can do this directly after the Main activity, as homework, or as the focus of a separate mathematics session to help students consolidate their learning and build fluency.

Students create a pattern of their own, which includes reflections and translations. They should ensure that they draw their pattern accurately, reflecting or translating each part of the pattern precisely. They use colours in their pattern, which they should also use in each reflected or translated shape. Note that the Practice Book says that, if they want to, they can also use rotational symmetry in their pattern.

| Differentiated outcomes | |
| --- | --- |
| **All students** | should use simple transformations with support. |
| **Most students** | will use examples of reflections and translations. |
| **Some students** | may use a wider range of transformations and name them. |



### Student Book page 177

Check that students have identified the different possible reflections and translations.

### Practice Book page 141

Check that students have coloured the pattern so that it contains reflections and translations.

Stretch zone: Check that students have sketched a pattern that includes translations.

---

# 10B Translations and reflections

## Explore  Student Book pages 178–179 • Practice Book page 142

### Specific learning focus

- Identify what happens to a polygon when it is translated and reflected on a coordinate grid.

### Global skills

- **Creative skills:** investigating

### Key vocabulary

- coordinates, translate, reflect, slide

### Resources

- prepared 'maze' using chalk or small cones in a large open space
- coloured card, scissors
- mini whiteboards and markers

### Language support

Ask students to give instructions. This helps you to work with individuals on pronunciation and accuracy of vocabulary. Model useful instructions, for example:

- *Walk forward five steps.*
- *Turn clockwise through 90 degrees.*

 ## Introductory activity

Before the activity, prepare a 'maze' on the floor with chalk. Include left and right turns through 90°.

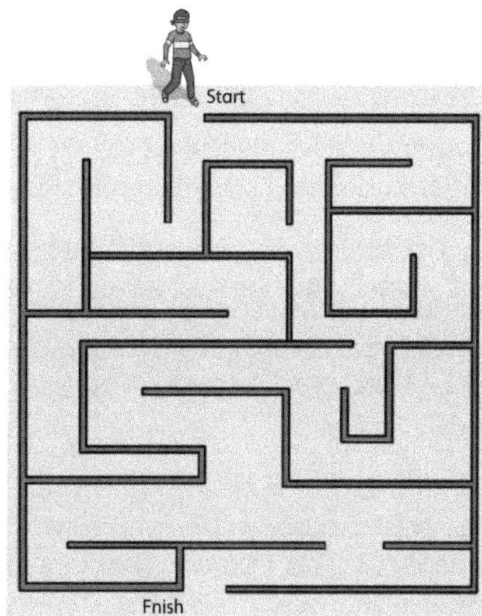


Ask a student to stand at the entrance to the maze. Blindfold them. Ask the other students to take it in turns to give the blindfolded student instructions to guide them through the maze. These instructions may be, for example:

- *Walk forward 4 steps.*
- *Turn 90° anti-clockwise.*

You could create two or three mazes and turn this activity into a competition. For example, the winner could be the team who completes the maze in the quickest time.

Give each student a piece of card and scissors and ask them to cut out a polygon of their choice. Before students start, tell them that their shape should fit on the grid on page 178 of the Student Book so that its vertices sit on intersections on the grid. Encourage students to use shapes that are not simple quadrilaterals. It could be a regular or irregular shape, perhaps a star or an arrow.

Ask students in pairs to complete the activities on pages 178–179 of the Student Book. They should draw different shapes and carry out the transformations individually and then check one another's answers.

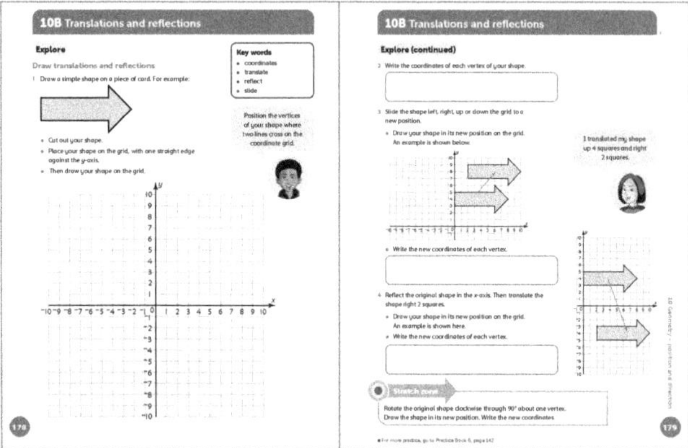

## Differentiation

**Supporting:** Help students to translate their shape and name the coordinates of the new points.

**Consolidating:** Ask students to describe the transformations using the coordinate grid.

**Extending:** Ask students to carry out multiple transformations and describe the effect.

**Stretch zone:** *Rotate the original shape clockwise through 90° about one vertex. Draw the shape in its new position. Write the new coordinates.*

Students use their cut-out shape. Show them how to keep one vertex in the same place and rotate the rest of the shape 'around this point' 90°. They then draw the shape in its new position and write the new coordinates of all the vertices.

## Reflection time

Draw a set of axes and a coordinate grid on the board. Ask two students to come to the front of the class. Ask one student to draw their initial shape on the grid. Then tell the second student that you want them say something quietly, so that only you and the other student can hear. Ask what their translation was for question 3 on page 179 of the Student Book. Ask the first student to translate their shape using these instructions and to draw their new shape on the grid. Ask the rest of the class to write the translation on their whiteboards.

After one minute, ask all students to show you their whiteboards at the same time. Show students with an incorrect translation what the effect of their translation would be, so that they can see their mistake.

**Practice Book:** Students complete Practice Book page 142. They can do this directly after the Main activity, as homework, or as the focus of a separate mathematics session to help students consolidate their learning and build fluency.

Students reflect a simple shape three times so that the same shape is drawn in all four quadrants of a coordinate grid. They write the new coordinates of the vertices each time they **reflect** the shape. *Is the final shape you drew a reflection of the first shape you drew? In which axes is this reflected?*

| Differentiated outcomes | |
|---|---|
| **All students** | should plot coordinates accurately and name the different transformations with support. |
| **Most students** | will use more complex shapes and name coordinates of points accurately after the transformations. |
| **Some students** | may use complex shapes, name their properties and name coordinates of points accurately after the transformations. |

## Answers

### Student Book pages 178–179

Answers will vary because students choose where to place their shape on the coordinate grid and where to move it to. While students are working, observe how they record the coordinates of their shape. Note who can reflect and **slide** their shapes correctly and who needs more practice.

### Practice Book page 142

Answers will vary because students draw their own shape on a coordinate grid. Check that they have reflected the shape correctly in the ways required.

Stretch zone: Students may notice that the final shape can be obtained from the original just by reflecting in the *y* axis.

# 10 Geometry – position and direction

## Connect  Student Book page 180

### Big idea

I can use coordinates to describe positions on a grid. I can describe the positions of objects in all four quadrants of a grid.

### Global skills

- **Creative skills:** exploring
- **Interpersonal skills:** teamwork

### Key vocabulary

- symmetry, line of symmetry, axis of symmetry, vertex, vertices, parallel, perpendicular, mirror line, reflect, reflection, rotate, rotation, translate, translation, transformation, clockwise, anti-clockwise

### Resources

- examples of logos such as those shown on page 180 of the Student Book as well as well-known logos of sportswear brands or restaurant chains
- squared paper

### Language support

As groups design their logos, engage in discussions with each group to consolidate the language from the unit. Ask, for example:

- *What shape is that?*
- *Where is that shape reflected?*
- *Is the shape translated in the logo? Can you describe the translation?*

 **Introductory activity**

Look together at page 180 of the Student Book. Display on the IWB, if possible. Together, look at the examples of logos. Discuss and display other well-known logos such as sportswear brands, restaurant chain logos or toy manufacturers. Ask students, in groups, to list the elements of an effective logo.

Take feedback from groups. Make a list of criteria for effective logos. This might include symmetry, simplicity or repetition of colours. Students might also spot shapes in logos that have been reflected, translated or rotated.

Explain that they will use these criteria to assess the logos at the end of the lesson.

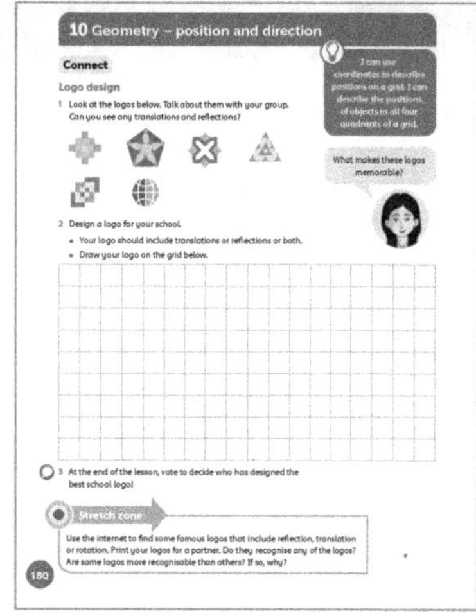

 **Main activity**

For this activity, organise the class to work in mixed-attainment groups of four students. Ask groups to read through the instructions on page 180.

Each member of the group designs their own logo for the school. Students in the group look at each logo in turn. They list the positives and what they would change. They pick their favourite logo and work together to improve it.

### Differentiation

Students work in mixed-attainment groups to support one another in working through the investigation.

**Stretch zone:** *Use the internet to find some famous logos that include reflection, translation or rotation. Print your logos for a partner. Do they recognise any of the logos? Are some logos more recognisable than others? If so, why?*

Students should find a wide range of logos and rank them according to the effectiveness criteria from the Introductory activity. Students could use their criteria to design alternative logos for their favourite brands.

 **Reflection time**

Ask each group to present their logo. The rest of the class assess each group's logo. The class then use the criteria from the Introductory activity to decide on the best logo.

| Differentiated outcomes | |
| --- | --- |
| **All students** | should use translation and reflections to design a logo with support. |
| **Most students** | will use translation and reflections to design a logo. |
| **Some students** | may design a logo using translation, rotations and reflections and describe the properties of the logo. |

## Student Book page 180

Answers will vary because students design their own logos. Check that students can describe correctly the translations, reflections and rotations they used for their logos.

# 10 Geometry – position and direction

## Review   Student Book page 181 • Practice Book page 143

## Global skills

• **Self-development skills:** reflecting on learning

## Student Book

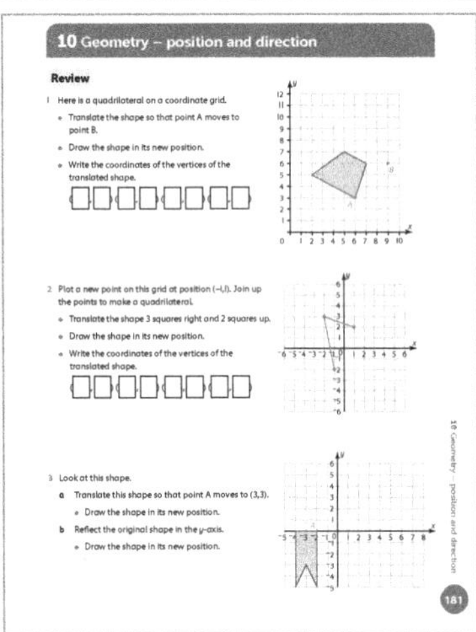

With young students, assessment activities are most effective when carried out as an everyday classroom activity. Students should have pencils and squared paper available to support them. Watch as they draw and reflect shapes using coordinates, identify lines of symmetry and describe translation of shapes, and observe their strategies for working them out.

Students should complete the Review activity on page 181 of the Student Book as individuals as it is a summative assessment.

## Answers

### Student Book page 181

**1**  (5, 8), (9, 6), (8, 10), (10, 9)

**2**  (1, 5), (2, 0), (2, 3), (4, 4)

**3 a**  (1, 3), (3, 3), (3, ⁻2), (2, 0), (1,⁻2)

  **b**  (2, 0), (4, 0), (4, ⁻5), (3, ⁻3), (2, ⁻5)

## Practice Book

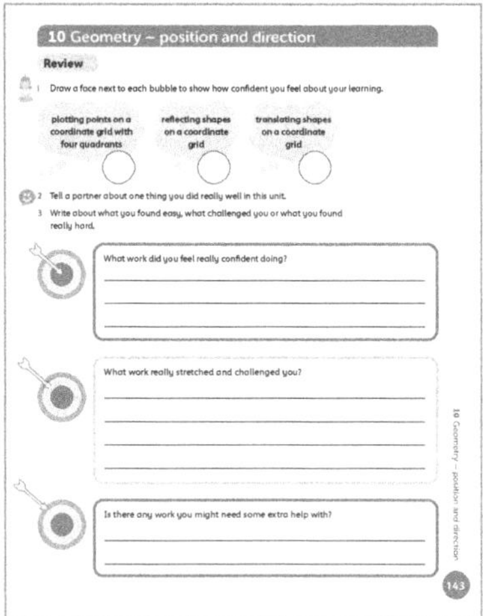

It is appropriate to complete this Practice Book Review as a whole-class discussion. You may choose to keep a record of the class discussion or a copy of the Review page for your own records. The Review provides an opportunity for students to reflect on their learning from the unit, to discuss any areas of mathematics that they feel went particularly well, and any areas that they feel less confident about. Ensure that all students have a copy of the Student Book as a reminder of the areas of mathematics that they have worked on in this unit.

Allow students plenty of time for discussion before asking them to complete the Practice Book page individually, and then, if appropriate, to share their responses with the rest of the class. If students complete this self-assessment at home, encourage them to discuss this with adults. Make a note of areas that students still feel unsure about, for example plotting coordinates correctly, not confusing the *x*-axis and *y*-axis and being able to translate and reflect shapes accurately across all four quadrants. Give students plenty of opportunities to plot coordinates and to translate or reflect coordinates and shapes.

## Additional material

There are additional end-of-unit assessments available on the *Oxford Owl for School* website.

# 11 Statistics

## Overview

### Big idea

This unit deals with three distinct areas of mathematics.

The first is looking at averages, specifically the mean. The second area dealt with is that of probability. The Big idea in this area is that of 'equally likely outcomes'; that is, that we can only calculate probabilities of events that are equally likely.

Finally, handling data is explored. This unit highlights that handling data is about making decisions about what data represents. Often, students do not know how to interpret data. They learn how to collect and organise the data but do not realise that there may be alternative interpretations of the data. It is important that students see the big picture from the beginning. They need to understand that interpretation is just as important as collection and representation of data.

The data handling cycle is:

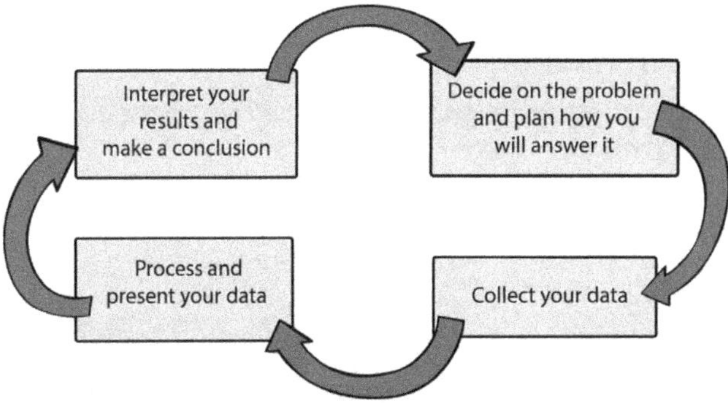


This diagram emphasises the importance of both problem posing and interpretation. It is important to realise that the data handling cycle is continuous. To emphasise this, the handling data part of this unit is not broken down into separate sub-units as previous units were. Instead, students undertake a single unbroken investigation in which they follow data handling through one cycle.

### Look out for

- **Students who do not realise that the mean of a set of data must lie within the range of the data.** Calculation errors may produce a result outside of the range and students do not check that the answer is reasonable in the context of the problem.
- **Students who think that some equally likely outcomes are not equal.** From experience of playing games, students may think that some numbers are harder to get when being determined randomly. Looking at all possibilities will help overcome this.

### Possible misconceptions

- **Students may be used to simply answering questions based on data and so will not realise that there can be alternative interpretations of the data.** This unit asks students to take a critical view of the data throughout. It also encourages them to make their own decisions about which charts to use to represent and interpret the data.
- **Students may confuse the mean with other averages they may have heard of.** If students have read about the mean or mode, they may be confused about the difference between them, but in this unit only the mean is considered.

### Key vocabulary

- average, mean, probability, equally likely outcomes, event, likely, possible, unlikely, most/least likely
- outcome, certain, impossible, equal chance, fair, unfair, data, survey, questionnaire, tally
- graph, line graph, pictogram, bar chart, table, frequency table
- Carroll diagram, Venn diagram
- question, collect, present, interpret, interrogate
- most/least popular, most/least common, interpret

## Coverage in lessons

| Learning objective | E | 11A | 11B | 11C | C | R |
|---|---|---|---|---|---|---|
| Interpret and construct pie charts and line graphs and use these to solve problems. | ✓ | | | ✓ | ✓ | ✓ |
| Calculate and interpret the mean as an average. | | ✓ | | ✓ | ✓ | ✓ |
| Use the language associated with probability to describe and compare possible outcomes. | | | ✓ | | | |

# 11 Statistics

## Engage  Student Book page 182

### Big question

- How can we find out the answers to some interesting questions and how can we present our findings?

### Global skills

- **Creative skills:** exploring
- **Real-world skills:** interpreting information
- **Interpersonal skills:** communication
- **Self-development skills:** reflecting on learning

### Key vocabulary

- data, survey, questionnaire, tally
- graph, line graph, pictogram, bar chart, table, frequency table
- Venn diagram
- most/least popular, most/least common, mean

### Resources

- range of magazines and newspapers containing different ways of representing data
- scissors, glue
- large sheets of paper
- Resource sheet 11.1: the data handling cycle

### Language support

As a class, read through the posters from the Introductory activity to model key vocabulary. These posters should be available for the whole unit.

 Introductory activity

Organise the class into mixed-attainment groups of three or four students. Groups find different examples of representing data in magazines and newspapers. They should cut out the examples they find and create a poster, labelling the different types of graphs and charts. Alternatively, you could look on the internet for various examples.

Each group should report to the whole class about the examples they have found. Discuss, using questions such as: *Why do you think a pie chart was used to display this data? Why is a bar chart a good way to display this data? What information does this line graph tell us?* Groups spend more time adding to the vocabulary on their posters as a result of other groups reporting back.

 Main activity

Look together at page 182 of the Student Book. Display on the IWB, if possible. Look at each example of a chart. *What does each one tell us? What information is missing from each one?* Read out some of the questions in the speech bubbles. Ask students to discuss the speech bubble questions, in pairs, using the posters from the Introductory activity to support them. After they have had time for discussion, ask pairs to feed back their ideas for each question. Ask the rest of the class, *Do you all agree? Can you explain why, or why not?*

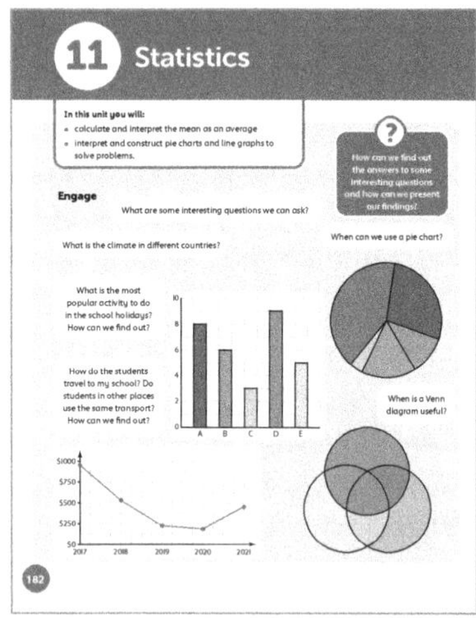

### Differentiation

**Supporting:** Help students to name the graphs and charts they are looking at.

**Consolidating:** Ask students to name the graphs and charts they are looking at and describe how they are used.

**Extending:** Challenge students to name the graphs and charts they are looking at and describe how and why they are used.

 Reflection time

Take the question from one of the speech bubbles on page 182 of the Student Book: *How do the students travel to my school?*

Show how to answer this question by:

1 constructing a frequency chart

2 constructing a graph (students can decide which type would be best and why)

3 interpreting the graph to answer the question.

Give students Resource sheet 11.1 (the data handling cycle) to show how they have completed the cycle: specified the problem, collected the data, represented the data and interpreted the data.

# 11A Averages



## Specific learning focus

- Calculate and understand the mean of a set of data.

## Global skills

- **Creative skills:** exploring, problem solving
- **Real-world skills:** interpreting information

## Key vocabulary

- average, mean

## Resources

- tape measures
- counters, cubes or similar items to count
- calculators

## Language support

Help students get a sense of the concept of an **average** to represent a group of data. Reinforce the idea that the **mean** is a measure of the data and is always within the range of the data, with some data being smaller than the mean and some being greater.

Ask questions that put the word 'mean' into context to support students' understanding, for example:

- *How do you find the mean of these numbers?*
- *Which set of figures has the greater mean?*
- *What is the mean height of this group of people?*

 **Introductory activity**

Ask five students to come to the front and ask them each to take a handful of counters (or similar) from a bag. Ask each student how many counters they have in their hand and record the numbers on the board, for example 5, 7, 5, 8 and 5. Ask them what they would say if they were asked to give one number that approximately expresses how many counters each person has.

Explain that the mean is a way of calculating how many each person would have if they put all the counters together and shared them equally. *What is the total number of counters held by the five students? How many would each student have if they were shared out equally?*

Write these calculations on the board:
5 + 7 + 5 + 8 + 5 = 30 and 30 ÷ 5 = 6.

The mean is 6 counters, and although no student took 6 counters from the bag, it represents a typical amount that represents the range of the data.

 **Main activity**

Give each pair of students a tape measure and ask them to measure the circumference of their heads around the forehead. They can help each other measure. List the measurements on the board. Using calculators, ask students to total the measurements and then divide the total by the number of students in the class to find the mean head size. Then ask, *How many students have a head circumference smaller than the mean? How many have a head circumference greater than the mean? Are there any students whose head circumference is the same as the mean?*

Ask students to complete the activities on page 183 of the Student Book. They should work in small groups as they need to gather data about each other.

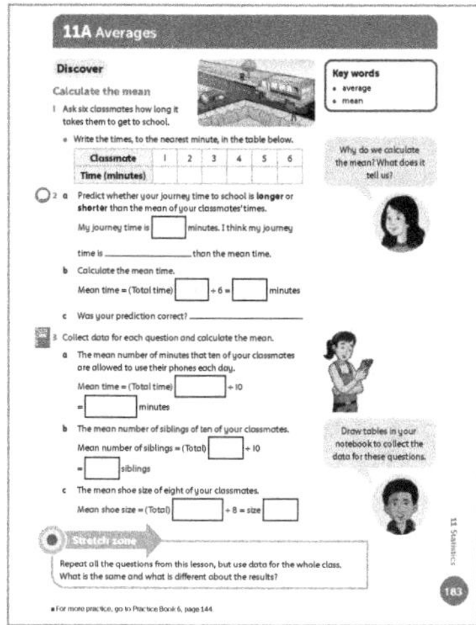

## Differentiation

**Supporting:** Help students to calculate the mean from a set of data.

**Consolidating:** Ask students to explain how the mean is calculated.

**Extending:** Challenge students to describe what the mean of a set of data represents.

**Stretch zone:** *Repeat all the questions from this lesson but use data for the whole class. What is the same and what is different about the results?*

Students should develop their understanding of the mean as a measure of a typical data value within a set of data.

 **Reflection time**

Ask each group to share their findings from their data collection and calculation of the mean. Ask whether their means represented the data and seemed like a typical example of the data.

Ask students to calculate the mean of these amounts: 1, 1, 1, 1, 1, 1, 1, 1, 1, 100.

The total is 109 and so the mean is 10.9. *Does 10.9 seem to represent the data? Why?*

**Practice Book:** Students complete Practice Book page 144. They can do this directly after the Main activity, as homework, or as the focus of a separate mathematics session to help students consolidate their learning and build fluency.

Students ask five people a series of questions. They then work out the mean of each data set.

| Differentiated outcomes | |
| --- | --- |
| **All students** | should calculate the mean of a set of data with support. |
| **Most students** | will calculate the mean of a set of data. |
| **Some students** | may calculate the mean of a set of data and interpret it in the context of a problem. |

# 11A Averages

## Explore
Student Book page 184 • Practice Book page 145

### Specific learning focus

- Calculate mean times to compare Olympic sprint data.

### Global skills

- **Creative skills:** investigating
- **Real-world skills:** interpreting information

### Key vocabulary

- average, mean

### Resources

- calculators
- newspapers or magazines containing articles that mention averages or means
- large sheets of paper

### Language support

Ask questions about the mean to support students' understanding in different contexts, for example:

- *What is the mean time for the women's 100 m sprint in 2016?*
- *Do women have the faster mean time in any year?*
- *What do the mean times in each year tell you about athletes' speeds?*

## Answers

### Student Book page 183

All answers depend on data collected by students about their classmates. Check that for the data collected they have found the mean where appropriate and interpreted what it tells them about the data.

### Practice Book page 144

All answers depend on data collected by students about their classmates. Check that for the data collected they have found the mean where appropriate and interpreted what it tells them about the data.

Stretch zone: Accept all sensible comparisons of the data using the mean values.

## Introductory activity

Ask four students to come to the front of the class. Time them for 60 seconds to see how many times they can write their name. Write the results on the board, for example 18, 24, 25 and 17. Now ask students to calculate the mean number of times these four students wrote their names in 60 seconds: $18 + 24 + 25 + 17 = 84$, and $84 \div 4 = 21$.

*If the mean number of times to write their names in 60 seconds is 21, why might some students write their name more times or fewer times than this?* Students might suggest that people with longer names will write it fewer times in a minute, or it could depend on each student's handwriting speed.

Use this as an example of the fact that the mean may give an idea about the data but the data still needs to be interrogated to understand what it means and why the data is varied.

## Main activity

Explain to students that sport is a rich source of data for calculating averages such as the mean number of goals scored, or the mean length jumped, or the mean speed for a race.

Ask students to look at the activity on page 184 of the Student Book, which lists the top four times for the 100 m sprint by men and women at three different Olympic Games. Read the text at the top of the page so that students understand the question they are being asked to investigate: Are women's times getting closer to men's times in 100 m races? *What patterns will we be looking for in the data in order to answer this question?*

Students should work in pairs to calculate the mean time for each group to run the 100 m at each Olympic Games. They then use the means to make comparisons about the data and to answer the investigation question, then to make predictions for future races.

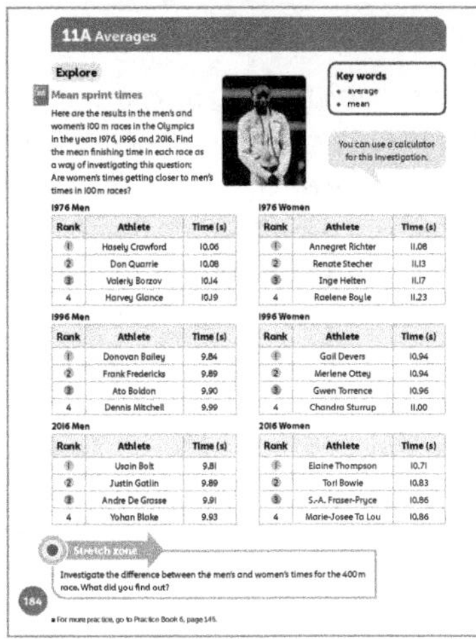

### 11A Averages

**Explore**

**Mean sprint times**

Here are the results in the men's and women's 100 m races in the Olympics in the years 1976, 1996 and 2016. Find the mean finishing time in each race as a way of investigating this question: Are women's times getting closer to men's times in 100 m races?

**Key words**
- average
- mean

You can use a calculator for this investigation.

**1976 Men**

| Rank | Athlete | Time (s) |
| --- | --- | --- |
| 1 | Hasely Crawford | 10.06 |
| 2 | Don Quarrie | 10.08 |
| 3 | Valeriy Borzov | 10.14 |
| 4 | Harvey Glance | 10.19 |

**1976 Women**

| Rank | Athlete | Time (s) |
| --- | --- | --- |
| 1 | Annegret Richter | 11.08 |
| 2 | Renate Stecher | 11.13 |
| 3 | Inge Helten | 11.17 |
| 4 | Raelene Boyle | 11.23 |

**1996 Men**

| Rank | Athlete | Time (s) |
| --- | --- | --- |
| 1 | Donovan Bailey | 9.84 |
| 2 | Frank Fredericks | 9.89 |
| 3 | Ato Boldon | 9.90 |
| 4 | Dennis Mitchell | 9.99 |

**1996 Women**

| Rank | Athlete | Time (s) |
| --- | --- | --- |
| 1 | Gail Devers | 10.94 |
| 2 | Merlene Ottey | 10.94 |
| 3 | Gwen Torrence | 10.96 |
| 4 | Chandra Sturrup | 11.00 |

**2016 Men**

| Rank | Athlete | Time (s) |
| --- | --- | --- |
| 1 | Usain Bolt | 9.81 |
| 2 | Justin Gatlin | 9.89 |
| 3 | Andre De Grasse | 9.91 |
| 4 | Yohan Blake | 9.93 |

**2016 Women**

| Rank | Athlete | Time (s) |
| --- | --- | --- |
| 1 | Elaine Thompson | 10.71 |
| 2 | Tori Bowie | 10.83 |
| 3 | S.-A. Fraser-Pryce | 10.86 |
| 4 | Marie-Josee Ta Lou | 10.86 |

**Stretch zone**

Investigate the difference between the men's and women's times for the 400 m race. What did you find out?

184

■ For more practice, go to Practice Book 6, page 145.

## Differentiation

**Supporting:** Help students to calculate the means for each group and explain what the mean represents in each case.

**Consolidating:** Ask students to explain how the mean figures can explain trends in the data over the years.

**Extending:** Challenge students to make predictions about the times in future Olympic Games.

**Stretch zone:** *Investigate the difference between the men's and women's times for the 400 m race. What did you find out?*

Students can extend the use of means to other sports for comparison between years or between men and women.

 **Reflection time**

Ask each pair to share something about their findings and how they were calculated.

Ask students to find examples of mean averages in newspapers or magazines and use them to create a poster illustrating how the mean is used in different contexts. These could be sports contexts or something different.

**Practice Book:** Students complete Practice Book page 145. They can do this directly after the Main activity, as homework, or as the focus of a separate mathematics session to help students consolidate their learning and build fluency.

Students look at the mean life expectancy in two different groups of countries. They look at where the countries are on a map. Then they interrogate the data and explain what they found out.

| Differentiated outcomes | |
| --- | --- |
| **All students** | should calculate the mean of a set of data with support. |
| **Most students** | will calculate the mean of a set of data. |
| **Some students** | may calculate the mean of a set of data and interpret it in the context of the data. |

## Answers

### Student Book page 184

| | |
| --- | --- |
| 1976 men: 10.12 | 1976 women: 11.15 |
| 1996 men: 9.91 | 1996 women: 10.96 |
| 2016 men: 9.89 | 2016 women: 10.82 |

The data suggest that both men and women are getting faster, and that women's race times are closing on the men's times.

### Practice Book page 145

**1** 83.52 years

**2** 56.55 years

**3** 26.97 years

**4** Students may notice that the countries with the lowest life expectancy are in Africa.

Stretch zone: Check that students have used accurate data and commented on the mean and differences between their country and five others.

# 11B Probability



## Specific learning focus

- Use the language associated with probability to discuss events, to assess likelihood and risk, including those with equally likely outcomes.

## Global skills

- **Creative skills:** investigating
- **Real-world skills:** interpreting information

## Key vocabulary

- probability

## Resources

- dice
- counters

## Language support

Use questions such as:

- *Are the outcomes equally likely? How do you know?*
- *How many possible ways can you score 7?*
- *What is the probability of an impossible event?*

 **Introductory activity**

Ask students, in pairs, to discuss the following statements and say how likely they are, using the vocabulary of likelihood: impossible, unlikely, equally likely, likely and certain.

- It will snow in July.
- A rocket will land on Mars in the next 10 years.
- I will go to school tomorrow.
- I will roll an even number on a dice.

Ask pairs to describe how they decided how likely each statement is.

Ask them to think of events for each likelihood and then share them as a class.

 **Main activity**

Arrange for the class to play 'The Great Horse Race Game' in mixed-attainment groups of four to six students. It is important that they play the game without any prior instruction about how the probabilities work.

Each group will need:

- two dice
- 12 different-coloured counters to represent the horses
- the game board on page 185 of the Student Book.



*Game instructions:*

*Players take turns to roll the two dice, then move the horse (counter), whose number is the sum of the scores on the dice, forward one space on the board. For example, if a 6 and a 4 are thrown on the dice, then horse 10 is moved one space forward. The horse that reaches the winning post first wins. Note: students are not assigned horses to move; they just take it in turns to move the relevant horse, depending on the roll of the dice.*

Let students play the game three or four times until they begin to notice patterns in the results, and identify which horses are most likely/least likely to win. When they have noticed a pattern, allow the groups to discuss the questions in the speech bubbles on page 185 of the Student Book.

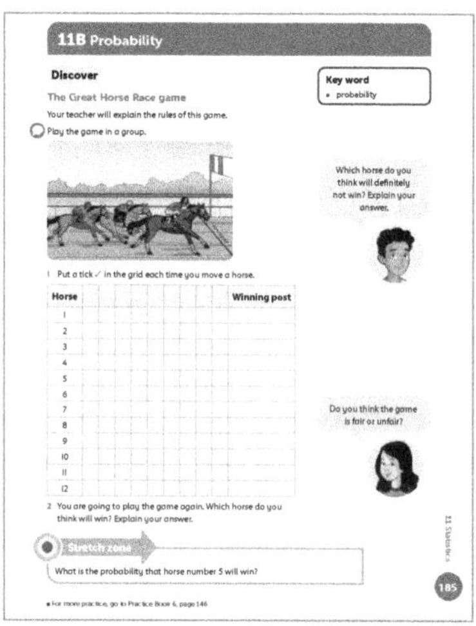

## Differentiation

**Supporting:** Ask students, after the playing the game a few times, why they think a particular horse will win.

**Consolidating:** Ask students why the game is unfair. Which horse is most likely to win? Why?

**Extending:** Ask students to calculate the probabilities of each horse winning.

**Stretch zone:** *What is the probability that horse number 5 will win?*

Discuss with students how to work out the probability of each number horse winning the race. There are 36 different combinations of totals using two dice, for example, 1 and 1; 1 and 2; 2 and 1; 2 and 2; 2 and 3; 3 and 2 (and so on). The number 5 horse has a 4/36 probability of winning as it could be thrown either with a 1 and a 4, a 4 and a 1, a 2 and a 3 or a 3 and a 2.

 **Reflection time**

Write the following table on the board. Ask students to complete the table. (The first three rows are completed as an example.) Use the table to remind students that the probability of an event happening is the number of possible ways of that event happening divided by the total number of possible outcomes. Remind students that the individual outcomes must be equally likely, and that the sum of the probabilities of all the possible outcomes is 1 (or 36/36).

| Dice score | Possible ways of making the score | Number of possible ways | Probability |
|---|---|---|---|
| 1 | Impossible | 0 | $\frac{0}{36}$ |
| 2 | (1, 1) | 1 | $\frac{1}{36}$ |
| 3 | (1, 2), (2, 1) | 2 | $\frac{2}{36} = \frac{1}{18}$ |
| 4 | | | |
| 5 | | | |
| 6 | | | |
| 7 | | | |
| 8 | | | |
| 9 | | | |
| 10 | | | |
| 11 | | | |
| 12 | | | |

**Practice Book:** Students complete Practice Book page 146. They can do this directly after the Main activity, as homework, or as the focus of a separate mathematics session to help students consolidate their learning and build fluency.

Students read statements that focus on the language of probability. All the statements are incorrect and students have to explain why each one is incorrect.

| Differentiated outcomes | |
|---|---|
| **All students** | should play the game and notice which horses are most likely to win. |
| **Most students** | will notice and explain the patterns – for example, it is impossible for horse 1 to win. |
| **Some students** | may make links to probability. |

## Answers

### Student Book page 185

Answers will vary because students play the game with other students.

### Practice Book page 146

1 All numbers on a fair dice are equally likely.

2 The numbers of boys and girls does not mean you can predict whether any particular baby will be a boy or girl. Each baby born is equally likely to be a boy or a girl.

3 It cannot be certain, only very likely that it will rain each day in the rainy season.

4 The outcomes are not equally likely as the teams may be of different strengths and one team will have a home advantage.

5 The next game also has a 1 in 10 chance, so you are still more likely to lose.

6 The runners will have different abilities and so their chances of winning are different.

Stretch zone: Look to see that students have made a reasonable attempt to approach how to calculate this, considering likelihoods and runners' abilities.

# 11B Probability

## Explore
Student Book pages 186–187 • Practice Book page 147

### Specific learning focus
- Use the language associated with probability to discuss events, to assess likelihood and risk, including those with equally likely outcomes.

### Global skills
- **Creative skills:** investigating

### Key vocabulary
- equally likely, outcomes, event

### Resources
- mini whiteboards and markers
- selection of magazines that include stories using probability
- pot containing lollipop sticks, each with a student's name on it

### Language support
Use questions such as
- *Are the **outcomes equally likely**? How do you know?*
- *How many possible ways can you score 7 with two dice?*
- *What is the probability of an impossible event?*
- *What does a probability of 1 mean?*

Write the following phrases on the board:
- It is very likely that …
- It is certain that …
- There is a fifty–fifty (or equal) chance that …
- It is impossible for …
- It is unlikely that …

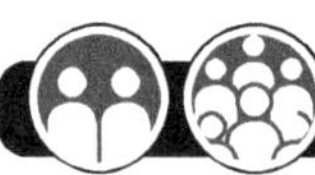 **Introductory activity**

Give each pair a mini whiteboard. Ask them to complete, on their whiteboard, the phrases you have written on the board. Tell pairs to write down at least two responses for each statement: one response for each student. Take feedback from students. Make sure that others in the class agree, especially with the statements 'It is unlikely that …' and 'It is very likely that …'. Talk through 'how likely' something has to be before it is 'very likely'. Remind students of the meaning of fifty–fifty or equal chance. Talk about their examples – are they exactly 50% or 1 in 2 chances?

 **Main activity**

Remind students that probability is 'Required outcomes'/'Possible outcomes'. For example, if I want to roll an even number on a six-sided dice:

required outcomes are: 2, 4 or 6

possible outcomes are: 1, 2, 3, 4, 5, 6.

Ask students to work in pairs on the activities on pages 186–187 of the Student Book. This allows them to discuss the equally likely outcomes and to check answers. As they work on the examples, ask individual students, *How are you working out the probabilities?*

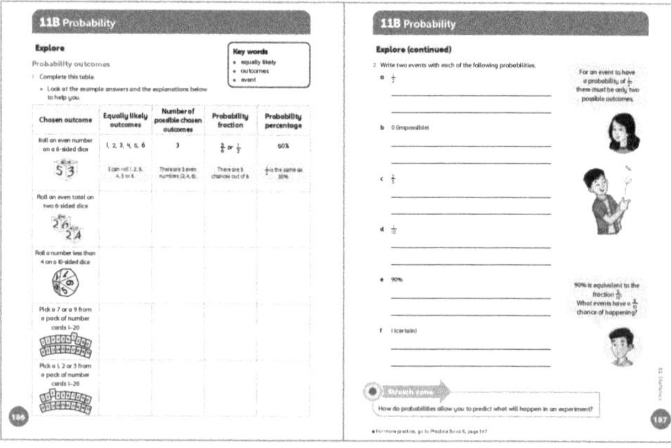

## Differentiation
**Supporting:** Ask students to describe all the equally likely outcomes.

**Consolidating:** Ask students to tell you how to find a probability.

**Extending:** Challenge students to find probabilities as fractions, percentages and decimals.

**Stretch zone:** *How do probabilities allow you to predict what will happen in an experiment?*

Students should come to understand that when tossing a coin, for example, the probability of getting a 'head' is 0.5. So if the coin is tossed 100 times, you would expect approximately 50 heads.

As an additional activity, ask students to explore the language of probability in the media. They use newspapers or magazines to find stories that include the use of probability. You could ask students to assess whether the use of probabilities is accurate.

 **Reflection time**

Take lollipop sticks from the pot to select some students to come to the front of the class. Ask each student to give one of their chosen **events** from question 2 in the Main activity. Ask other students to decide what is the probability of this event happening, to check that the answer is correct.

| Differentiated outcomes | |
| --- | --- |
| **All students** | should find equally likely outcomes. |
| **Most students** | will find probabilities as fractions. |
| **Some students** | may find probabilities as fractions, percentages and decimals. |

## Answers

### Student Book pages 186–187

**1**

| Chosen outcome | Equally likely outcomes | Number of possible chosen outcomes | Probability fraction | Probability percentage |
| --- | --- | --- | --- | --- |
| Rolling an even number on a 6-sided dice | I can roll 1, 2, 3, 4, 5 or 6 | There are 3 even numbers (2, 4, 6) | $\frac{3}{6} = \frac{1}{2}$ | 50% |
| Rolling an even total on two 6-sided dice | There are 36 different pairs of numbers. | 18 | $\frac{18}{36} = \frac{1}{2}$ | 50% |
| Rolling a number less than 4 on a 10-sided dice | 1, 2, 3, 4, 5, 6, 7, 8, 9, 10 | 1, 2 or 3 so 3 possible required outcomes | $\frac{3}{10}$ | 30% |
| Picking a 7 or a 9 from a pack of number cards 1–20 | 20 cards in a pack | 7 or 9, so 2 possible required outcomes | $\frac{2}{20} = \frac{1}{10}$ | 10% |
| Picking a 1, 2 or 3 out of a pack of number cards 1–20 | 20 cards in a pack | 1, 2 or 3 so 3 possible required outcomes | $\frac{3}{20}$ | 15% |

**2** Answers will vary. Check that students' events correctly match the probabilities.

### Practice Book page 147

**1**

| Chosen outcome | Number of equally likely outcomes | Number of possible chosen outcomes | Probability fraction |
| --- | --- | --- | --- |
| Rolling an odd number on a 1–6 dice | There are 6 possible scores | 3 (1, 3 or 5) | $\frac{3}{6} = \frac{1}{2}$ |
| Rolling a total of 9 on two dice | There are 36 possible outcomes | 4 (3 and 6; 4 and 5; 5 and 4; 6 and 3) | $\frac{4}{36} = \frac{1}{9}$ |
| Rolling an even number on a 1–10 dice | There are 10 possible scores | 5 (2, 4, 6, 8 or 10) | $\frac{5}{10} = \frac{1}{2}$ |

Stretch zone: Check that students have correctly identified an event with a probability of $\frac{1}{3}$, for example rolling a 1 or a 2 on a 1–6 dice.

# 11C Handling data extended project

## Activities 1–4
**Student Book pages 188–191 · Practice Book pages 148–151**

These teacher's notes follow a different pattern from others in this book. The 'Handling data' section is written to run as one continuous activity over several sessions. You may choose to allocate a lesson to each part of the process, but it is better to allow groups to work at different rates. These notes are divided up to show you how to support students in each section of the process. This is the last time that students will study data handling in the programme. The aim is that they draw on all their knowledge from other stages to follow the data handling cycle as a single continuous experience.

### Specific learning focus

- Explore how data is used in everyday life.
- Solve a problem by representing, extracting and interpreting data in tables, graphs, charts and diagrams.
- Find the mean of a set of data.

### Global skills

- **Creative skills:** investigating
- **Real-world skills:** presenting information
- **Interpersonal skills:** teamwork
- **Self-development skills:** reflecting on learning

### Key vocabulary

- question, survey, data, research, collect, present, interpret, pie chart, line graph

### Resources

Ideally students need access to the internet for research. If this is not possible, give them access to a large number of reference books, newspapers and magazines. Students also need large sheets of paper and access to the best technology available in school to prepare their presentations. Scrap paper, squared paper and coloured pencils will all be used.

### Language support

Write the key vocabulary on a flipchart. Display these words throughout the series of activities. As you work with students during this section of the unit, model the use of these words.

When students plan their presentations, remind them that they need to use this key vocabulary.

## Activity 1: Define the question

Look together at page 188 of the Student Book. Display on the IWB, if possible. Read through the data handling cycle, ensuring that students understand what happens at each stage of the cycle. Explain that Activity 1 is the first part of the data handling cycle: 'Decide on the problem and plan how you will answer it'.

Ask students to look at the speech bubbles on Student Book page 188. Tell them to work individually and to list all the **questions** from the page that they are interested in finding out about, or to come up with their own ideas. They should list these in the Student Book on page 188. An alternative is to ask students to look at magazines and newspapers and to find all the stories that use data. You can discuss how the journalists might have researched these stories. The following are common areas that students often like to explore.

- How can we improve school dinners?
- Is the school environmentally friendly (including recycling and travel to school)?
- How does the climate in our country compare with that in other areas?
- What programmes are good for a school television/ radio station?
- Are students at our school getting better at mathematics?

When students have prepared their lists of questions, ask them to work in pairs to decide on the three questions that they think are best.

Talk about practicality at this point. Encourage students to choose questions that they are genuinely interested in finding an answer to.

Then ask students to work in groups of four, five or six to decide on a list of three questions for their group. At this point, list all the questions on the board. Ask individual students to pick a question to research.

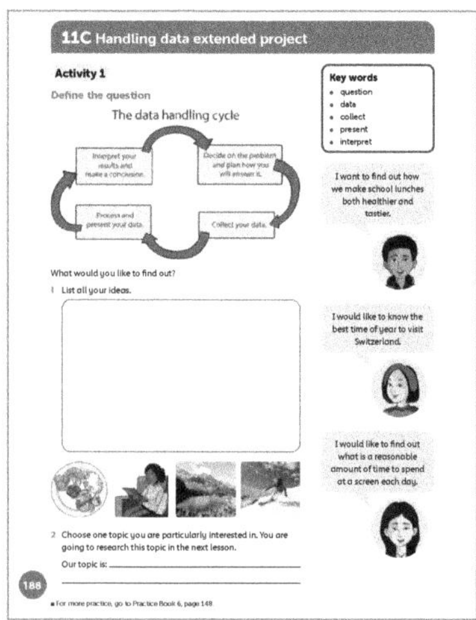

## Activity 2: Research the topic

Students stay in their groups of four, five or six for the rest of this section of the unit. In this part, ask them to research all the data they can find about the question they are asking. Encourage them to use the internet, newspapers and information texts to carry out this research. For example, a group exploring school dinners can find out what dinners schools in different countries provide, how school dinners have changed over a specified time in their own country, and what the state or government policy is in relation to the provision of school dinners. Students record their information on page 189 of the Student Book.

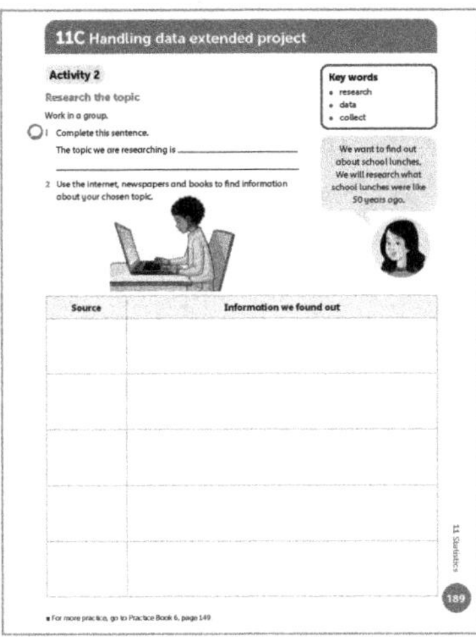

## Activity 3: Carry out a survey

The next stage is 'to **collect** data from a range of sources.' This involves either carrying out a survey or collecting data that already exists to help with the research question. For example: the school dinners group may put together a questionnaire that asks a range of questions related to satisfaction with school dinners and another questionnaire about what food is available. The group need to make sure that they ask these questions to everyone who has an interest in the answer. This includes parents, students, the dinner staff and teachers.

Groups looking at areas such as climate or athletic performance can collect appropriate data from other sources such as weather websites or world athletics championship results.

Students should plan their survey using page 190 of the Student Book and may use a tally chart to collect the data.

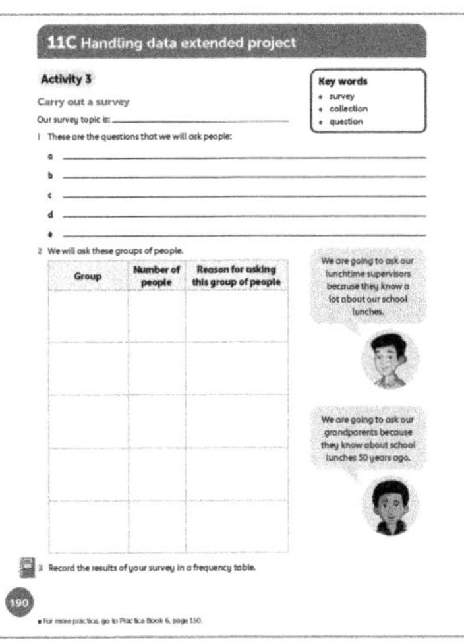

## Activity 4: Present the data

Students need to draw together and analyse their results. It is likely that a frequency table will be most appropriate representation to initially show the most/least popular choices of answer. For some questions, however, different graphs or charts may be most useful. The Student Book asks students to draw a line graph or a pie chart to represent their data. For example, athletics results over time can be **presented** effectively as a line graph, showing the change over time. Differences in opinions can be represented in a pie chart. It is also sometimes appropriate to show the data in a graph such as a block graph, pictogram or bar chart. Venn diagrams and Carroll diagrams are also useful. Direct students to complete page 191 of the Student Book and encourage them to explain why they have represented the data in their chosen way. If neither a pie chart nor a line graph is appropriate to represent their data, students must be able to explain why these representations are not suitable and select a more suitable chart. Alternatively, you could agree on two sets of 'class data' that they must represent using a pie chart or line graph.

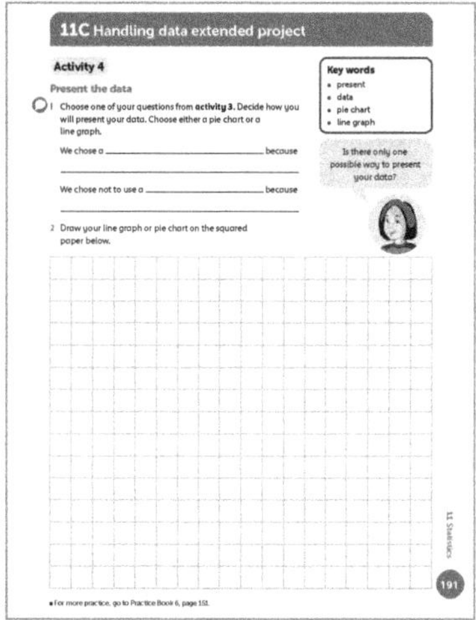

### Differentiation

**Supporting:** Help students with formulating purposeful research questions and guide them to discover where they can source some of the information they need.

**Consolidating:** Ask students to describe the different stages of the process and explain why they have chosen the methods they have.

**Extending:** Challenge students to interrogate and interpret their data to generate solutions to their research question.

### 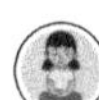 Reflection time

Allow each group time to present their study, so they can outline the choice of topic, data collection methods and choices of presentation methods. This may take a whole lesson to give every group ample time to share what they have done so far and to answer questions from the other groups. The **interpretation** phase of the cycle is done in the Connect lesson.

**Practice Book:** Students complete Practice Book pages 148–151. These pages can be done directly after the Main activity, as homework, or as the focus of a separate mathematics session to help students consolidate their learning and build fluency.

Students consolidate their learning from the extended project on handling data, including carrying out more research before collecting data, representing data in pie charts or line graphs and interpreting and interrogating the data.

| Differentiated outcomes | |
| --- | --- |
| **All students** | should take part in group research to answer an identified question and represent it using a suitable chart. |
| **Most students** | will support the design of a survey, collection of data and representation of the data on different tables and charts. |
| **Some students** | may analyse the data and use it to demonstrate how their research question can be answered. |

## Answers

### Student Book

**Page 188**

Students will record their topic ideas and identify which one they will focus on for their study. Check that they have chosen a topic that is manageable.

**Page 189**

Students will use this page to record the information about their topic that they find from the internet, books and newspapers.

**Page 190**

Check that students have selected appropriate survey questions for their topic and identified suitable groups to ask.

**Page 191**

Students will choose a method of presentation for their data, along with explanations for their choices.

### Practice Book

**Page 148**

Students choose their own countries and then research data on the mean temperature in those countries. Check that they have completed the table of temperatures correctly and used it to write four facts about the data.

Stretch zone: Students should identify something they find surprising about the data and explain why.

**Page 149**

Students draw a line graph to compare the date from their three countries. Check that they have completed the line graph correctly.

Stretch zone: Check that students can identify advantages and disadvantages of a line graph and a table, for example the line graph gives an easy visual representation of the data.

**Page 150**

Students design survey questions with multiple options as responses. Check that their questions and responses are suitable.

Stretch zone: The survey questions are used with another group of ten people. Students should be able to compare the data from the responses and describe any similarities and differences.

**Page 151**

Students draw pie charts to represent the data from their survey questions responses on page 150. Check that they have correctly completed the pie charts and keys.

Stretch zone: Students should be able to compare the data from the pie charts and describe any similarities and differences.

# 11 Statistics

## Connect  Student Book page 192

### Big idea

We can use surveys and research to find out the answers to some interesting questions. We can use pie charts and line graphs to present our findings.

### Global skills

- **Creative skills:** investigating
- **Real-world skills:** presenting information
- **Self-development skills:** reflecting on learning

### Key vocabulary

- data, mean, interpret, interrogate

### Resources

- students' data and charts from their surveys

### Language support

As students present the results of their research, ask them questions about their data collection and recording, for example:

- *How did you decide on your topic for research?*
- *How did you plan to collect your data?*
- *What type of charts did you choose to represent your data and why?*
- *What did your data tell you and did it help answer your original question?*

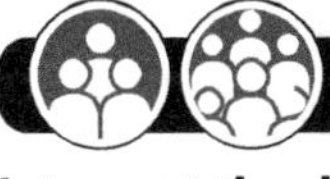 **Introductory activity**

**Interpret the data and plan a presentation**

The final part of the cycle is to **interpret** and discuss the data. As the groups analyse the data they have collected, remind them of how to calculate the mean if this is an appropriate measure for their data.

Quickly remind students how to calculate the mean by carrying out a survey of shoe sizes of the class.

Complete a class frequency table, then calculate the mean by totalling all the shoe sizes and dividing this by the number of students. You could also use this data to calculate the probability of someone in their class having a given shoe size.

 **Main activity**

Students should demonstrate what they have learned about carrying out an investigation using the data handling cycle.

Ask groups to share their final representations from the extended project. Students should, individually, complete Student Book page 192, describing the stages that they carried out during the data handling cycle. They are given the beginning of each sentence, which they need to complete, in order to present their results and some interpretation of what the results tell them. In addition, you could ask each student to write five facts that they can interpret from their information.

Finally, ask each group to write three to five questions, which can be answered by looking at their graph. They swap their questions with another group to see whether the other group can answer these questions from their graph.

*Look back at your original question. Did you answer this question? What else did you find out? Was there anything that surprised you?*

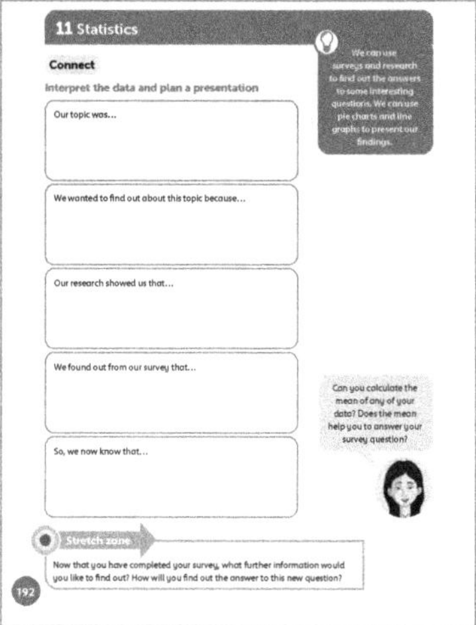

## Differentiation

Students work in mixed-attainment groups for this activity. More-confident students can help others with the data interpretation. As you ask questions while they work, explain to students that you might ask anyone in the group and they should all be able to say what they found out.

**Stretch zone:** *Now that you have completed your survey, what further information would you like to find out? How will you find out the answer to this new question?*

###  Reflection time

Bring the groups together as a class to discuss the research studies and compare graphs and charts. Ask each group in turn to say briefly what they found out and whether they answered their original question or not. They share the key headlines from their study. Groups can discuss the merits of each other's studies.

| Differentiated outcomes | |
| --- | --- |
| **All students** | should take part in group presentation of their research study. |
| **Most students** | will explain the process of carrying out the data handling cycle for their research study. |
| **Some students** | may analyse the data and use it to demonstrate how their research question can be answered and developed further. |

## 11 Statistics

**Review** Student Book page 193 · Practice Book page 152

### Student Book

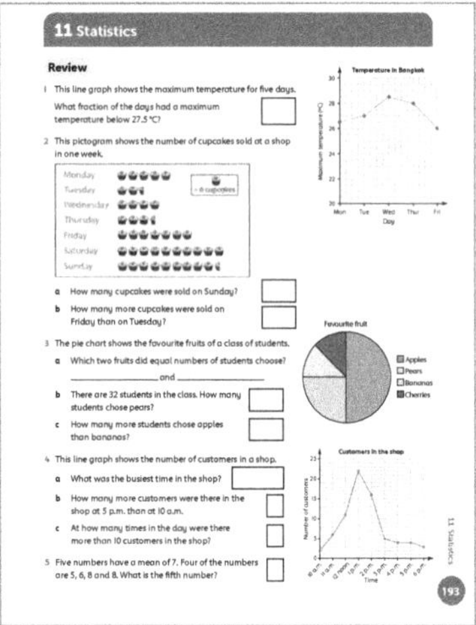

With young students, assessment activities are most effective when carried out as an everyday classroom activity. Students should have access to examples of different types of charts, so they can refer to these to support them. Give students opportunities to create questions for which they can collect and organise data. Extend by asking a range of questions about the data and using it to answer their original question, including interpreting the data by calculating the mean of sets of numerical data.

### Answers

**Student Book page 193**

**1** $\frac{3}{5}$

**2 a** 57 **b** 27

**3 a** bananas and cherries **b** 8 **c** 12

**4 a** 1 p.m. **b** 2 **c** 3

**5** 8

### Answers

**Student Book page 192**

Students will summarise their research study using the boxes and sentence stems given. Check that these are appropriate for their study and that they understand how the data handling cycle has helped them answer a question about a topic of interest.

### Practice Book

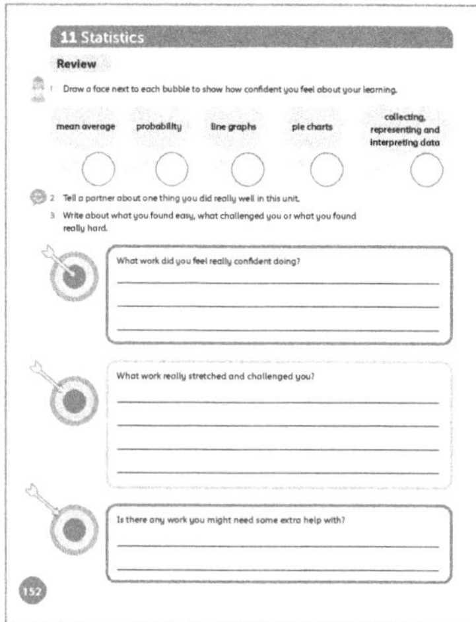

It is appropriate to complete this Practice Book Review as a whole-class discussion. You may choose to keep a record of the class discussion or a copy of the Review page for your own records. The Review provides an opportunity for students to reflect on their learning from the unit, to discuss any areas of mathematics that they feel went particularly well, and any areas that they feel less confident about. Ensure that all students have a copy of the Student Book as a reminder of the areas of mathematics that they have worked on in this unit.

Allow students plenty of time for discussion before asking them to complete the Practice Book page individually, and then, if appropriate, to share their responses with the rest of the class. If students complete this self-assessment at home, encourage them to discuss this with adults. Make a note of areas that students still feel unsure about. As students are working, remind them to use the displays. Reassure them that they will become used to the vocabulary as they collect data, represent it on different types of chart and use tables, charts and measures such as the mean to interpret the data. Students will also be growing more accustomed to representing likelihood of events, or probabilities, both verbally and numerically.

### Additional material

There are additional end-of-unit assessments available on the *Oxford Owl for School* website.

# Glossary

| | |
|---|---|
| **algebra** | the use of letters to represent unknown values in calculations |
| **associative law** | the law that says that when you add or multiply any three real numbers, the grouping (or association) of the numbers does not affect the result, e.g. $5 + 2 + 3 = 10, 5 + 5 = 10$ |
| **average** | a number that best represents a set of numbers. There are different ways of determining a representative number (average): mean, median and mode |
| **BIDMAS** | an aid to help remember the order of operations. B (brackets), I (indices), D and M (divide and multiply), A and S (add and subtract) |
| **brackets** | in an equation, brackets tell you to complete the operation within the brackets first |
| **centilitre** | one hundredth of a litre. There are 10 millilitres in 1 centilitre. |
| **circumference** | the distance all the way round a circle; the perimeter of a circle |
| **common denominator** | a common multiple of the denominators of various fractions |
| **common factor** | numbers that are factors to more than one other numbers, e.g. the common factors of 18 and 24 are 2, 3 and 6 |
| **common multiple** | a number that is in two or more multiplication tables, e.g. 15 is a common multiple of 3 and 5, as it is in the 3 times table and the 5 times table |
| **commutative law** | the law that says you can add or multiply numbers in any order, e.g. $4 + 7 = 11, 7 + 4 = 11$ |
| **composite number** | any positive number greater than 1 that is not a prime number. A composite number is divisible by at least one number that is not 1 or itself |
| **constant** | a number whose value is fixed in an expression, e.g. in $3x \times 8 = 17$, 8 is a constant. |
| **conversion** | changing something from one thing into another. You can use conversion graphs and tables when converting between units |
| **coordinate grid** | the grid used to plot coordinates |
| **coordinates** | the numbers on both axes that combine to give a particular point, e.g. (2,5) |
| **cross-section** | a cut straight across a 3D shape; a slice of an object |
| **decade** | a period of 10 years. A century is divided into 10 decades, e.g. the 20th century is divided into the 1900s, 1910s,1920s, 1930s and so on |
| **decimal fraction** | tenths, hundredths, thousandths and so on. They have digits to the right of the decimal point |
| **decimal place** | 4.235 has 3 decimal places. Rounded to 1 decimal place, it is 4.2 |
| **degree of accuracy** | e.g. 2.657 is 2.7 to a degree of accuracy of one decimal place |
| **diameter** | the distance from one side of a circle to the other, going through the centre of the circle |
| **distributive law** | a law that says that multiplying a number by a group of numbers added together is the same as doing each multiplication separately, e.g. $3 \times (3+2) = 3 \times 3 + 3 \times 2$ |
| **dividend** | the quantity that has to be divided |
| **divisor** | a number that is divided into another number |
| **dodecahedron** | a 3D shape that has 12 flat faces. A regular dodecahedron has 12 regular pentagonal faces |
| **elevation (side elevation, front elevation)** | a particular side of a shape, e.g. front elevation |
| **enlarge, enlargement** | changing a shape so that it is larger but remains the same shape as previously |
| **equal chance (even chance, fifty-fifty chance)** | any two events that are equally likely. They have a probability of 50% or $\frac{1}{2}$, e.g. if we toss a coin, it is equally likely to come down heads or tails |
| **equally likely** | any events that have the same probability, e.g. if you have 2 yellow balls, 2 blue balls and 2 black balls in a bag it is equally likely that you will pick out a yellow ball, a blue ball or a black ball |

| **equation** | two expressions that are equal and joined together by the = sign |
| **formula (plural formulae)** | a rule that can be written in(plural formulae) words or using letters and symbols |
| **inch** | (plural inches) an imperial unit used to measure length. Twelve inches measure the same as one foot. An inch measures about $2\frac{1}{2}$cm |
| **integer** | any whole number. An integer can be a positive or a negative number. Zero is also an integer |
| **interior angle** | an angle inside a shape: a triangle has three interior angles; a pentagon has five interior angles |
| **line graph** | a graph that shows continuous data over a period of time |
| **linear** | progressing from one stage to another in a series of sequential steps |
| **mean** | a kind of average. To find the mean, total the quantities and then divide by the number of quantities |
| **millisecond** | a unit used to measure a very small amount of time. There are 1000 milliseconds in 1 second |
| **mixed number** | a number made up of a whole number and a fraction |
| **net** | a 2D plan of a 3D shape |
| **order of operations** | the order in which you perform mathematical operations |
| **order of rotational symmetry** | the number of times a shape can be rotated so that it looks the same in the rotation as it did in the original |
| **ounce (oz)(plural ounces)** | an imperial unit of weight or mass. 1 oz is about 25 g |
| **parallelogram** | a four-sided shape in which opposite sides are parallel to each other |
| **percentage** | out of a hundred. A percentage is another way of writing a fraction that has a denominator of 100. The symbol for per cent is % |
| **percentage decrease, percentage increase** | the amount by which a percentage has gone down or up over time |
| **perimeter** | the distance all the way round a shape |
| **perpendicular height** | the height of (e.g. a triangle) when measured at a right angle to the base |
| **pie chart** | A circular chart that shows results of a survey in wedges of the circle, e.g. as percentages. |
| **plan** | a diagram showing what an object looks like |
| **prime factor** | a factor of a number that is also a prime number |
| **prime number** | a number that has only two factors, which are 1 and the number itself. The number 1 is not a prime number because it has only one factor |
| **proper fraction** | a fraction with a numerator that is smaller than its denominator. It is a fraction that is worth less than 1 |
| **protractor** | a piece of equipment used to measure the size of angles |
| **proportion** | the number of items of a particular type in a group of items, e.g.10 counters, 4 are yellow, 6 are green. The proportion of yellow counters is $\frac{4}{10}$ |
| **quadrant** | one of four parts of a coordinate grid |
| **quotient** | the product of a division problem |
| **radius** | the distance from the centre of a circle to the edge of the circle |
| **ratio** | a way of comparing one quantity to another |
| **rectilinear** | a rectilinear shape is one consisting of or bounded by straight lines |
| **reflect, reflection** | a mirror image of a shape across a mirror line |
| **reflex angle** | an angle larger than 180° and smaller than 360° |
| **rhombus** | a 2D shape with four equal sides. The opposite sides are parallel. Rhombus is the correct name for a diamond shape |
| **rotational symmetry** | the property a shape has when it looks the same after a rotation by a partial turn |
| **scale factor** | when enlarging a shape, the amount by which the shape has been enlarged |
| **simplify** | to make more straightforward. Fractions can be simplified by dividing the denominator and numerator by the same number, $\frac{10}{40} = \frac{5}{20} = \frac{1}{4}$ |

**symmetry**  a shape is symmetrical if a line can be places across it and one half of the shape is an exact reflection of the other half

**tangram**  a puzzle where several shapes are placed together to form a larger shape

**time zones**  parts of the world that have the same time of day, e.g. when it is 11 a.m. in New York, it is 5 p.m. in Rome.

**tonne**  a metric unit used to measure mass or weight. A tonne equals 1000 kilograms. A tonne is sometimes called a metric ton

**transformation**  a way of moving a shape or an object, e.g. turn (rotation), slide (translation) or flip (reflection)

**translate, translation**  the movement of a shape without turning it, e.g, moving it 3 squares right and 7 squares up on a grid

**trapezium**  a four-sided shape that has one pair of parallel sides and one pair of sides that are not parallel

**variable**  a symbol that works as a placeholder for expressions or quantities that are likely to vary or change

**volume**  the amount of space taken up by a solid shape. Volume is measured in cubic units such as $cm^3$ and $m^3$